A CHEMICAL HISTORY TOUR

A CHEMICAL HISTORY TOUR

Picturing Chemistry from Alchemy to Modern Molecular Science

by

ARTHUR GREENBERG

Professor and Chair
Department of Chemistry
University of North Carolina at Charlotte
Charlotte, NC 28223

Current address
Dean
College of Engineering and Physical Sciences
University of New Hampshire
Durham, NH 03824

A JOHN WILEY & SONS, INC., PUBLICATION

New York • Chichester • Weinheim • Brisbane • Singapore • Toronto

This book is printed on acid-free paper. ∞

Published simultaneously in Canada.

For ordering and customer service, call 1-800-CALL-WILEY.

Library of Congress Cataloging-in-Publication Data:

Greenberg, Arthur.
A chemical history tour : picturing chemistry from alchemy to modern molecular science / by Arthur Greenberg.
p. cm.
A Wiley-Interscience publication.
Includes index.
ISBN 0-471-35408-2 (alk. paper)
1. Chemistry—History I. Title.
QD11.G74 2000
540′.9—dc21 99-38865
CIP

Printed in the United States of America.

10 9 8 7 6 5 4 3

"Joy, Joy, Joy, Tears of Joy"

—from translation of a manuscript found sewn into the lining of an item of Blaise Pascal's clothing

Now shall my inward joy arise,
And burst into a Song;
Almighty Love inspires my Heart,
And Pleasure tunes my Tongue

—from *Africa*, composed by William Billings

William Billings of Boston, from
"The Singing Master's Assistant" (1778)

CONTENTS

PREFACE

Amiable reader, the purpose of this book is to treat you to a light-hearted tour through selected highlights of chemical history. I hope to provide an entertaining and attractive, but informative resource for chemistry teachers, practicing professionals in science and medicine, as well as the lay public interested in science and appreciative of artwork and illustration. This book is designed as a picture book with sufficient text to explain details and context, and like any tour, it is idiosyncratic in the highlights that it chooses to show the tourist.

A Chemical History Tour is meant to be skimmed as well as read. It begins with the practical, medical and mystical roots of chemistry, and traces in pictures and words chemistry's evolution into a modern science. Our tour starts with the metaphorical frontispiece in the 1738 edition of *Physica Subterranea*, the last edition of the text by Becher that introduced the phlogiston concept, and ends with a "quantum corral" of 48 iron atoms placed one-by-one using an atomic force microscope. The book's postscript is comprised of the images in three short works of the modern poet Seamus Heaney. We trace the evolution of alchemical concepts into phlogiston and examine critical steps in the development of modern chemistry. Our coverage starts to thin out in the nineteenth century and is, deliberately, very sparce through the twentieth century.

One important theme of this tour reflects our very human need to pictorialize matter: four elements, three principles, platonic solids such as the cube, corpuscles of atoms with and without hooks, two-dimensional "clumps" of atoms, two-dimensional molecules, three-dimensional molecules, fairies linking arms, "ball-and-stick" and "space-filling" models, solar-system atoms, cubic atoms with electrons at the corners, resonating structures, atoms hooked together by springs, atomic and molecular orbitals, electron-density contours on computer screens. Such images will recur throughout the book.

Chemistry has the most interesting history of any of the sciences. Its origins are both mystical and practical. What is the origin of matter? Did it arise from "nothingness" or from sub-atomic particles or from some fundamental primary matter (*prima materia*)? What about the underlying structure of matter? Is it infinitely divisible? Does it reach fundamental limits of division? Even today, one may question whether a single atom of gold is really gold: How many gold atoms are required to make a submicroscopic cluster that conducts electricity and has luster? Do "atoms" of acid have sharp corners? Are "corpuscles" of water smooth and slippery? Does matter evolve? What is the Elixir of Life? The Fountain of Youth? Do we combat poison (and sickness) with stronger poison or do we try to neutralize it? On the one hand, stomach acid upset can be neutralized with alkaline bicarbonate of soda. On the other, venereal diseases may be treated with toxic metallic substances just as we treat cancer with toxic medications that attack our DNA.

The light coverage of the twentieth century will certainly draw the attention of some not-so-amiable reviewers. I would defend this admitted weakness by noting that the exponential explosion of information during modern times would overwhelm the contents in this book. Moreover, the modern findings that continue to pass muster are included in the current chemistry texts. *A Chemical History Tour* is meant to supplement and enliven the coverage in a modern course. It makes no pretense of completeness. Nevertheless, we include the discoveries of subatomic structure, x-ray crystallography, the Kossel–Lewis–Langmuir picture of bonding based on the octet rule, development of the quantum mechanics—the underlying basis of the Periodic Table, as well as resonance theory. The DNA double helix is included because it is a triumph of structural chemistry and its structure immediately explained its function. Indeed, DNA's function, duplication, implied that its structure would likely have "two-ness."

We conclude the pictorial tour with two examples of *fin-de-millenium* structural chemistry. One of these is the synthesis of nanoscopic polyhedra using preconstructed "toy parts" (linear or bent bifunctional molecules; trigonal planar or trigonal pyramidal trifunctional molecules). These are simply mixed in the right ratio to form the polyhedra in almost 100% yield in about 10 minutes. The trick is to copy nature by employing weak bonds that form–break–reform until the system self-anneals to the desired structure. A nanoscopic dodecahedron thus obtained recalls the Pythagoreans' view of the heavenly, or fifth, element ("ether") postulated some 2500 years ago. The other example is the use of the scanning tunneling microscope (STM and its modifications) to view individual atoms and even to pick them up and move them one-by-one. In this regard, another theme in our tour is the resistance to the reality of atoms that continued for over 100 years after Dalton's theory was enunciated in 1803. Indeed, in the "minutes" before its universal acceptance shortly after the start of the twentieth century, Ludwig Boltzmann committed suicide due in part, it is believed, to his failure to convince all physicists and chemists of the reality of atoms. The 48 individual iron atoms "lassoed" together by the atomic force microscope to form the "quantum corral" even provide a direct STM "picture" of wave–particle duality.

I anticipate justified criticism of this idiosyncratic tour due to the numerous sites not visited. I freely admit that there are countless other paths through chemical history, and I apologize in advance for discoveries omitted or given short shrift. However, I want this book to be useful and to fulfill this mission it must be read and enjoyed by nonspecialists. A more thorough or encyclopedic approach will not help to achieve this goal. Although I have attempted to recognize contributions beyond those of Western culture, I am aware of and apologize for the weak coverage given to early science in Chinese, Indian, African, Moslem, and other cultures. This is really more an artifact of the availability of printed books rather than intent.

Although our tour is meant to be both light-hearted and light reading it tackles some of the important issues that are often too lightly or confusingly broached in introductory courses and are difficult to teach. We do, however, try our hand at humor and some of the earthiness so evident in the Renaissance works of Chaucer and Rabelais. Why not include Van Helmont's recipe for punishment of anonymous "slovens" who leave excrement at one's doorstep? By providing such vignettes, I hope to reengage chemists and other scientists in the history of our field, its manner of expressing and illustrating itself and its engagement with the wider culture. I hope to provide teachers in introductory chemistry courses with some guidance through difficult teaching areas and a few anecdotes to lighten the occasional slow lecture. And if a few students are caught snickering over a page of Rabelaisian chemical lore or some bad puns, would that be such a bad thing?

Arthur Greenberg
Charlotte, North Carolina

COLOR PLATE CAPTIONS

FIGURE 32 ■ Black-and-white reproduction of the color photograph of the 1671 oil painting, *The Alchemist*, by Hendrick Heerschop, in the Collection of Isabel and Alfred Bader. The author expresses his gratitude to Dr. Bader for permission to reproduce the image and also for his helpful discussion of the Bush drawing (see Fig. 30).

FIGURE 112 ■ This beautiful hand-colored figure (see color plates) is from the 1857 edition of Edward Youmans' *Chemical Atlas* (New York, first published in 1854).

FIGURE 114 ■ Plate from Youmans' *Chemical Atlas* (Fig. 112). The organization of the "Primeval Forest" of organic chemistry by Laurent and Gerhardt included the concept homology. The units of homology are CH_2 rather than C_2H_2 as shown. The confusion was the result of discrepancies in the atomic weight of elements and assumed formulas. These would be cleared up very shortly in the Karlsruhe Congress of 1860.

FIGURE 115 ■ Plate from Youmans' *Chemical Atlas*. Although Wöhler and Liebig discovered that silver fulminate and silver cyanate were isomers (term coined by Berzelius in 1830) and it was suspected that the origin was the different arrangement of atoms, the concept of valence remained to be discovered. In this plate Youmans depicts isomers as different arrangements of atoms. But he postulates that their chemical history plays a role. For example, since the atomic arrangements in the carbon allotropes graphite and diamond are different (presumably different charcoals are also allotropes), then hydrocarbon isomers (butane and isobutane, for example) maintain the different carbon arrangements of the allotropes from which they were (presumably) derived.

FIGURE 116 ■ Plate from Youmans' *Chemical Atlas* (original in color) depicting the three prevailing theories of organic chemistry structure in reactivity prior to Karlsruhe.

FIGURE 129 ■ An ebullient flame from the 1857 edition of Youmans' *Chemical Atlas* (see Fig. 112; the errors in formulas such as HO for water are discussed in the text).

FIGURE 136 ■ Pages from Maxfield Parrish's beginning notebook. Courtesy of the Quaker Collection, Haverford College Library.

COVER ILLUSTRATION ■ The artwork on the cover of this book is from an egg tempera painting (original in full color; author's private collection) signed in 1845 and is a version of a seventeenth-century work by Davied Teniers the Younger (J Read, *Prelude to Chemistry*, The MacMillan Co. New York, Plate 29; J. Read, *The Alchemist in Life, Literature and Art*, Thomas Nelson and Sons Ltd, London, 1947, Plate 21 and pp 72–79). It has some mischief in it: the leg of the table has a mouth and an eye reminiscent of a tortoise or dragon—both potent alchemical symbols. The painting is signed "las voy" ("les noy" or similar) with some symbols and we do not know the identity of the artist.

ACKNOWLEDGMENTS

I believe that my idea for this book was stimulated by the book *Chemistry Imagined* (Smithsonian, 1993), written by Roald Hoffmann in collaboration with the artist Vivian Torrence. I am grateful to Professor Roald Hoffmann for his kind response to my partial manuscript and his generous activities on behalf of it, as well as to Dr. Jeffrey Sturchio for early encouragement on this project. I thank my daughter, Rachel, for her meticulous scanning of most of the figures in this book and for her healthy skepticism that added to my own motivation. The artistic endeavors of my son David were another stimulus, and I thank my loving wife Susan for a lifetime of support and tolerating and commenting on early-morning readings of my essays. I am grateful for the comments and suggestions of my long-time friend and colleague, Professor Joel F. Liebman, and my newer Chemistry Department colleague Professor Daniel Rabinovich at the University of North Carolina at Charlotte; as well as those of my father, Murray Greenberg, who enjoyed a long, productive career in industrial chemistry. Art Department colleague Rita L. Shumaker provided two original works of interpretive artwork for this book and helpful readings. Professor Susan Gardner of the English Department renewed my confidence in this project and has been my imagined reader. She has gently encouraged my attempts to write creatively, added to the multicultural perspective of this book, edited me lightly yet firmly, alerted me to errors and unintended neologisms in English and French and co-authored the essay "Lavoisier in Love." I am also grateful to Dr. Barbara Goldman at John Wiley and Sons for early suggestions on this project and her faith in it, as well as to Jill Roter at Wiley and Ron Lewis at UNC at Charlotte who have been joys to work with. Unless otherwise noted, the figures are from books or artwork in my own collection.

SUGGESTIONS FOR FURTHER READING

I am not a chemical historian. Fortunately, there are a number of truly wonderful books treating chemical history. The most authoritative is the inspirational four-volume reference work, *A History of Chemistry* (McMillan, 1961—1964), by James R. Partington—it is rigorous, amply referenced, engagingly written, and nicely illustrated, and it has been a major source of information and insight for me. I have also relied heavily on the book by Aaron J. Ihde, *The Development of Modern Chemistry* (Harper & Row, 1964), and the more recent book by William H. Brock, *The Norton History of Chemistry* (Norton, 1993). Both of these are admirably suited to be textbooks in a course on the history of chemistry although Ihde's book is over 35 years old. I am particularly fond of the 1927 book *Old Chemistries* (Van Nostrand, 1927) by Edgar Fahs Smith. I feel that I am in Professor Smith's den on a cold winter's night as he shows me his antiquarian book collection and gently reads selected passages as we are warmed by the fireplace. And how I wish that I could have met the erudite and ebullient John Read. His trilogy, *A Prelude to Chemistry* (McMillan, 1937), *Humour And Humanism in Chemistry* (G. Bell, 1947), and *The Alchemist In Life, Literature and Art* (Thomas Nelson and Sons, 1947), provides the reader with healthy doses of laughter and learning. In *Humour and Humanism*, Read gives us the "box-score" of an Alchemical Cricket Match of All-Stars from the Bible (Noah, Moses), Greek and Roman mythology (Jupiter, Neptune, Aphrodite), ancient cultures (Cleopatra, Aristotle), the Renaissance (Paracelsus, Maier), and the early history of our science (Boyle, Lavoisier). The puns are deliciously low. He also writes a one-act play, "The Nobel Prize" ("A Chemic Drama in One Act") and happily treats us to the bawdier moments in Ben Jonson's 1610 play *The Alchemist*. Professor Read also stimulated the first performance of Michael Maier's seventeenth-century alchemical music (by the "Chymic Choir" at St. Andrews College in 1935). I also recommend two other recent books: *Women In Chemistry: Their Changing Roles From Alchemical Times to the Mid-Twentieth Century* by M. Rayner-Canham and G. Rayner-Canham (American Chemical Society and Chemical Heritage Foundation, 1998) and *A Dictionary of Alchemical Imagery* by Lyndy Abraham (Cambridge University Press, 1998). Omissions from my very brief list are not meant as criticisms and may simply reflect my own limited studies.

A CHEMICAL
HISTORY TOUR

SECTION I
PRACTICAL CHEMISTRY, MINING, AND METALLURGY

WHAT FRESH HELL IS *THIS*?[1]

What does this allegorical figure (Fig. 1) represent? This bald, muscular figure has the symbols of seven original metals arrayed around (and possibly including) the head. The all-too-perfect roundness of the head appears to correspond to the perfect circle that represents gold.

The elements, also including antimony and sulfur, are also buried in the intestines of the figure—literally its bowels—and now we have a hint of its nature. Any attempts at further interpretation are in the realm of psychology rather than science, and indeed the famous psychologist C.G. Jung owned a valuable collection of alchemical books and manuscripts and wrote extensively on the subject.[2]

At its heart, alchemy postulated a fundamental matter or state, the *Prima Materia*, the basis for formation of all substances. The definitions[2] of the *Prima Materia* are broad, partly chemical, partly mythological: quicksilver, iron, gold, lead, salt, sulfur, water, air, fire, earth, mother, moon, dragon, dew. At a more philosophical level, it has been defined as Hades as well as Earth.[2] Another figure from a seventeenth-century book on alchemy was identified by Jung as the *Prima Materia*—a similar muscular Earth shown suckling the "son of the philosophers."[2] This figure also has the breasts of a woman; this hermaphroditic being is reminiscent of the derivation of Eve from Adam and the subsequent seeding of the human species.

Let us cling to the Earth analogy because it seems to help in understanding the presence of the elements in its bowels. The small figure in the upper abdomen may be considered to be a type of Earth Spirit nurturing the growth of living things (see vegetation below it) and "multiplication" of the metals. The unique positions of gold (the head as well as the highest level in the intestines) implies *transmutation*—the conversion of base metals into noble metals. The figure holds a harp, representing harmony, and an isosceles triangle, representing symmetry. It is a metaphor for the unity that the true alchemists perceived between their art and nature.

This plate is the frontispiece from the book *Physica Subterranea* published by the German chemist and physician Georg Ernst Stahl in 1738.[3] It is the last edition of the famous book published by Johann Joachim Becher in 1669. Becher evolved chemistry's first unifying theory, the Phlogiston Theory, from alchemical concepts and it was subsequently made useful by Stahl. So in this plate are themes of alchemical transmutation, spiritual beliefs, and early chemical science that will begin our Tour.

FIGURE 1 ■ Frontispiece from the final edition of *Physica Subterranea* by Johann Joachim Becher (Leipzig, 1738). The figure may represent the Primary Matter (*Prima Materia*).

1. With apologies to the writer Dorothy Parker.
2. N. Schwartz-Salant, *Jung on Alchemy*, Princeton University Press, Princeton, NJ, 1995, pp. 25–30; 44–49.
3. A different interpretation of this figure, namely as Saturn, is to be found in C.A. Reichen, *A History of Chemistry*, Hawthorne Books, New York, 1963, p. 8.

THE ESSENCE OF MATTER: FOUR ELEMENTS (OR FIVE); THREE PRINCIPLES (OR TWO) OR THREE SUBATOMIC PARTICLES (OR MORE)

The ancient Greek philosophers were not scientists. They were, however, original thinkers who attempted to explain nature on a logical basis rather than by the whims of gods and goddesses. The father of this movement is considered to be Thales of Miletus, and during the sixth century B.C., he conceived of water as the essence of all matter. (We note later in this book that, in the mid-seventeenth century, Van Helmont had a somewhat similar view.) Thales is reputed to have predicted the total solar eclipse of 585 B.C., said to have occurred during a naval battle—although there is no basis for him having the knowledge to make such a prediction.[1] One of his successors in the Milesian School was Empedocles of Agrigentum (ca. 490–430 B.C.).[1] Empedocles is said to be the first to propose that all matter is composed of four primordial elements of equal

FIGURE 2 ■ The four elements of the ancients: Fire, Air, Earth, and Water from St. Isidore, *De Responsione Mundi Et Astrorum Ordinatione* (Augsburg, 1472) (courtesy of The Beinecke Rare Book and Manuscript Library, Yale University).

importance,[2,3] although similar ideas appear to have formed in Egypt, India, and China (five elements) around 1500 B.C.[2] Figure 2 depicts the four earthly elements. It appears in *De Responsione Mundi et Astrorum Ordinatione* (Augsburg, 1472), a book derived from the writings of Saint Isidorus, Bishop of Seville, during the seventh century A.D.[4]

Although Empedocles wrote about the actual physical structure of matter, it was only during the fifth century B.C. that two philosophers of the Milesian School enunciated a coherent atomic cosmology. None of the writings of Leucipus remain, but he is widely accepted as real and some of the writings of Democritus (ca. 460–ca. 370 B.C.),[1] his student, are known. For these scholars there were two realities in nature: Atoms (*atomos*, meaning not cuttable) and Void (derived from *vacuus*, meaning empty).[3] Void was considered to be as real as Atoms. Atoms of water were thought to be smooth and slippery; those of iron were jagged with hooks.

Aristotle (384–322 B.C.) is considered to be one of the two greatest thinkers of ancient times, the other being Plato.[1] Aristotle proposed a kind of primordial, heavenly element, "ether," and to each of the four earthly elements attributed two pairs of opposite or contrary "qualities" (wet versus dry; hot versus cold). The relationships between the elements and their qualities are depicted in a square that nicely places contrary qualities on opposite edges. The square is one of the fundamental symbols that often appear in alchemical manuscripts and books even as late as the eighteenth century. Thus, a liquid (rich in water) is cold and wet while its vapor (rich in air) is hot and wet. To vaporize a liquid, simply add heat—move from the cold edge to the hot edge of the square. To dissolve a solid (rich in earth), add wet; to burn the solid, add hot. Fire was not solid, liquid, or gas but a form of internal energy—perhaps related to the eighteenth-century concept of "caloric" propounded by Lavoisier.[2]

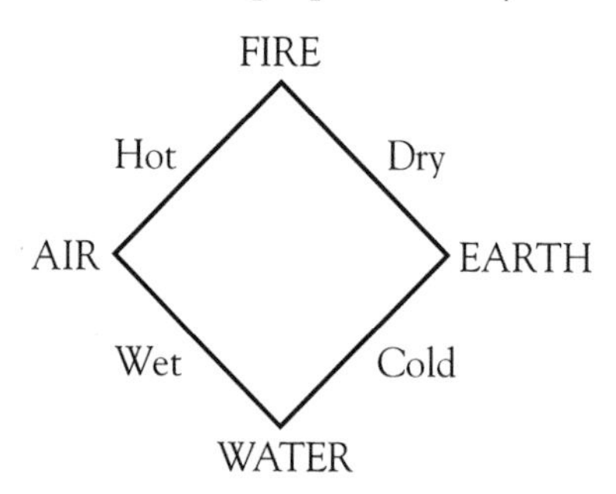

Aristotle was an anti-Atomist, in part, because he did not believe that space could be empty. This view was adopted by the great mathematician and philosopher Rene Descartes (1596–1650) who envisioned only two principles in matter (extent and movement) and rejected the four Aristotlean qualities. The idea of *extent* led him to reject the idea of finite atoms and the concept of void he considered ridiculous ("Nature abhors a vacuum"[5]). Thus, in the seventeenth and eighteenth centuries we have intellectual conflict between the Cartesians (school of Descartes) and the Corpuscular school (corpuscles were similar in concept to atoms), which included Robert Boyle and Isaac Newton.[6]

During the Renaissance, the classical Greek views of nature were finally challenged by the likes of Paracelsus.[7] Paracelsus extended an earlier view of matter that held that it was a union between an exalted sulfur of the philosophers ("Sophic Sulfur"—characterized often as male) and an exalted mercury of the philosophers ("Sophic Mercury"—characterized often as female). These are not related to the chemical elements we now recognize as sulfur and mercury. To these Paracelsus added Salt as the third Principle. Now, Mercury is Spirit, Sulfur is Soul, and Salt is Material Body. The relationship is depicted as a triangle, the other great metaphor found in alchemical manuscripts and books

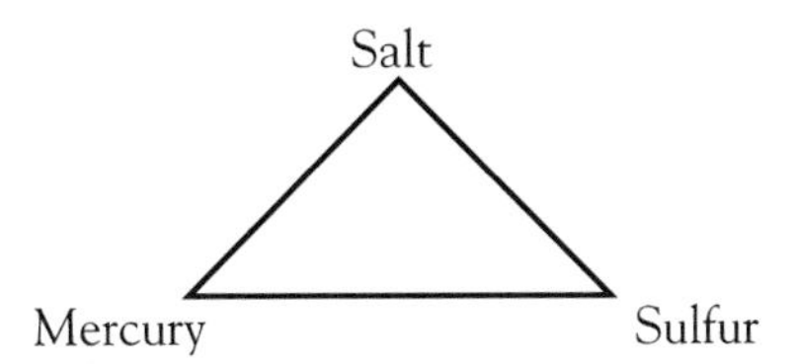

through the eighteenth century. All matter is composed of these three principles in various proportions. Later in this book (Fig. 43) we see two such symbolic triangles in Oswald Croll's *Basilica Chymica*. Croll presented Paracelsan alchemy —the bottom triangle presents Life, Spirit, Body (or Fire, Air, Water or Animal, Vegetable, Mineral). Symbols of triangles and squares abound in alchemy. The Sioux view the circle as their high ideal: "circle of Life," the tipi, the campfire.[8] In his nineteenth-century satire *Flatland*, Edwin Abbot portrays increasing perfection through each successive generation as a triangle begets a square, which begets a pentagon, and so on. A *megagon* is close to the perfection of a circle —a kind of generational transmutation.

The modern view of the atom is that it *is* divisible and that the fundamental particles making up all atoms of all elements are protons (positive charge), neutrons (zero charge), in an unimaginably dense nucleus occupying a miniscule fraction of the atom's volume, and electrons (negative charge).[9] The positive nucleus and the negative electrons are our modern "contraries." (Incidentally, it was Benjamin Franklin who introduced the negative–positive nomenclature in the context of electricity.[10]) The electrons are considered to be fundamental particles of infinite lifetime and are actually one of six subatomic particles called leptons. Protons and neutrons are not considered fundamental and are two of a very complex class of subatomic particles called hadrons. Outside of the nucleus, a free neutron has a half-life of only 17 minutes and decays into a proton, an electron (β particle), and an antineutrino—another lepton.[9] So, based upon this modern view, we can draw a Paracelsan-style triangle, but not equilateral in the sense that the neutron can give rise to the other two. The modern *Prima Materia* could be a dense neutron star.

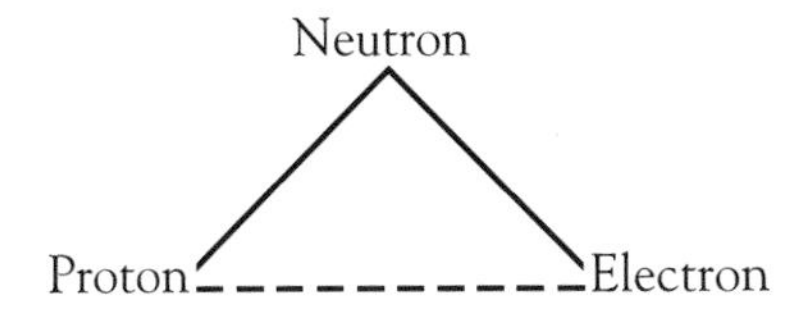

1. *Encyclopedia Brittanica*, 15th ed. Vol. 11, Chicago, 1986, p. 670.
2. J. Read, *Prelude to Chemistry*, MacMillan, New York, 1937, pp. 8–11.
3. B. Pullman, *The Atom in the History of Human Thought*, Oxford University Press, New York, 1998, pp. 2–47.

4. I. MacPhail, *Alchemy and The Occult*, Yale University Library, New Haven, 1968, Vol. 1, pp. 3–4.
5. Two sources for quotations simply refer this phrase(*Natura abhorret vacuum*) to a Latin proverb [B. Evans, *Dictionary of Quotations*, Delacorte Press, New York, 1968, p. 720, and *Dictionary of Foreign Quotations*, R. Collison and M. Collison (eds.), Facts on File, New York, 1980, p. 241]. One source attributes it to *Gargantua* in 1534 but from an ancient Latin source [A. Partington (ed.), *The Oxford Dictionary of Quotations*, 4th ed., Oxford University Press, New York, 1992, p. 534; *Bartlett's Familiar Quotations*, 16th ed., J. Kaplan (ed.), Little, Brown, Boston, 1992, p. 277] attributes the phrase to Spinoza in 1677. Just thought you'd want to know this one for the next Happy Hour.
6. B. Pullman, op. cit., pp. 140–142, 157–163.
7. J. Read, op. cit., pp. 21–30.
8. J. Lame Deer and R. Erdoes, *Lame Deer Seeker of Visions*, Simon and Schuster, New York, 1972, pp. 108–118.
9. B. Pullman, op. cit., pp. 343–353.
10. J.R. Partington, *A History of Chemistry*, MacMillan, London, 1962, Vol. 3, p. 66.

UNIFYING THE INFINITE AND THE INFINITESIMAL

It is human nature to try to harmonize our universe—to attempt to unify the infinite with the infinitesimal. Pythagoras and his followers developed a purely mathematical conception of the universe. As Pullman notes:[1] "Indeed, the Pythagoreans held that numbers are the essence of all things. Numbers are the source of what is real; they themselves constitute the things of the world."

Mendeleev developed the periodic table roughly 2400 years after Pythagoras died. He could not possibly have understood the origin of its order. But in 1926, the new quantum mechanics of Schrödinger explained the periodic table on the simple basis of four quantum numbers (n, l, m_1, and m_s) that students now learn in high school. Pythagoras would have been pleased but not surprised.

Figure 3 is from Johannes Kepler's *Harmonices Mundi* (1619). The fanciful drawings on the middle right depict the five platonic solids—polyhedra whose faces are uniformly composed of triangles, squares, or pentagons. The Pythagorean Philolaus of Tarentum (480 B.C.–?) is generally credited with equating the four earthly elements to these polyhedra.[1] Starting from the top center and moving counterclockwise, we have the tetrahedron (fire), octahedron (air), cube (earth), and icosahedron (water). Plato added the fifth solid, the dodecahedron, to represent the universe (similar to Aristotle's ether). The tetrahedron is the sharpest of these polyhedra, and fire is, thus, the "most penetrating" element. The dodecahedron is most sphere-like, most perfect. Its pentagons are also unique—you cannot tile a floor with pentagons as you can with triangles, squares, and hexagons. Plato further imagined that the four earthly elements were themselves composed of fundamental triangles—an isosceles right triangle A (derived from halving the square face of the cube) and a right-triangle B (derived from halving the equilateral triangular face of the tetrahedron, octahedron, or icosahedron). Earth was composed of triangle A. Air, fire, and water were composed of triangle B and could therefore be interconverted.[1]

In his 1596 book *Mysterium Cosmographicum*, Kepler proposed a solar system that placed the orbits of the six known planets on concentric spheres inscribed within and circumscribed on these five polyhedra arranged concentri-

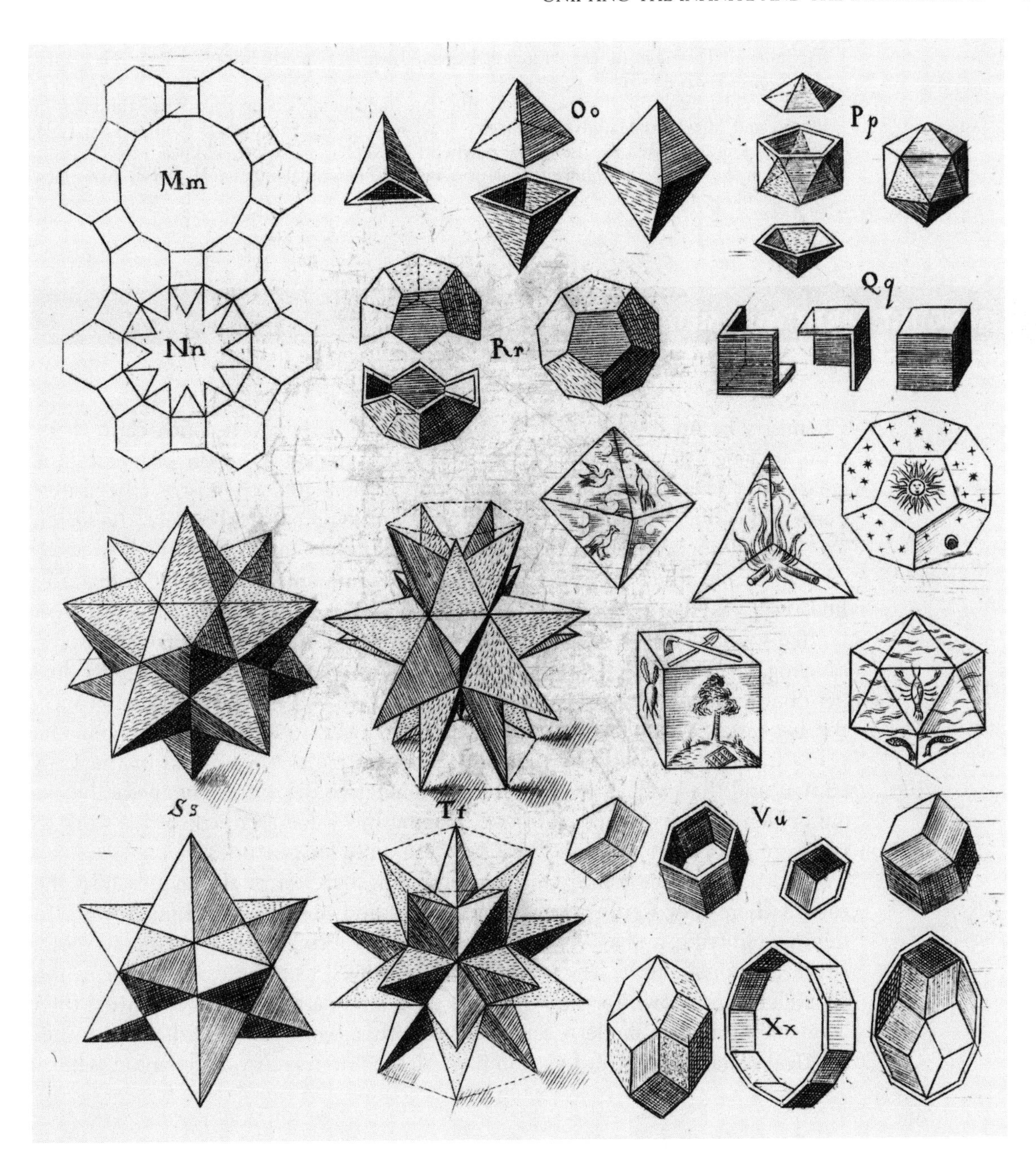

FIGURE 3 ■ Polyhedra in Johannes Kepler's *Harmonices Mundi* (Linz, 1619). Note the five Platonic solids on the middle right of this figure representing the four earthly elements Air, Fire, Water, and Earth as well as the fifth (heavenly) element Ether (courtesy of Division of Rare and Manuscript Collection, Carl A. Kroch Library, Cornell University).

cally.[2] In the words of Jacob Bronowski:[3] "All science is the search for unity in hidden likenesses." He states further: "To us, the analogies by which Kepler listened for the movement of the planets in the music of the spheres are far-fetched. Yet are they more so than the wild leap by which Rutherford and Bohr in our own century found a model for the atom in, of all places, the planetary system?"

1. B. Pullman, *The Atom In The History of Human Thought*, Oxford University Press, New York, 1998, pp. 25–27, 49–57.
2. Kepler's polyhedral model is beautifully illustrated and described on page 95 of the book by Istvan and Magdolna Hargittai, *Symmetry—A Unifying Concept*, Shelter, Bolinas, CA, 1994. This book also inspired my use of the polyhedra in Kepler's *Harmonices Mundi*.
3. J. Bronowski, *Science and Human Values*, revised ed., Perennial Library Harper & Row, New York, 1965, pp. 12–13.

SEEDING THE EARTH WITH METALS

Chemistry began to emerge as a science in the early seventeenth century. Its roots included practical chemistry (the mining and purification of metals, the creation of jewelry, pottery, and weaponry), medicinal chemistry (the use of herbs and various preparations made from them), and mystical beliefs (the search for the Philosopher's Stone or the Universal Elixir).

Figure 4 is the frontispiece from the final German edition (1736) of Lazarus Ercker's book *Aula Subterranea . . .* , which was first published in Prague in 1574. Unlike so many books of the sixteenth century, this important treatise on ores, assaying, and mineral chemistry was clearly and simply written by an individual personally experienced in the mining arts. For this reason (and for its beauty) the book was reprinted in numerous editions over a period of 160 years. The plates in this 1736 edition are made from the original blocks used in the 1574 edition and the gradual, but slight and cumulative deteriorations in the blocks are evident in the various editions.[1] Imagine the value ascribed to this work to motivate printers to preserve the blocks carefully for centuries.

This handsome plate depicts the seeding by God of the metals into the earth (where they were thought to multiply) and the laborious human work in mining, purifying, and assaying them. Although we recognize seven metals (gold, silver, mercury, copper, lead, tin, and iron) as well as arsenic and sulfur as the nine elements known to the Ancients, they were certainly not recognized then as elements in the modern sense. Instead they were considered to be rather mystical combinations of, for example, salt, sophic mercury, and sophic sulfur.

1. A.G. Sisco and C.S. Smith, *Lazarus Ercker's Treatise on Ores and Assaying* (translated from the German Edition of 1580), The University of Chicago Press, Chicago, 1951.

FIGURE 4 ■ Frontispiece from the final edition of *Aula Subterranea* by Lazarus Ercker (Frankfurt, 1736) depicting God seeding the earth with metals and their harvesting and refining by people. (The first edition of this book was published in 1574; the original blocks were employed to strike the plates in all subsequent editions.)

CHYMICALL CHARACTERS

This table of chemical symbols (see Fig. 5) is found in the book titled *The Royal Pharmacopoea, Galenical and Chymical, According to the Practice of the Most Eminent and Learned Physitians of France, and Publish'd with their Several Approbations*, the English edition published in 1678. The author, Moses Charas, fled religious persecution in France to join the enlightened intellectual environment in the England of Charles II, who chartered the Royal Society. Its membership included Robert Boyle, Robert Hooke, and Isaac Newton.

The elements listed in the table include the nine ancient elements described previously and a few others readily separable. Gold, of course, being "inert," is commonly found in an uncombined state and its high density (about 9 times denser than sand) allows it to be panned. Actually, we now also know that inert gases such as helium, neon, argon, krypton, and xenon are also found uncombined in nature, but they are colorless and odorless. In any case, we are suddenly over 200 years ahead of ourselves and apologize to the reader for getting carried away by our enthusiasm.

The association of elements with planets and their symbols, evident in Figure 5, appears to have been adopted from the ideas of Arab cultures during the Middle Ages. Association of gold with the sun is too obvious. The others are more subtle. For example, of the planets, mercury appeared to the Ancients to move most rapidly in the sky and was most suited as a messenger. Mercury's wings nicely represent the metal's volatility. In contrast, Saturn was the most distant of planets observed by the Ancients (Uranus, Neptune, and Pluto were discovered in the eighteenth, nineteenth, and twentieth centuries, respectively). The apparent slow movement of this planet through the skies was likened to Saturn, the god of seed or agriculture, who is sometimes depicted with a wooden leg. Lead was dense, slow . . . leaden. A person who is *saturnine* is sluggish or gloomy (not to be confused with a person who is *saturnalian*—riotously merry or orgiastic after the Roman holiday Saturnalia).

But let's return to a modern use of metaphor, based upon the toxic element lead, and visit the book *The Periodic Table*, by Primo Levi,[1] who used 21 elements as metaphors in 21 stories. For example:

> My father and all of us Rodmunds in the paternal line have always plied this trade, which consists in knowing a certain heavy rock, finding it in distant countries, heating it in a certain way that we know, and extracting black lead from it. Near my village there was a large bed; it is said that it had been discovered by one of my ancestors whom they called Rodmund Blue Teeth. It is a village of lead-smiths; everyone there knows how to smelt and work it, but only we Rodmunds know how to find the rock and make sure it is the real lead rock, and not one of the many heavy rocks that the gods have strewn over the mountain so as to deceive man. It is the gods who make the veins of metals grow under the ground, but they keep them secret, hidden; he who finds them is almost their equal, and so the gods do not love him and try to bewilder him. They do not love us Rodmunds: but we don't care.

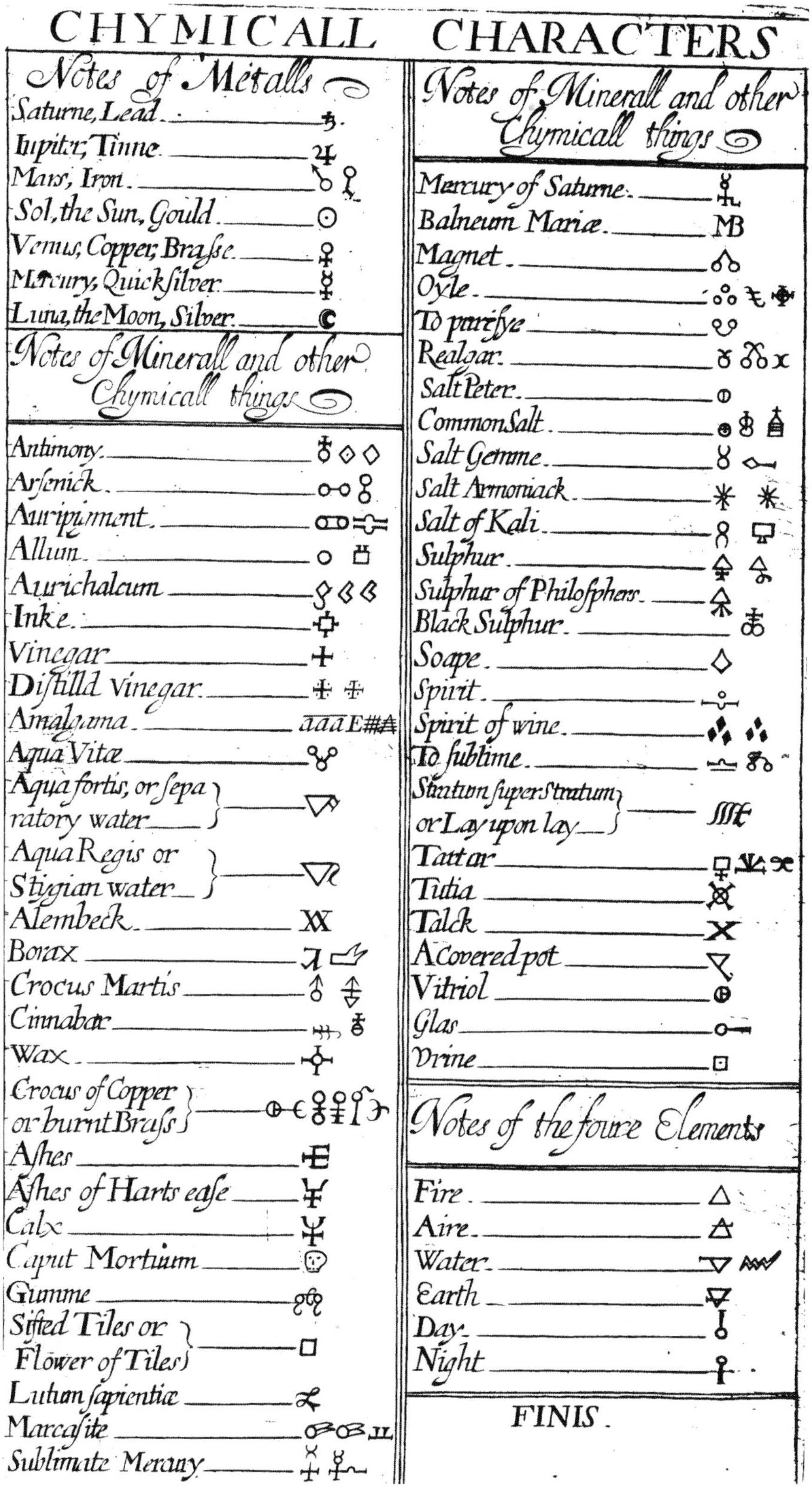

CHYMICALL CHARACTERS

Notes of Metalls

Saturne, Lead.
Iupiter, Tinne.
Mars, Iron.
Sol, the Sun, Gould.
Venus, Copper, Brasse.
Mercury, Quicksilver.
Luna, the Moon, Silver.

Notes of Minerall and other Chymicall things

Antimony.
Arsenick.
Auripigment.
Allum.
Aurichalcum.
Inke.
Vinegar.
Distilld Vinegar.
Amalgama.
Aqua Vitæ.
Aqua fortis, or separatory water.
Aqua Regis or Stigian water.
Alembeck.
Borax.
Crocus Martis.
Cinnabar.
Wax.
Crocus of Copper or burnt Brass.
Ashes.
Ashes of Harts ease.
Calx.
Caput Mortuum.
Gumme.
Sifted Tiles or Flower of Tiles.
Lutum sapientiæ.
Marcasite.
Sublimate Mercury.

Notes of Minerall and other Chymicall things

Mercury of Saturne.
Balneum Mariæ.
Magnet.
Oyle.
To purifye.
Realgar.
Salt Peter.
Common Salt.
Salt Gemme.
Salt Armoniack.
Salt of Kali.
Sulphur.
Sulphur of Philosphers.
Black Sulphur.
Soape.
Spirit.
Spirit of wine.
To sublime.
Stratum super Stratum or Lay upon lay.
Tartar.
Tutia.
Talck.
A Covered pot.
Vitriol.
Glas.
Urine.

Notes of the foure Elements

Fire.
Aire.
Water.
Earth.
Day.
Night.

FINIS.

FIGURE 5 ■ Chemical symbols from *The Royal Pharmacopoea* by Moses Charas (London, 1678).

> All the men have resumed their former trades, but not I: just as the lead, without us, does not see the light, so we cannot live without lead. Ours is an art that makes us rich, but it also makes us die young. Some say that this happens because the metal enters our blood and slowly impoverishes it; others think instead that it is a revenge of the gods, but in any case it matters little to us Rodmunds that our lives are short, because we are rich, respected and see the world.
>
> So, after six generations in one place, I began traveling again, in search of rock to smelt or to be smelted by other people, teaching them the art in exchange for gold. We Rodmunds are wizards, that's what we are: we change lead into gold.

With the naked eye, ancient people could discern that the planet Mars is red, just as is the calx of iron ("rust"). Associating Mars—the god of war—with iron—the stuff of weapons, as well as with blood—is intuitively reasonable. Late twentieth-century business executives wore red "power ties" to meetings. But in an almost too wonderful confirmation of ancient intuition, the findings of the NASA Viking Mission, which landed two spacecraft on Mars in 1976, indicated a red surface composed of oxides of iron: eyeball chemical analysis by the Ancients at over 30 million miles—not bad!

But let us take irony one or two steps further. As of this writing, it appears that Mars sent its own messenger to Antartica 13,000 years ago in the form of Meteorite ALH84001.[2] Comparison of the carbon isotope content in the carbonate globules of the meteorite with Viking data indicated its Martian origin. Among the fragments of chemical evidence, the finding of iron(II) sulfide coexisting with iron oxides suggests a biogenic origin since these two are essentially incompatible under abiotic conditions. The electrifying, if perhaps premature, conclusion of the scientists[2]:

> Although there are alternative explanations for each of these phenomena taken individually, when they are considered collectively, particularly in view of their spatial association, we conclude that they are evidence for primitive life on early Mars.

1. P. Levi, *The Periodic Table* (English translation of the Italian text), Schocken Books, New York, 1984 (see pp. 80–81 for the three quotations employed here).
2. D.S. McKay, E.K. Gibson, Jr., K.L. Thomas-Keprta, H. Vali, C.S. Romanek, S.J. Clemett, X.D., F. Chillier, C.R. Maechling, and R.N. Zare, *Science*, **273**(5277):924–930, 1996.

PRACTICAL CHEMISTRY: MINING, ASSAYING, AND REFINING[1]

Figure 6 depicts the inside view of an assay laboratory of the late sixteenth century. Figures 6 to 16, like Figure 4, are from the 1736 edition of Ercker's *Aula Subterranae* . . . and were printed using plates from the 1574 edition. Figure 7 depicts a machine washing alluvial gold ores. The great density of gold, 19.3

FIGURE 6 ■ A sixteenth-century assay laboratory (Ercker; see Fig. 4).

FIGURE 7 ■ A sixteenth-century machine washing alluvial gold ores (Ercker, see Fig. 4). Gold's great density (19.3 g/cm^3) permits its ready separation from other, lighter minerals.

FIGURE 8 ■ Making cupels from calcined, crushed bones ground into a paste with beer and molded. The oxides of baser metals such as iron are absorbed into the cupel while molten gold or silver remain on its surface (Ercker, see Fig. 4).

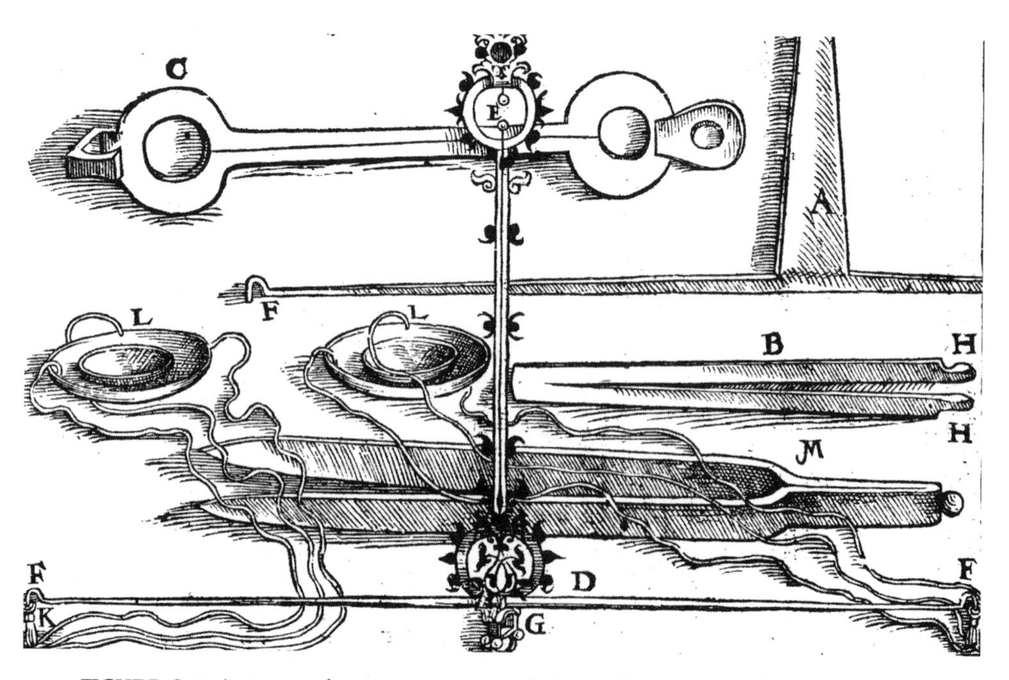

FIGURE 9 ■ A sixteenth-century assayer's balance (see text; Ercker, see Fig. 4).

FIGURE 10 ■ Use of mercury to dissolve gold in ore concentrates. The gold amalgam is then heated and mercury distills (Ercker, see Fig. 4).

g/cm^3 (the density of water, 1.0 g/cm^3; mercury "only" 13.6 g/cm^3), allows its ready separation from sand and other minerals. Figure 8 depicts the operations in making cupels. Cupellation was a technique for purifying gold or silver in ores. Cupels were cuplike objects made of ground bones in which ground ores were placed. The ores were principally sulfides and heating in air roasted the sulfides and formed oxides of the less noble (more reactive) metals while melting gold or silver. The oxides were absorbed into the cupel while a droplet of gold or silver remained on its surface.

To make cupels, calf or sheep bones are calcined (heated in open air), crushed, and ground to the texture of flour and the "ash" is moistened with strong beer. The ash is then placed in cupel molds (see A and C, Fig. 8) and coated with facing ashes, best obtained according to Ercker, from the foreheads of calves' skulls. The molded ash is then pounded and shaped (see H, man pounding cupels), removed from the molds (see B and D and the stack of cupels E), and allowed to dry. In Figure 8, G depicts a man washing ashes and F is a ball of washed ashes.

Figure 9 depicts an assayer's balance including: (A) forged balance beam, (B) shackle, (C) half of shackle, (D) filed assay beam with half of shackle, (E) two little beads—upper end of shackle and pointer, (F) ends, (G) how the beam is suspended, (H) sleeves of shackle, (K) knots by which strings are hung, (L) pans of the balance, and (M) assay head forceps.

Figure 10 depicts the amalgamation of gold concentrates and recovery of mercury by distillation of the amalgam. One of the earliest precepts of chemistry is *like dissolves like*, which explains why oil floats on water while alcohol freely

FIGURE 11 ■ Distillation of *aqua regia* (3:1 HCl/HNO_3) (Ercker, see Fig. 4). This "kingly water" is capable of dissolving gold. (See essay in Section VIII "The Chemistry of Gold is Noble But Not Simple.")

mixes with water. Mercury, being a liquid metal, dissolves other pure metals and forms alloys called amalgams. Relatively mild heating of the amalgam frees the volatile mercury from the metal of interest. However, mercury does not dissolve salts (calxes or oxides, sulfides) of metals. Thus, crushed ore was treated by Ercker with vinegar for 2 or 3 days and then washed and rubbed into mercury by hand and then with a wooden pestle by the amalgamator depicted in Figure 10(F). (*Note*: Elemental mercury is very toxic. It caused nerve damage in workers who made hats in England during the 19th century—this was "Mad Hatters' Disease"—the source of the madness of the tea party in Alice in Wonderland. There has been some concern late in the twentieth century that amalgams used to make tooth fillings give off a steady stream of mercury vapor.) The mercury itself was purified by squeezing through a leather bag [see (L) and (G) in Fig. 10]. Distillation of mercury from the amalgam employed a large furnace called an athanor (A), which supplied uniform and constant heat, side chambers (B), an earthenware receiver (C) and a still head (D), a blind head through which

FIGURE 12 ■ The use of "parting acid" (mostly HNO_3) to separate silver from gold. Silver is soluble and gold is not soluble in this acid (Ercker, see Fig. 4).

FIGURE 13 ■ A sixteenth-century self-stoking cementation furnace (see text; Ercker, see Fig. 4).

FIGURE 14 ■ Smelting of bismuth ore in open wind; freshly formed molten bismuth flows into the pans (Ercker, see Fig. 4).

FIGURE 15 ■ Steps in the leaching and concentration (by boiling) of saltpetre obtained from old sheep dung (Ercker, see Fig. 4).

water can be poured for cooling purposes (E), and an iron pot [lower part (H); upper part (K)] to contain the amalgam to be heated. Also depicted (M) is a man who remelts gold using bellows.

Aqua regia (three parts hydrochloric acid to one part nitric acid) had the valuable property of dissolving gold and allowing its ready recovery (see our later discussion of this subtle chemistry). Figure 11 shows the distillation of *aqua regia* involving the athanor (A) and a chamber (B) for the flask, situated as in (C). (D) is the glass distillation head and (E) the receiver.

Figure 12 depicts the use of parting acid to separate gold and silver. Parting acid (essentially nitric acid) "dissolves" silver but not gold and is obtained by melting pure saltpetre (potassium nitrate, KNO_3) with vitriol, $FeSO_4$, adding a small amount of water and distilling.

Figure 13 shows a self-stoking furnace for cementation—a process having some similarities to cupellation for purifying gold. The "cement" is made by taking four parts of brick dust, two parts of salt, and one part of white vitriol (zinc sulfate, $ZnSO_4$), grinding the mixed solid, and moistening the powder with urine or sharp wine vinegar. One-finger thickness of the cement is used to cover the bottom of the pot and upon this layer are placed thinly hammered strips of less pure gold, moistened with urine, for further purification. Then follows al-

FIGURE 16 ■ Pans and tubs from crystallizing concentrated leachate for saltpetre (see Fig. 15). One hundred pounds of the concentrate yields about 70 pounds of saltpetre (Ercker; see Fig. 4).

ternating layers of cement and gold strips finishing with a top layer, one-half-finger thick, of cement. The furnace is applied for 24 hours at a temperature lower than gold's melting point. At the conclusion, the powder is cleaned off and the resulting gold is said to be 23 carat. Pure gold is 24 carat.

Figure 14 depicts the smelting of bismuth in open air with the aid of a very stylized wind. Walnut-sized pieces of ore are placed in pans such that wind-blown fire will smelt the ore and cause liquid bismuth to flow in the pans.

Although saltpetre was used to make nitric acid (for research?) on a small scale, its largest demand was for its use in manufacturing gunpowder. Figure 15 depicts steps in the leaching and concentration, by boiling, of saltpetre. First, the best "earth" for obtaining saltpetre was said by Ercker to come from old sheep pens (which contain the remains of excrement and rotted building matter). Part (A) depicts the "earth" to be leached and (B) shows pipes containing water to run into the vats. The vats are continuously drained into gutters (C) that run the leachate into a sump (D). Part (E) depicts a little vat from which the leachate runs into a boiler, and (F) to (L) depict parts of the furnace. The boilers distill off considerable water to make a concentrated "liquor."

Figure 16 shows pans (F) and tubs (G) for crystallizing concentrated leachate. One hundred pounds of this concentrate yield about 70 pounds of crystalline saltpetre upon standing.

1. The translations and interpretations used here were obtained from A.G. Sisco and C.S. Smith, *Lazarus Ercker's Treatise on Ores and Assaying* (translated from the German Edition of 1580), The University of Chicago Press, Chicago, 1951.

SECTION II
SPIRITUAL AND ALLEGORICAL ALCHEMY

THE PHILOSOPHER'S STONE CAN NO LONGER BE PROTECTED BY PATENT

John Read's wonderful trilogy, *Prelude to Chemistry*,[1] *Humour and Humanism in Chemistry*,[2] and *The Alchemist in Life, Literature and Art*,[3] include many choice gems. For example, in *Prelude* we see publicly and plainly disclosed for the first time ever, and therefore no longer patentable, the recipe for The Philosopher's Stone[1] (also known as *Lapidus Philosophorum*, The Red Tincture, The Quintessence, The Panacea, The Elixir of Life, Virgins Milke, Spittle of Lune, Blood of the Salamander, The Metalline Menstruall, and hundreds of other straightforward names).[1] In *Humour* Read produces the box score for a cosmic cricket match between a timeless all-star team led by Hermes Trismegistos (223 runs) and another team captained by Noah (210 runs).[2] The game was umpired by Solomon and Ham and scored by the Bacon boys (Roger and Francis). For the winners, Aristotle contributed 4 runs (earthly elements) and Paracelsus 3 runs (the *tria prima* of sulfur, mercury, and salt)—it only gets worse!

In any case, and without further ado, here is the recipe for The Philosopher's Stone ("quicksilver" is the real element mercury):[1]

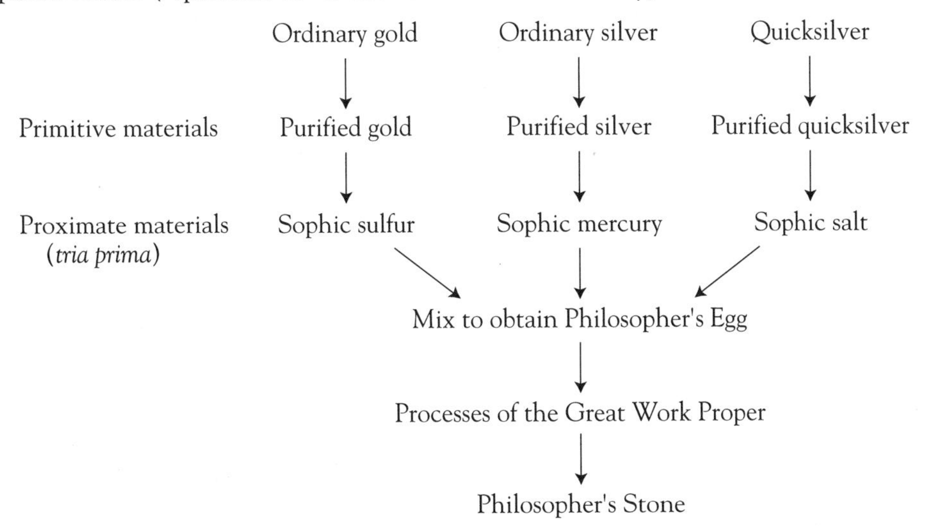

The Twelve Keys of Basil Valentine [Figs. 17 to 20, including Fig. 20(c) for the "Abstract"] depict the Processes of The Great Work in the days before patent attorneys. The images are almost as obscure as legalese and clearly meant to protect his venture capital. Some of them refer to specific processes (twelve is not uncommon—one for each sign of the Zodiac).[1] Each process is best done under the appropriate sign (e.g., distillation under Virgo; digestion under Leo —what else?; the actual use of The Stone, projection, under Pisces). That means, optimally, one year for manufacturing each Quintessence and lots of "aging" space—inefficient use of time and commercial square-footage that will certainly vex the company's accountants.

1. J. Read, *Prelude To Chemistry*, MacMillan, New York, 1937, pp. 127–142.
2. J. Read, *Humour and Humanism in Chemistry*, Bell, London, 1947, pp. 12–14.
3. J. Read, *The Alchemist in Life, Literature and Art*, Nelson, Edinburgh, 1947.

THE TWELVE KEYS OF BASIL VALENTINE: FIRST KEY, THE WOLF OF METALS AND THE IMPURE KING

The best evidence indicates that Basil Valentine (the Valiant King, a Benedictine cleric monk said to be born in 1394) never existed. Books attributed to him such as the ever-popular *Triumphal Chariot of Antimony*, first published in 1604, are generally attributed to the publisher Johann Thölde who, in turn, had perhaps "improved upon" earlier manuscripts that had come into his hands.[1] Nevertheless, they contain interesting images and even some useful information.

The Twelve Keys of Basil Valentine provide images pertaining to the Great Work and have been analyzed by many authors including John Read.[2] In Figure 17(a), we see the First Key. The wolf in this picture is generally considered to represent antimony (Sb or *stibium*), although, until recent centuries, that term really meant its ore stibnite (Sb_2S_3). Antimony had been referred to in the alchemical literature as *lupus metallorum* ("wolf of metals").[2] Actually, it is a *metalloid.* One of its forms (allotropes) is metallic—a brittle gray substance with relatively poor thermal and electrical conductivity, rather unlike typical metals. In his delightful book, Venetsky quips: "As if in revenge for the unwillingness of other metals to accept it in their family, molten antimony dissolves almost all of them."[3] In modern terms, we recognize a metal capable of dissolving other metals ("like dissolves like") but also something akin to a nonmetal, capable of oxidizing other metals.

We see the wolf near a figure of Saturn. [Remember the old man with the wooden leg in the essay on chemical symbols (Fig. 5)? Antimony had also been called the "lead of the philosophers"[2]; the ancients described antimony as the progeny of lead through heating.] Now, if impure gold is heated in the fire (three times—the queen holds three flowers—pretty obvious, eh?), the king will emerge. The king is gold (perhaps seed of gold or sophic sulfur[2]). The queen represents purified silver from which is derived sophic mercury. The first purification gives the "primitive materials" of the Stone—derived from gold, silver, and mercury.[2] This picture actually represents the following chemistry occurring in the fire:

$$Au + Ag + Cu + Sb_2S_3 \rightarrow Au/Sb + Ag_2S + CuS$$

FIGURE 17 ■ Depictions of the Twelve Keys of Basil Valentine (from Basil Valentine, *Letztes Testament . . .*, Strasburg, 1667). See text for interpretations. (a) First Key; (b) Second Key; (c) Third Key; (d) Fourth Key.

(a) (b)

(c) (d)

$(Au + Ag + Cu)$ represents impure gold here; Au/Sb is gold alloy.

$$Au/Sb + O_2 \rightarrow Au + Sb_2O_3 \text{ (vapor)}$$

Similar chemistry occurs in the purification of antimony from stibnite through heating with metallic iron. It is also noteworthy that when metallic ("red") copper and antimony are alloyed (e.g., 6% Sb), the resulting metal looks very much like gold. The following is from a Syriac manuscript dating from the Crusades:[4]

> Throw in with red copper some antimony roasted in olive oil and it will become gold-like.

1. J.R. Partington, *A History of Chemistry*, MacMillan, London, 1961, pp. 183–203.
2. J. Read, *Prelude to Chemistry*, MacMillan, New York, 1937, pp. 196–211.
3. S.I. Venetsky, *On Rare and Scattered Metals—Tales About Metals*, (translated from the Russian by N.G. Kittell), Mir, Moscow, 1983, p. 83.
4. M.L. Dufresnoy and J. Dufrenoy, *Journal of Chemical Education*, **27**:595–597, 1950.

RATSO RIZZO AND THE POET VIRGIL AS TRANSMUTING AGENTS?

Most of the remaining eleven keys of Basil Valentine have been analyzed by Read[1] and I will rely primarily upon his insights. The *Second Key* [Fig. 17(b)] is said to represent the operation *separation*, a purifying of watery matter from its dregs. The matter appears to be volatile quicksilver which has undergone a kind of molting, under the influence of Sol (Sophic Sulfur) and Luna (Sophic Mercury), en route to Sophic Salt. The rooster (cock) held on the left is a male symbol implying the need for conjunction. The *Third Key* [Fig. 17(c)] includes a winged dragon, fox, pelican, and cock. Dragons appear to have been used for many symbols and are sometimes used interchangeably with snakes. Winged dragons sometimes represent Sophic Mercury, sometimes Proximate Material.[2] Read did not discuss this key. My reading of the 1671 English edition indicates that this key pertains to purification of gold to Sophic Sulfur. The *Fourth Key* [Fig. 17(d)] clearly refers to *putrefaction*, the necessary blackening that starts the operation. The symbols of the crow and skeleton are clear here. It was widely known from the work of the third-century alchemist Mary Prophetissa (or Maria the Jewess, for whom the hot-water bath or *Bain Marie* for controlled heating was named) that heating of a lead–copper alloy with sulfur produced a black mass.[3] So too did heating together of the four base metals lead, tin, copper, and iron.[3]

Putrefaction, as the first step toward transmutation to gold, is loaded with religious symbolism. The idea here is total abasement before salvation can begin. Impurities and imperfections must similarly be removed from metals in order for them to transmute to gold. Humans must remove their imperfections to achieve a state of grace. In *The Divine Comedy*, Dante Alighieri must first be guided by the classical Roman poet Virgil through Hell before he can enter Purgatory and thence to Paradise.[4] In the 1969 movie *Midnight Cowboy*, Joe Buck (Jon Voight) leaves the earthly sphere (rural Texas) and must first experience Hell (Times Square in the late 1960s New York City), in the company of Ratso Rizzo (Dustin Hoffman). He discovers his inner gentility and true nature and achieves salvation through a similar journey, by Greyhound bus, from colder to gentler climes (Florida as Paradise?). Virgil, who died before the birth of Jesus, was of course a nonbeliever in Christianity and could never enter Paradise. Ratso Rizzo dies of consumption on the bus and also never enters Paradise.

The *Fifth Key* [Fig. 18(a)] is said by Read to represent the operation *solution*. The *Sixth Key* [Fig. 18(b)] represents *conjunction*—the marriage of the King (Sophic Sulfur) and the Queen (Sophic Mercury), the conjunction of the Sun (Sol) and Moon (Luna), the fiery two-headed man and rain, condensation, and fertility. This operation is an excuse for lots of naughty pictures in alchemical manuscripts and texts—the golden seed follows coitus. The *Seventh Key* [Fig. 18(c)] implies the four earthly elements, the Heavens (chaos is quite a complex concept), and the three Paracelsan principles. The double circle can symbolize the interaction between earthly and heavenly spheres. The stem at the top of the sphere appears again in the summary figure [Fig. 20(c)] and Read interprets this figure [and implicitly the vessel in Fig. 18(c)] as the Philosopher's Egg in which the Proximate Materials are joined: The stem is a kind of placenta. The Philosopher's Egg is hermetically sealed and may be placed for long periods in a special furnace termed an athanor—a kind of uterus. The *Eighth Key* [Fig. 19(a)], another graveyard scene, is said by Read to represent *fermentation*.

The *Ninth Key* [Fig. 19(b)] is a marvelous representation referring, in part, to the color changes that occur during the Great Work.[1,5] The falling figure is Saturn (base metals, notably lead); the rising figure is perhaps Luna (Sophic Mercury). The outer four-sided figure represents the four elements and the three snakes represent the *tria prima* (sulfur, mercury, salt). The four birds represent color changes. At the top, the crow represents blackness, then counterclockwise, the swan represents white, the peacock is multicolored, sometimes simply citrine, and the phoenix represents The Red Tincture (The Stone). It is interesting that the Sioux Nation recognizes four "true colors"—black, red, yellow, and white —the same as in the Ninth Key. Red is also the most important, representing earth, pipestone, and blood. The colors correspond to the four compass directions: west, north, east, and south.[6]

The *Tenth Key* [Fig. 19(c)] represents the *tria prima*. The three symbols near the corners of the triangle (clockwise from top left) are: gold, silver and mercury. These are the three elements that are purified to make Sophic Sulfur, Sophic Mercury and Sophic Salt, respectively. The double borders of the circle (heavenly perfection) and the triangle represent the duality of the earthly and heavenly spheres. The German phrases translate thusly[7]: "From Hermogenes I was born" (top); "Hyperion has chosen me" (right); "Without Jamsuph I am lost" (left). In Gnostic mythology, Hermogenes developed the doctrine of the eternity of matter.[8] In Greek mythology, Hyperion was a Titan recognized as the Father of the Sun (Helios), the moon (Selene) and the dawn (Eos).[9] Jamsuph, from the Kabbalists, refers to the Red Sea—a sign of God's power—the parting of the Red Sea may refer to the splitting of matter.[10] Translation of the Hebrew has been more elusive and is possibly Kabbalistic in nature.[11] The *Eleventh Key*

(a)

(b)

(c)

FIGURE 18 ■ (a) Fifth Key of Basil Valentine; (b) Sixth Key; (c) Seventh Key (see Fig. 17).

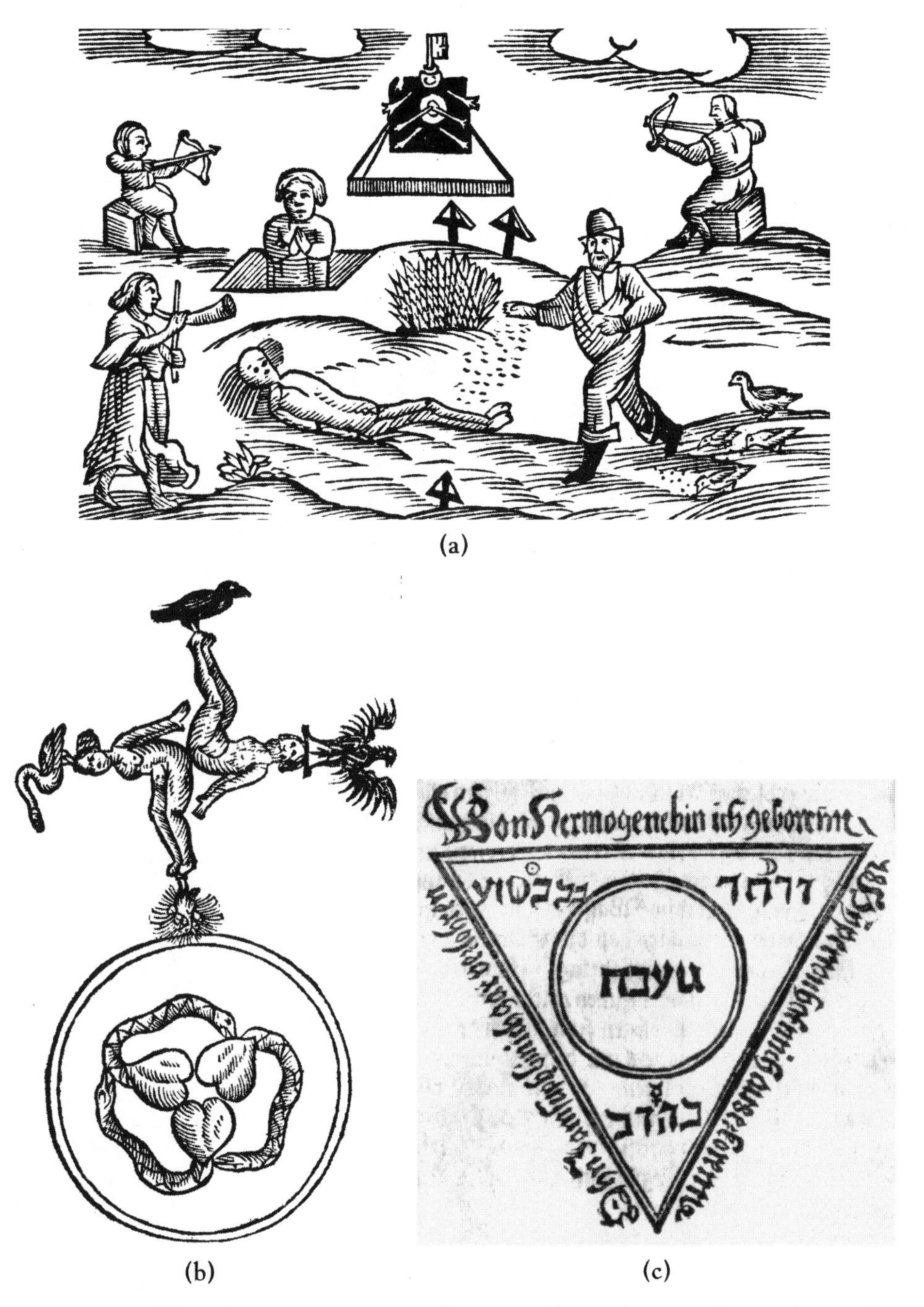

FIGURE 19 ■ (a) Eighth Key of Basil Valentine; (b) Ninth Key; (c) Tenth Key (see Fig. 17).

[Fig. 20(a)] shows lion whelps and depicts the *multiplication* achievable by the Stone. The two vessels represent the Philosopher's Egg (Vase of Hermes) where *conjunction* takes place [also see the double pelican and its symbolism in Fig. 27(c)]. Read has described the *Twelfth Key* [Fig. 20(b)] as representing *calcination* (whitening, drying) with the lion and snake as fixed and volatile principles and

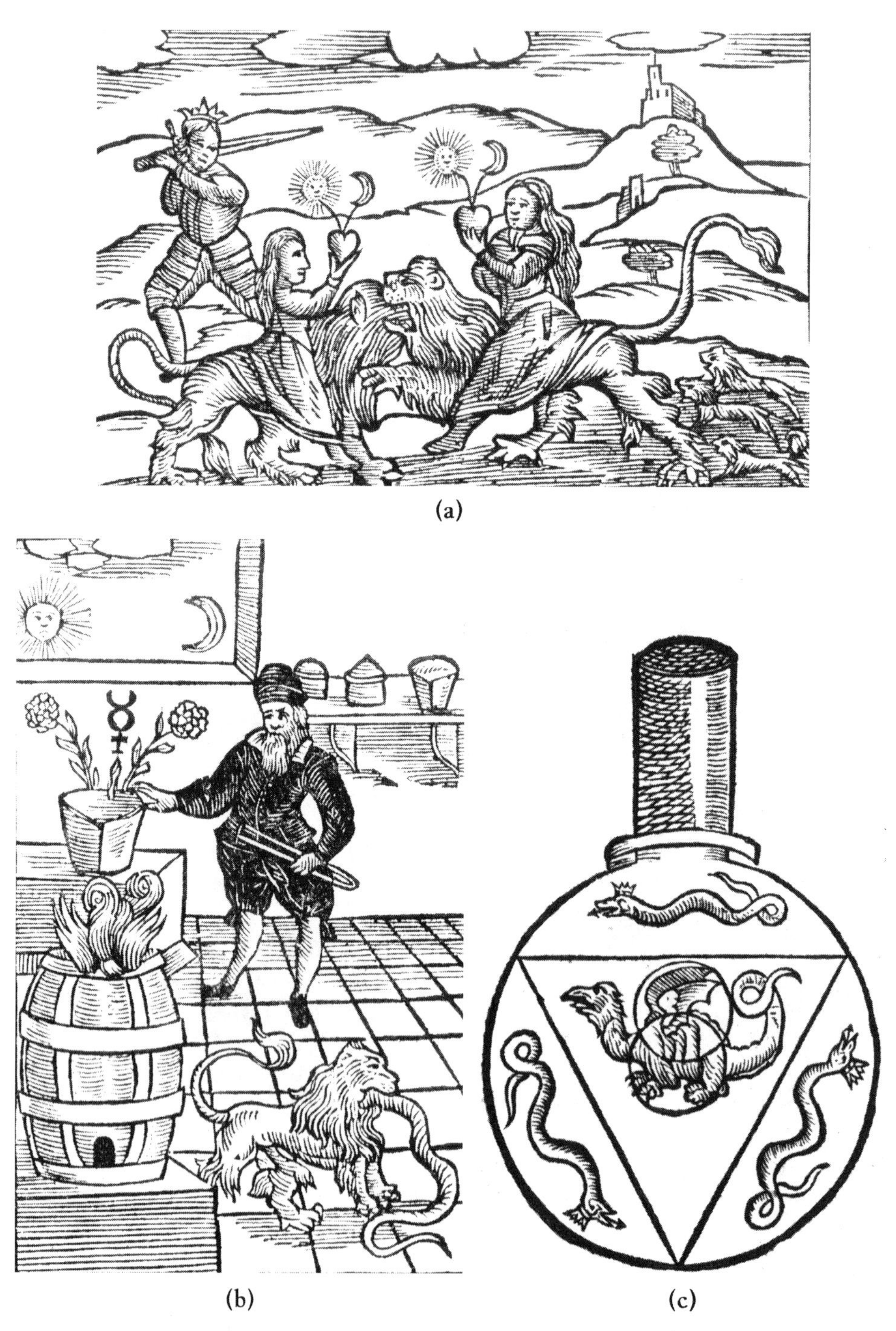

FIGURE 20 ■ (a) Eleventh Key of Basil Valentine; (b) Twelfth Key; (c) Summary image for the Twelve Keys of Basil Valentine (see Fig. 17).

the flowers as purified noble metals. The dragon here is said to represent the Proximate Material of the Stone; the circles around its wings and paws are the volatile and fixed principles. Figure 20(c) is the summary (literature Abstract?) of the work.

1. J. Read, *Prelude To Chemistry*, MacMillan, New York, 1937, pp. 196–211, 260–267.
2. J. Read, op. cit., pp. 106–108, 208, 269–272.
3. J. Read, op. cit., pp. 13–17.
4. I am grateful to Professor Susan Gardner for this discussion.
5. J. Read, op. cit., pp. 145–148.
6. J. Lame Deer and R. Erdoes, *Lame Deer Seeker of Visions*, Simon & Schuster, New York, 1972, pp. 116–117.
7. I am grateful to Professor Ralf Thiede for this translation.
8. W. Doniger, Mythologies Compiled by Yves Bonnefoy, University of Chicago Press, Chicago, 1991, Vol. 2, p. 677.
9. W. Doniger, op. cit., Vol. 1, pp. 371, 375.
10. I am grateful to Professor Laura Duhan Kaplan for her help in interpretation.
11. R. Patai, *The Jewish Alchemists*, Princeton University Press, Princeton, 1994.

CATAWBA INDIAN POTTERY: FOUR COLORS AND A MIRACLE OF SURVIVAL

The Ninth Key of Basil Valentine [Fig. 19(B)] describes four colors of transmutation in the Great Work: black, white, citrine (a yellow) and ultimately red, symbolized by the crow, swan, peacock, and phoenix, respectively. It is interesting that these are the four characteristic colors of earthenware fabricated for thousands of years by aboriginal peoples in diverse lands.

The Catawba Indians located in South Carolina spoke a Siouan language.[1] They were a once powerful nation that alternately coexisted and fought with the Cherokees in the Carolinas. However, as of June 1908, only nineteen houses and ninety-eight Catawbas were counted on the reservation and in its surroundings in York County.[2] Although pre-Columbian Catawba pottery was largely utilitarian (cooking pots, water jugs), starting in the eighteenth century it became a source of hard currency for the Indians. They began to fashion *objets d'art* in addition to traditional pieces. These were often taken to the port city of Charleston, South Carolina, traded, and sold. The very survival of Catawba culture came to depend to a significant extent on the sale and trade of pottery, largely fabricated by women. This is elegantly stated by former Catawba Tribal Historian Tom Blumer:[3]

> [T]he Catawba pottery tradition has survived for over 4,500 years. That it has done so is a tribute to the tenacity of the people who make up the Catawba Nation and the power of pottery, as an art form, to define that Nation and help it endure. It is a miracle of survival that will take the Catawba to the Third Millenium and beyond.

Figure 21 (left) shows a mostly reddish-brown headed bowl with three running legs made by Master Potter Sara Ayers (b. 1919).[3] The legs are off

FIGURE 21 ■ Two pieces of Catawba Indian pottery: left, two-headed fluted bowl with three "running" legs by master Potter Sara Ayers; right, two-headed, fluted bowl by young master Monty ("Hawk") Branham. Catawba pottery is still made essentially as it was 4500 years ago. [Photograph by Thomas W. ("Wade") Bruton.]

center, and the broken symmetry provides a wonderful dynamic to the piece. Also shown in Figure 21 (right) is a two-headed, fluted bowl made by young master Monty ("Hawk") Branham (b. 1961). The heads were ultimately derived from a mold made over 100 years ago by the great Martha Jane Harris. These pieces are made almost the same way they were in prehistoric times. Clay is dug from holy and secret sites along the Catawba River which contain rich deposits of kaolinite, sifted, mixed, and dried in the sun and rolled and pounded to remove air pockets. Clean kaolinite is fluffy and white. Pipe clay has organic matter and is heavier. The two are usually mixed to make a pot. Larger pots are built using layers of coils of clay that are shaped and smoothed, then allowed to dry. A pot may be incised with symbols just before being totally dry and, when dry, it is laboriously burnished with smooth river stones that have usually been passed between generations of women. The pots are then wood-fired in pits in the ground, removed, and allowed to slowly cool. Open-pit firing is considered to be low temperature (1200°C or 2200°F) or soft firing as opposed to hard firing (1450°C or 2650°F).[4] Air pockets in the clay and even slight wind gusts often cause a high degree of breakage. The high shine in the finished product is due to hours of burnishing rather than to glaze, which is never used. Clay pots, which are unglazed, are not considered suitable for holding water since they "sweat" and will stain furniture. However, one can imagine taking a water jug into the field—its sweating and vaporization from the surface will cool the bulk of the water inside. Furthermore, the frequent heating and decomposition of fat as well as protein from sinew and meat will coat the inside of a cooking pot and seal it.

The colors in this pottery are largely due to the iron so abundant in all clays.[4] Iron is the fourth most abundant element in the earth's crust. It is largely found in the iron(II) (ferrous) or iron(III) (ferric) oxidation states. Iron(II)oxide (FeO), iron(III)oxide (Fe_2O_3, hematite), and ferroferric oxide (Fe_3O_4), which contains both Fe(II) and Fe(III), are the three oxides of iron commonly encountered. The mottled coloring of the pot depends upon the degree of oxidation and also reflects the smoke and soot of the wood employed in firing since different woods burn at different temperatures and oxygen levels.[5] One of my former professors at Princeton University, Tom Spiro, called the color changes associated with "tweaking" the environments of transition metals, such as iron, "tickling electrons." Under oxygen-rich conditions, the dominant colors are "white" (really buff), and yellow and red and are due to a greater abundance of Fe(III). Oxygen-poor conditions can be achieved by "smother-burning" pots by surrounding and covering them with wood. The presence of carbon monoxide (CO) causes more reducing conditions conducive to enrichment in Fe(II). This is the way to deliberately produce a shiny, black pot; otherwise, coloring is left largely to the fates. Traces of manganese also help to blacken pots as will soot.[5] When removed from the fire, the pieces are usually dark and then lighten as they cool. Dynamic chemistry is occurring, for example, disproportionation of FeO to Fe_3O_4 and Fe although Fe will further oxidize.[6] Sometimes a greasy-looking area can be seen on the surfaces of the pots. This is probably due to local vitrification perhaps by a local concentration of feldspar or mica.[4]

1. J.H. Merrell, *The Indians' New World*, The University of North Carolina Press, Chapel Hill, 1989.
2. M.R. Harrington, *American Anthropologist*, 10:399–407, 1989.
3. T. Blumer in Pamphlet *Catawba Pottery: Legacy of Survival, 7 Master Potters*, South Carolina Arts Commission and Catawba Cultural Preservation Project, Columbia, 1995.
4. *Encyclopedia Brittanica*, 15th ed., Chicago, 1986, Vol. 17, pp. 101–103.
5. I am grateful for discussions with Professor Victor A. Greenhut.
6. F.A. Cotton and G. Wilkinson, *Advanced Inorganic Chemistry*, 5th ed., Wiley, New York, 1988, pp. 711–713.

DRAGONS, SERPENTS, AND ORDER OUT OF CHAOS

The sometimes wildly allegorical depictions of alchemical relationships are well illustrated by Figures 22 and 23, which come from *Della Tramutatione Metallica* (Brescia, 1599). It is a virtual reprint of the 1572 edition but with addition of the *Concordontia de filosifi*: a listing of alchemical works largely attributed to Arnold of Villanova.[1,2] The first edition (1564) contains a list of alchemists and alchemical works, which was expanded in the 1572 and 1599 editions.[2] The author, Giovanni Battista Nazari, is reported to have read widely in alchemy

FIGURE 22 ■ A depiction of the *tria prima* (Sophic Mercury, Sulfur, and Salt) from *Della Tramutatione Metallica* by Giovanni Battista Nazari (Brescia, 1599). A very similar figure, depicting Austrian physician Franz Anton Mesmer, appeared in an anti-Mesmer pamphlet published in 1784.[3]

over a 40-year period but is blamed "... for describing spurious operations, which possibly helped ruin the people who tried them ...."[2] The book includes several dream sequences including one in which the author converses with Bernhardus Trevisanus (born 1406 in Padua), who, starting at age 14, devoted the remainder of his life to the study of alchemy.[4] The psychologist C.G. Jung had a lifelong interest in dreams and alchemy and owned a copy of the 1599 edition.[1]

Figure 22 clearly represents the *tria prima.* Perhaps the old dragon is a representation of the ultimate source of these sophic elements—the *prima materia* or fundamental matter. Figure 23 is a depiction of the generation, starting from chaos, of the six lower metals (the six crowns) and ultimately gold (the King).

Figure 24 is a drawing, executed in 1999, by artist Rita L. Shumaker.[5] It depicts the male–female (gold–silver; sun–moon) relationship. The two entwined dragons also represent male and female (fixed and volatile) principles and, with the rod or central stem held by the male figure, form a caduceus—the familiar medical symbol. The central stem is said to consist of "the gold of the philosophers." The original form of the caduceus is said to have been a cross representing the four ancient elements.[6] The square in the background of Figure 24 represents these four elements. The drawing represents the *conjunctio,*

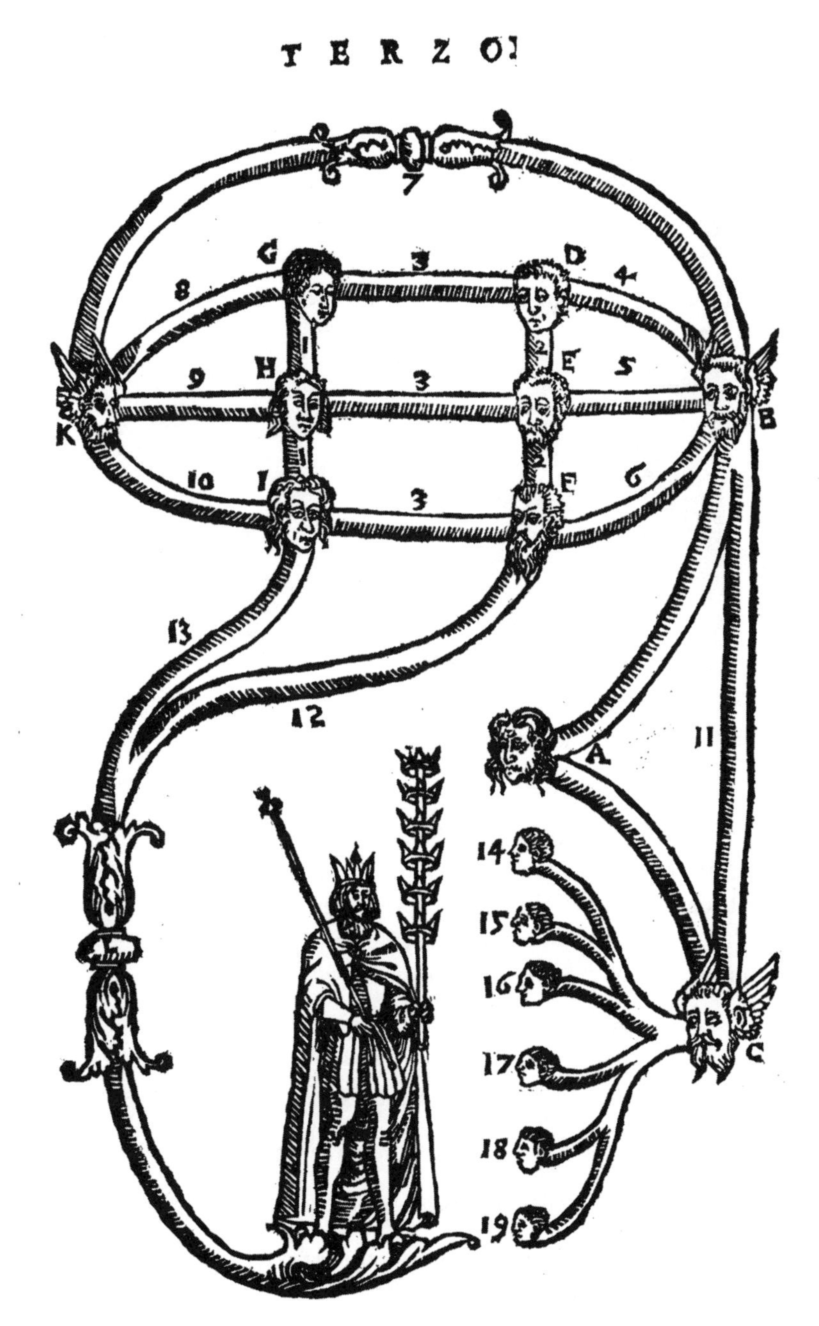

FIGURE 23 ■ A depiction of the birth and evolution of the six lower metals (six crowns) and Gold (the King) starting from Chaos (Nazari, see Fig. 22).

FIGURE 24 ■ Artist Rita L. Shumaker's rendition in 1999 of male and female allegorical images. The imagery of the caduceus is also evident in this drawing.

the alchemical wedding of male and female, spirit and body. We encourage you, gentle reader, to find the "chymicall characters" (see Fig. 5) in this figure. There are actually three dragons in this drawing representing the *tria prima* (salt, sulfur, mercury) as "metaphors for unconscious intuition and feeling, vital spirit or will, and the impulse to give creative form in matter."[5]

1. I. MacPhail, *Alchemy and the Occult*, Yale University Library, New Haven, 1968, pp. 178–181.
2. J. Ferguson, *Bibliotheca Chemica*, Derek Verschoyle, London, 1954 (reprint of 1906 ed.), Vol. II, pp. 131–132.
3. F. A. Pattie, *Mesmer and Animal Magnetism: A Chapter in the History of Medicine*, Edmonston, Hamilton, 1994, pp. 178–179.
4. J. Ferguson, *Bibliotheca Chemica*, Derek Verschoyle, London, 1954 (reprint of 1906 ed.), Vol. I, pp. 100–104.
5. The author thanks Ms. Rita L. Shumaker, a faculty member at the University of North Carolina at Charlotte, for this original drawing and its interpretation.
6. J. Read, *Prelude To Chemistry*, MacMillan, New York, 1937, pp. 105–116.

TODAY'S SPECIALS: OIL OF SCORPION AND LADY'S SPOT FADE-IN CREAM

Figure 25 is the frontispiece from the 1608 book *De Distillatione* depicting the author Giambattista Della Porta (1545–1615),[1] a polymath who authored books on plants, physiognomy, physics, chemistry, and mathematics, wrote "some of the best Italian comedies of his age," and published a design for a steam engine.[1,2] "This book is as rare as it is beautiful."[3] The dedications in the preface are set in Hebrew, Persian, Chaldaic, Illyrian, and Armenian typescripts attributed to the Vatican type foundry.[4]

Porta's book *Magia Naturalis*, first published in 1558, a compendium of popular science, was reprinted for over 100 years. A mixture of technical information and misinformation, it cites the procedure of the Greek physician and pharmacist Pedanius Dioscorides[1] (ca. 40–ca. 90 A.D.) for heating "antimony" [really stibnite—see Saturn and the wolf in Fig. 17(a)] into lead despite the fact that sixteenth-century practitioners knew they were different and could not be so interconverted.[5] *Magia Naturalis* includes a preparation of a cosmetic that will

FIGURE 25 ■ Frontispiece depicting the polymath Giambattista Della Porta in his beautiful book *De Distillatione Lib. IX* (Rome, 1608).

produce spots (a kind of fade-*in* cream for women)—a bit of Renaissance fraternity house humor perhaps.

De Distillationibus also exemplifies the playful wit of the Renaissance, likening chemical glassware to animals. Figure 26(a) depicts a *matrass*[6,7]: it has a round bottom and long neck like an ostrich (phials for rectifying alcohol had a similar appearance) and is part of a distillation apparatus called an *alembic*, which has a distilling head that could be attached to a receiver (see Figs. 72 and 73). The liquid to be distilled must be fairly volatile to make it to the top of the long neck. Figure 26(b) is a flat, stylized retort called a tortoise along with a rather stylized tortoise with a doglike head.

Could the hexagons with circles inside them on the tortoise's shell be a leap of about 330 years into the future to our modern structure for benzene? We suspect not since benzene would not be discovered for another 200 years. However, when we discover that Kekulé claimed in the 1860s to have dreamed of benzene's structure formed from three snakes biting tails in a circle, perhaps a subliminal message from another reptile 260 years earlier might not seem quite so strange.

The distillation apparatus in Figure 27(a) places the alembic head on top of a wide-mouth flask (a kind of *cucurbit*, a more squat version of a matrass). This apparatus would be more useful for a less volatile liquid. Figure 27(b) is a one-piece *pelican*. Note how the bird's neck forms a curved arm as it bites its chest. When closed at the top, the pelican was used for prolonged heating at the boiling point of the recirculating (refluxing) solvent. Figure 27(c) shows a *double pelican* in which the two wedded vessels exchange vapors and fluids for a prolonged period. We hesitate to provide further interpretation of the metaphor except to remind the reader that the book *was* printed in Rome seemingly with some degree of church assent.[4] Figure 28(a) shows a common retort. Figure 28(b) depicts a still capable of fractional distillation. The upper receivers are enriched in the more-volatile substances and the lower vessels are enriched in the less-

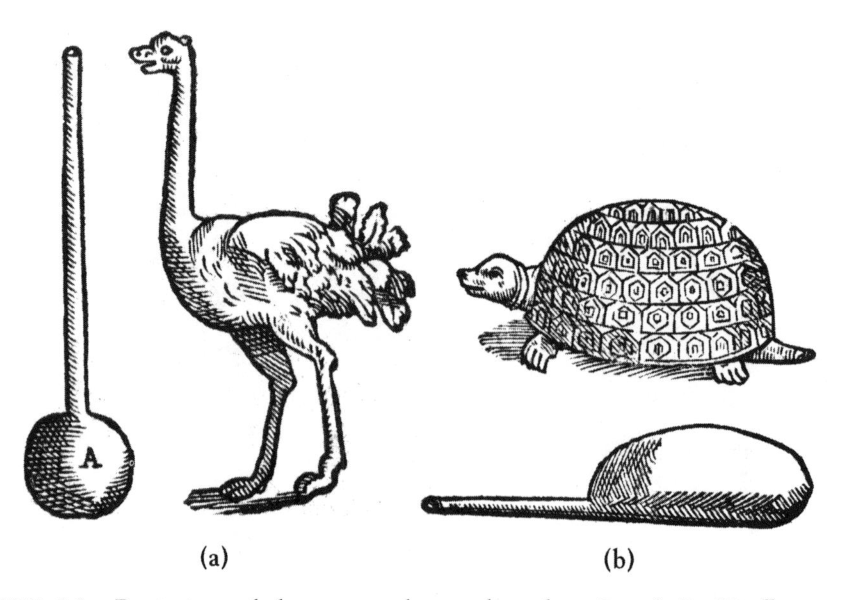

FIGURE 26 ■ Depictions of glassware and metaphors from Porta's *De Distillatione* (Fig. 25): (a) Matrass and ostrich; (b) Flat retortlike "tortoise" and tortoise (somewhat "dog-headed" methinks; is that benzene on the shell?).

volatile substances. Since fractional distillation is one of the first experiments in an introductory organic chemistry course, perhaps the seven-headed monster is a college sophomore's preconception of his or her laboratory instructor. Then again, perhaps not.

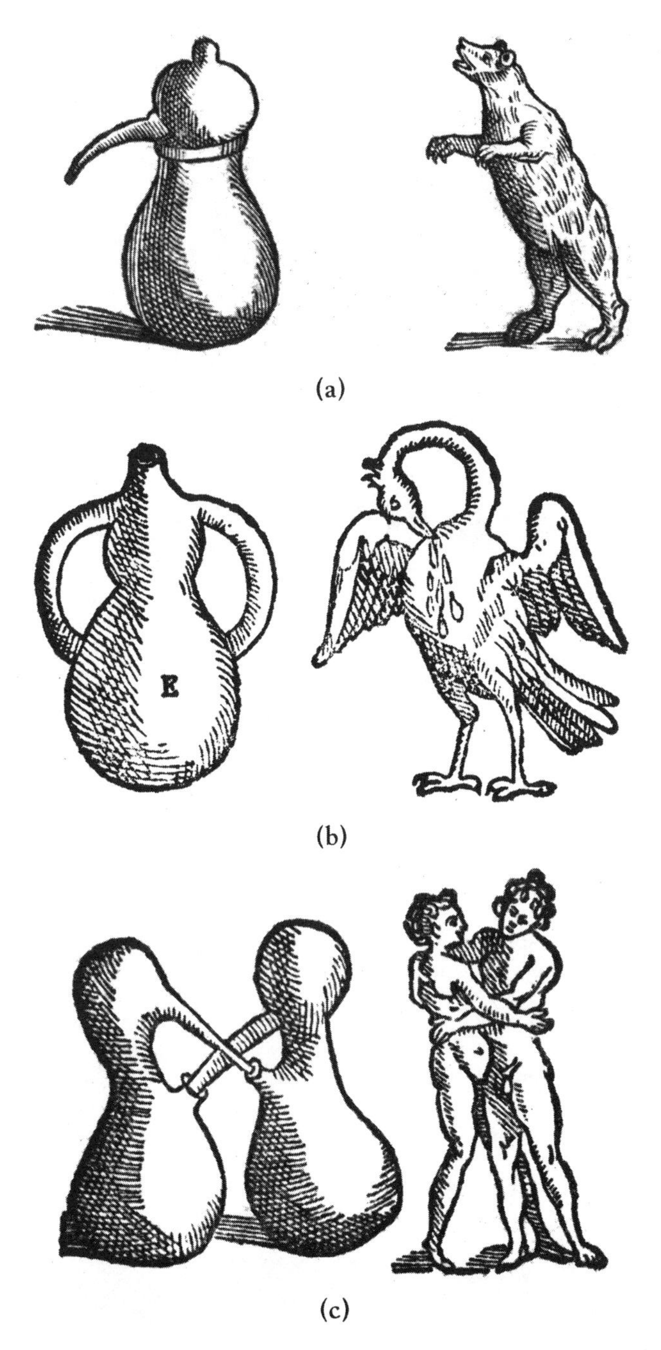

FIGURE 27 ■ Depictions of glassware and metaphors from Porta's *De Distillatione* (Fig. 25): (a) Distillation apparatus employing a distillation head (alembic) atop a wide-mouth flask (or cucurbit) along with matching bear; (b) one-piece pelican for refluxing a liquid and the pelican itself biting its chest—considered a blood of Christ symbol; (c) A double pelican for prolonged exchange of hot fluids and an interesting metaphor.

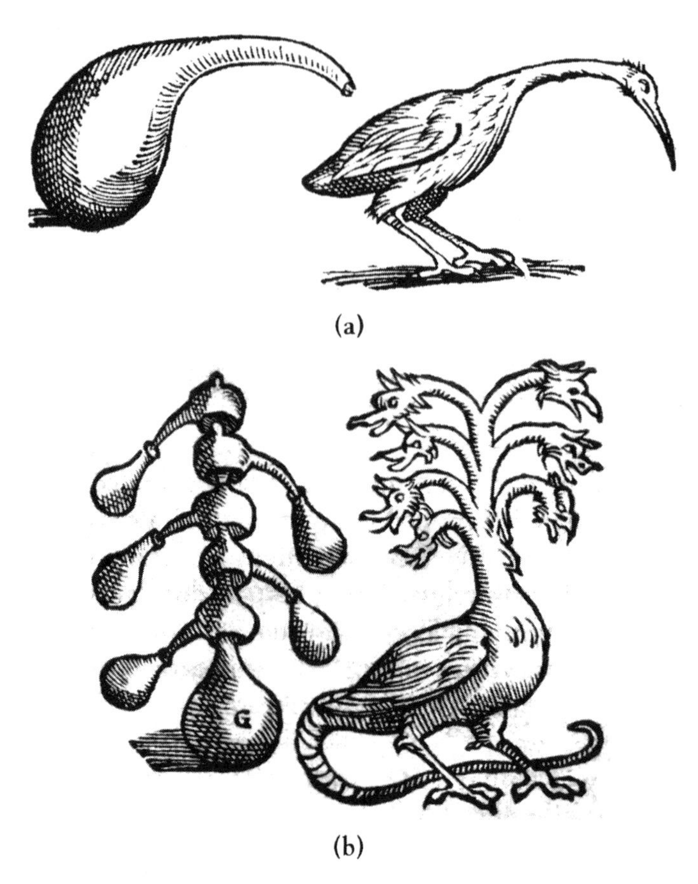

FIGURE 28 ■ Depictions of glassware and metaphors from Porta's *De Distillatione* (Fig. 25): (a) Common retort and appropriate bird; (b) fractional distillation apparatus and depiction of a seven-headed beast (or perhaps the Organic Chemistry Laboratory Instructor).

1. Also called Giambattista della Porta as well as Giovanni Battista Della Porta (see *Encyclopedia Brittanica*, 15th ed., 1986, Chicago, Vol. 9, p. 624, which lists his birthdate as "1535?").
2. J. Ferguson, *Bibliotheca Chemica*, Derek Verschboyle, London, 1954, Vol. II, p. 216.
3. D.I. Duveen, *Bibliotheca Alchemica et Chemica*, HES, Utrecht, 1986, p. 481.
4. I. MacPhail, *Alchemy and The Occult*, Yale University Library, New Haven, CT, 1968, Vol. 1, pp. 212–215.
5. J.M. Stillman, *The Story of Alchemy and Early Chemistry*, Dover, New York, 1960, pp. 349–352.
6. J. Eklund, *The Incompleat Chemist—Being An Essay on the Eighteenth-Century Chemist in the Laboratory With a Dictionary Of Obsolete Chemical Terms of the Period*, Smithsonian Institution Press, Washington, D.C., 1975.
7. F. Ferchl and A. Sussenguth, *A Pictorial History of Chemistry*, William Heinemann, London, 1939, pp. 73–75, 105–108.

"VULGAR AND COMMON ERRORS"

Why would a knowledgeable scholar like Porta reinforce incorrect information such as that heating of antimony produces lead? Scientific experimentation was still only in its infancy. Early writers such as Pliny often turned folklore into fact. In his book *Pseudodoxia Epidemica: Or, Enquiries Into Very Many Received Tenents, and Commonly Presumed Truths*[1] the physician Thomas Browne notes on page 83:

> And first we hear it in every mans mouth, and in many good Authors we reade it, That a Diamond, which is the hardest of stones, and not yielding unto steele, Emery, or any thing, but its own powder, is yet made soft, or broke by the bloud of a Goat;

Goat's blood softens a diamond so that it can be shattered? Browne refers to this "vulgar and common error" and notes that, while some scholars accepted it, diamond cutters, whom we can presume as unscholarly, knew it was not true. He traces the misconception to the notion that in order to produce such potent blood, some scholars wrote that goats must be fed certain herbs that were said to dissolve kidney stones in humans. Since kidney stones are also extremely hard and can be "broken," why not diamonds?

Browne further noted that "glasse is poyson, according unto common conceit." Yet he pointed out that glass is made from sand, which is not poisonous. He had also fed finely ground glass to dogs: "a dram thereof, subtilly powdered in butter or paste, without any visible disturbance." The confusion arises from the common and successful practice of adding "glasse grossely or coursely powdered" to bait in order to "destroy myce and rats." Clearly, it is internal bleeding caused by the coarse glass rather than the chemical nature of glass that is deadly.

1. Thomas Browne, *Pseudodoxia Epidemica: Or, Enquiries into Very Many Received Tenents, and Commonly Presumed Truths*, T.H. for Edward Dod, London, 1646. I thank my daughter Rachel Greenberg for bringing goat's blood and diamonds to my attention.

WHAT IS WRONG WITH THIS PICTURE?

"Distillatio" (Fig. 29) is an engraving[1] by Phillip Galle executed around 1580 from an oil by Stradanus (Jan van der Straet or Johannes Vander Straaten, 1523–1605), who painted other alchemical scenes as well.[2] This beehive of activity would cheer the heart of any modern-day research director or university professor. The alchemist is perhaps reading from the contemporary chemical literature and, in the manner of the recently born scientific method, trying to replicate a recent advance (Porta's 1558 work on the distillation of scorpion oil?).

What is wrong with this picture? Well, for starters, none of the graduate students, post-doctoral researchers, or technicians is wearing any eye protection. The professor's eyeglasses might pass muster through the middle twentieth century but not afterward as long as they lack protection on the sides. The ventilation system is antiquated to say the least; there is no evidence of fire extinguishers or a sprinkler system. A visit from the Fire Marshall should be in the offing. Admittedly, the large pestle hanging by an elastic band in the right front

FIGURE 29 ▪ *Distillatio*, an engraving (ca. 1580) by Phillip Galle after a painting by Stradanus. How many laboratory safety violations can you spot?

is a nice safety touch, although there should be a protective wire screen around it to keep it from swinging and conking an unsuspecting researcher.

The laboratory seems to be well equipped with the best the late sixteenth century has to offer. That large water bath with the multitude of stills shown in the center suggests a well-funded research operation. The two (!) hooded stills at the right front, one in operation and one idle (without hood), and the high-tech athanor, a furnace for incubating the Egg of the Philosophers, in the upper left, suggest that money is no object and that the alchemist is a good grantsman. And therein lies the greatest inconsistency. The professor is actually in "the trenches" doing science with his research group and not writing funding proposals, midsemester grade warnings, or explanations of low teaching evaluations by his students. Post-tenure reviews are still over 400 years away.

John Read describes this picture as a depiction of a late-sixteenth-century Italian laboratory bustling with "ordered and affluent activity."[2] This is in marked contrast to the poverty depicted in the 1558 "An Alchemist At Work" by Pieter Brueghel the Elder.[2] The sheaf of grain lying on the floor in the Stradanus "Distillatio" is said by Read to typify the "vital principle" although we would recognize it today as a fire hazard.

1. This engraving is from the author's private collection.
2. J. Read, *The Alchemist in Life, Literature and Art*, Thomas Nelson, London, 1947, pp. 66–68.

PROTECTING THE ROMAN EMPIRE'S CURRENCY FROM THE BLACK ART

Figure 30 is a whimsical eighteenth-century drawing[1] partly in the style of David Teniers, the Younger. There is a mysterious Arab, or possibly a Jew,[2] inappropriately garbed for a day in the laboratory. Then, there is the furtive figure peering in the doorway—a dark, sinister-looking cloaked character. From the expression of the alchemist, ancient magic is happening in the flask or perhaps there's a clue in the analysis of a woman's urine.[2]

Egyptian and Arabic cultures played crucial roles in the development of practical chemistry and alchemy.[3] Figures and ornaments of almost pure copper dating from around 4000 B.C. have been isolated from ancient Egyptian and Chaldean sites (Chaldea is now southern Iraq). Bronze alloys from Egypt dating to 2000 B.C., glass furnaces found at Tel-El Amarna dating back to 1400 B.C., evidence of pigments, cosmetics, and medicines all support the early and profound impact of these cultures.

While there is evidence suggesting the possibility of an even earlier origin in China,[3] the role of ancient Middle Eastern cultures in preserving western culture and in the beginnings of chemistry is undisputed. The origins of the word *alchemy* itself are very murky. *The Oxford English Dictionary*[3] cites the Arabic *alkimiya* derived from the Greek *chymeia*—itself related to a word meaning "to pour." A completely different origin[4] is also cited by this same august reference work referring to *Khem*, a word meaning "black," as the ancient name for Egypt

FIGURE 30 ▪ Pen-and-wash drawing of an alchemist (or a physician) signed F.P. Bush, 1769 from the author's personal collection (photograph by Dr. James Tait Goodrich, MD).

due to the blackness of its soil. Alchemy could have a double meaning here: referring to its Egyptian origins or its place as "a black art." And even here, "black" could refer to alchemy's dark and secretive nature *or* to the first step on the pathway to the Philosopher's Stone—the *Nigredo*—or initial conversion of matter to blackness. The father of alchemy, Hermes Trismegistus ("Hermes Thrice Magisterial"), who was said to predate Jesus by 2500 years, was an invention. 'Tis a mystery.

What is known is that the first verifiable person attached to an alchemical manuscript was Zosimos of Panopolis, who wrote in Alexandria, Egypt around 300 A.D. Alexandria was home to the greatest library of the classical world. Started in the third century B.C., it housed 400,000 to 500,000 books and manuscripts, mostly in Greek. The library was largely destroyed during civil wars toward the end of the third century A.D. and its "daughter library" sacked by Christians in 391 A.D.[5]

It is noteworthy that the Roman Emperor Diocletian,[6] who ruled from 285 to 305 A.D., was said to have ordered the destruction of alchemical books and manuscripts throughout the Roman Empire. As the story goes, he feared that transmutation of base metals to silver and gold would devalue the Empire's currency. (However, see the next essay, p. 46).

What manner of Emperor would destroy alchemical books and manuscripts merely to preserve the value of the Empire's currency? Being a bibliophile but not a scholar of antiquity, I tried to assess Diocletian as a politician from a *fin de siècle* (*actually, fin de millenium*) American perspective. Diocletian[6] stood for preservation of ancient virtues and the obligation of children to feed their parents in old age (no Social Security program here: a "social conservative"? a conservative Republican?). He also introduced a progressive income tax and the beginnings of the vast system of bureaucracy and technocracy that even today makes visits to state Departments of Motor Vehicles so memorable (a "tax-and-spend" liberal Democrat?). The coins he was trying to protect were inscribed *dominus et deus* ("ruler and god"). Does any reader out there know the Latin word for the Greek *hubris*?

In 1979, the Nobel Prize in Physics, awarded for a theory of unified weak and electromagnetic interactions between elementary particles, was shared by three scientists: Sheldon Glashow, Steven Weinberg, and Abdus Salam, two Jews born in New York and a Moslem born in Pakistan. Salam's Nobel Prize lecture[7] was particularly beautiful. He told of a young Scotsman named Michael who, almost eight centuries earlier, had traveled to study at the Arab Universities of Toledo and Cordova in Spain, centers for "the finest synthesis of Arabic, Greek, Latin, and Hebrew scholarship" and home to the Hebrew scholar Maimonides. Salam notes that Sarton's *A History of Science* credits the period 750 to 1100 A.D. to an unbroken period of intellectual dominance by Middle Eastern cultures. In contrast to wealthy countries with flourishing schools of research such as Syria and Egypt, Scotland, a poor but developing land, had little to offer Michael upon his return, and Salam says: "At least one of his masters counseled young Michael the Scot to go back to clipping sheep and to the weaving of woolen cloth." But it was around this time that scientific superiority began to shift to the West, and Salam continues:

And this brings us to this century when the cycle begun by Michael the Scot turns full circle, and it is we in the developing world who turn westward for science. As Al-Kindi wrote 1100 years ago: "It is fitting then for us not to be ashamed to acknowledge truth and to assimilate it from whatever source it comes to us. For him who scales the truth there is nothing of higher value than truth itself; it never cheapens or abases him."

1. This pen and wash drawing is signed "H.P. Bush, fecit 1769" and is from the author's personal collection. The author thanks Dr. James Tait Goodrich, M.D., James Tait Goodrich Antiquarian Books and Manuscripts for a photograph of this drawing.
2. The author thanks Dr. Alfred Bader, founder of Aldrich Chemical Company and renowned art collector, for his interpretation (personal correspondence). The stylized letters on the chemist's garb evoke both Arabic and Hebrew.
3. J.M. Stillman, *The Story of Alchemy and Early Chemistry*, Dover, New York, 1960, Chap. I.
4. *The Oxford English Dictionary*, Clarendon, Oxford, 1989, Vol. 1, p. 300.
5. *Encyclopedia Brittanica*, 15th ed., 1986, Vol. 1, Chicago, p. 251.
6. *Encyclopedia Brittanica*, 15th ed., 1986, Vol. 4, Chicago, pp. 105–106.
7. A. Salam, *Reviews of Modern Physics*, **52**(3):525–526, 1980. I am grateful to Professor Joel F. Liebman for making me aware of and suggesting Salam's Nobel Prize lecture.

GEBER AND RHAZES: ALCHEMISTS FROM THE BIBLICAL LANDS

The Diocletian story (see the previous essay) is a nice one. However, it seems that Arabic alchemy only reached the West (including Rome) around the eleventh century, so the story may be charitably termed "legendary."[1]

Most of our knowledge of Arabic alchemy derives from the writings of a mysterious eighth-century person named Jabir ibn Hayyan or Geber. Figure 31, the title page of *De Alchimia Libri Tres* (Of Alchemy in Three Books), published in Strasbourg in 1529, depicts a distillation furnace.[2] While alchemy also had origins in China and India, the cultures and languages of the biblical lands were more accessible to the Europeans who, starting in the fifteenth century, produced the first printed books.

Al-Razi (850–ca. 923) or Rhazes, a Persian physician, produced the text *Secret of Secrets*. It included a great deal of practical and useful chemistry. Brock suggests that the preparation of pure hydrochloric, nitric, and sulfuric acids by Europeans in the thirteenth century depended crucially on the technology described by Rhazes.[1] These incredibly powerful "biting serpents" played critical roles in opening up new chemical reactions: for example, the ability to "release phlogiston" from metals or (as we have understood for over 200 years) to oxidize the metal to its calx while reducing an aqueous acid so as to release hydrogen gas.

1. W.H. Brock, *The Norton History of Chemistry*, Norton, New York, 1993, pp. 19–24.
2. I. MacPhail, *Alchemy and the Occult*, Yale University Library, New Haven, CT, 1968, pp. 32–34. We acknowledge the Beinecke Library, Yale University for this figure.

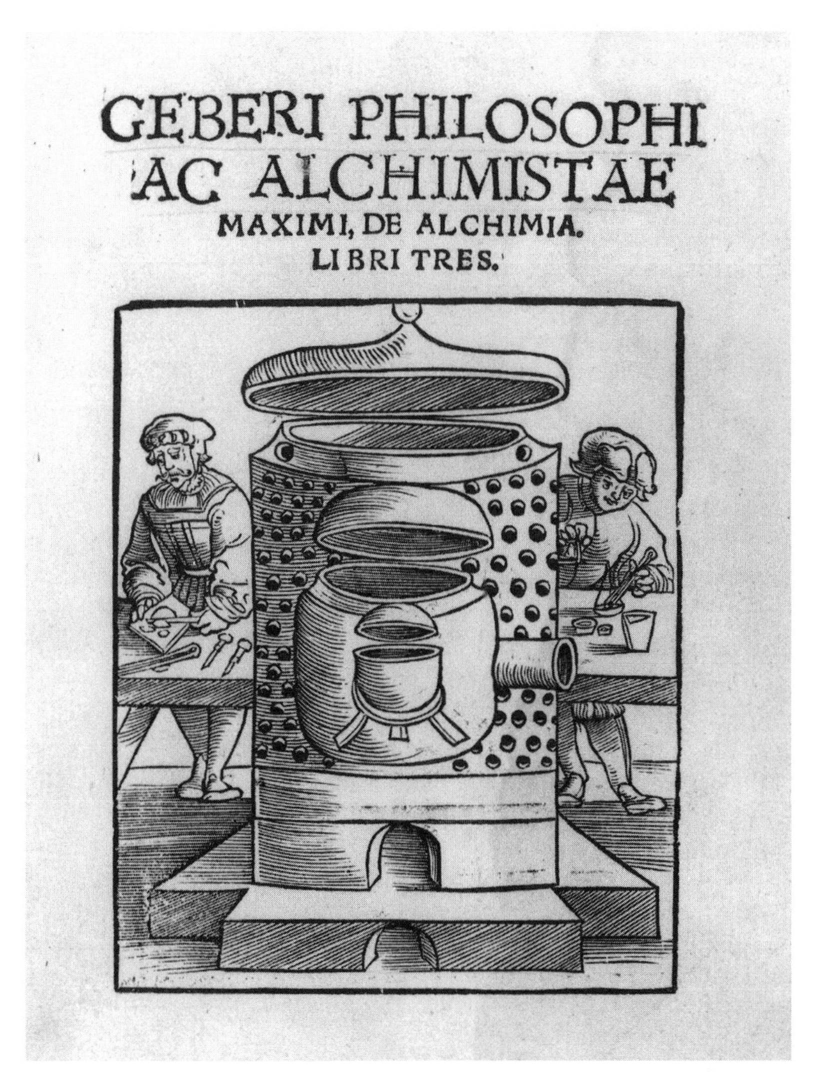

FIGURE 31 ■ Frontispiece from Geber, *De Alchimia Libri Tres* (Strasbourg, 1529) (courtesy of The Beinecke Rare Book and Manuscript Library, Yale University).

ALCHEMISTS AS ARTISTS' SUBJECTS

The sixteenth, seventeenth, and eighteenth centuries witnessed the painting of numerous masterworks of European art depicting alchemists and physicians at work. Two prominent American collections including such artwork are the Fisher Collection of Alchemical and Historical Pictures (now kept by Duquesne University, Pittsburgh, PA) and the Isabel and Alfred Bader Collection (Milwaukee, WI). Although there are suggestions that Albrecht Durer (1471–1528) understood alchemical imagery, he apparently never engraved an alchemist or a laboratory.[1] Two early masters who represented medieval alchemists were Hans Weiditz ("An Alchemist and his Assistant at Work"—executed around 1520) and Pieter Brueghel the Elder (1525–1569).[1] Brueghel's 1558 "An Alchemist at Work" achieved widespread fame due to the contemporary engraving of it by Hieronimus Cock.[1] Some other noted artists depicting alchemical scenes during this period were Stradanus (see Fig. 29), The De Bry, Adriaen van Ostade, David Teniers the Younger, Jan Havickz Steen, Cornelis Pitersz Bega, Hendrik Heerschop, Charles Meer Webb, Matheus van Hellemont, Bathasar van den Bosch, Franz Christophe Janneck, Fernand Desmoulin, Thomas Wijck, Wenzel von Brozik, William Pether, and David Ryckaert. Most of these were in the Dutch–Flemish school.[1] A notable painting by Englishman Joseph Wright ("The Discovery of Phosphorus," 1771) and a piece by Richard Corbould near the start of the nineteenth century began to depict the science rather than the art. In the nineteenth century, carricaturists James Gillray, Thomas Rowlandson, and George Cruikshank (see Figs. 126 and 127) took a shot at depicting chemical activity.[1]

Figure 32 was executed by Hendrick Heerschop in 1671 and is titled "The Alchemist." It is a black-and-white reproduction of a color photograph of a beautiful oil painting from the Isabel and Alfred Bader Collection.[2] The alchemist appears to smoke his pipe while watching a distillation. Hopefully, he is not distilling diethyl ether.

1. J. Read, *The Alchemist in Life, Literature and Art*, Thomas Nelson, London, 1947, pp. 56–91.
2. The author is grateful to Dr. Alfred Bader for making this photographic reproduction available and providing permission to reproduce it in black and white, as well as in color.

FIGURE 32 ■ Black-and-white reproduction of the color photograph of the 1671 oil painting, *The Alchemist*, by Hendrick Heerschop, in the Collection of Isabel and Alfred Bader. The author expresses his gratitude to Dr. Bader for permission to reproduce the image and also for his helpful discussion of the Bush drawing (see Fig. 30).

ALLEGORIES, MYTHS, AND METAPHORS

The sixteenth and seventeenth centuries witnessed the publication of many beautiful books that illustrated the alchemical art. For example, the title page of *Le Tableau des Riches Inventions* (Fig. 33), authored by Francesco Colonna, translated by Beroalde de Verville, and published in Paris in 1600,[1] depicts The Great Work beginning in chaos and culminating with the Stone (rising of the phoenix).[2] The tree stump in this figure represents the putrefaction (*Nigredo*) or the debasement at the beginning of the process.[3] The figure on the lower left is a Philosophical Tree[3] representing the complete work as well as the aspect of multiplication of gold. Fire is the transforming element. The winged dragon and wingless serpent represent the union of Sophic Mercury (winged means volatile) and fixed (nonvolatile) Sophic Sulfur. Their symbols are also shown. The psychologist C.G. Jung owned a copy of this book and wrote extensively on dreams and alchemical imagery.[4,5]

Michael Maier (ca. 1568–1622), whom John Read calls "a musical alchemist," was a physician, philosopher, alchemist, and classical scholar.[6] His extensive classical scholarship influenced his unification of alchemy and classical mythology.[6] Figure 34 is the title page of the 1618 book *Tripus Aureus* (The Golden Tripod).[7] It is a pun on the *tria prima* (mercury, sulfur, and salt), which "support" the synthesis of gold. The main objective is to present works by the "Three Possessors of the Philosopher's Stone." In Maier's own words:[7]

> Amiable reader, you behold three nurslings of the wealthy Art who by their studies have achieved the Stone. Cremerus in the middle, Norton himself on the left, Basil, lo, is seen on the right. Pray read their writings and search for the arms of Vulcan you who wish to pluck the apples of the Hesperian ground.

John Cremer, a fourteenth-century abbot who lived in Westminster, was reputed to have joined Raymond Lully in alchemical works in Westminster Abbey and the Tower of London.[7] Thomas Norton, author of the *Ordinall of Alchimy*, began writing this famous work in 1477.[7] The reputed Basil Valentine is mentioned later in our *Chemical History Tour* and also by Read.[8] Vulcan is the god of fire. "Arms of Vulcan" refers to fire as an instrument of chemical change.[7] The picture depicts an immediate proximity between a chemical laboratory and a chemical library—the duality of practice and theory. Although the American Chemical Society recommends location of a university chemical research library in the chemistry laboratory building, it is unlikely that they have quite this closeness in mind.

Maier's 1618 book *Atalanta Fugiens* (Atalanta Fleeing) contains 50 beautiful engravings and also includes music for about 50 three-voice (*tria prima*—get it?) fugues he composed.[9,10] The title page (Fig. 35) depicts the legend of Atalanta and the golden apples.[9] Atalanta, the fastest mortal, challenged any suitor to a foot race. If they lost, they died. She would wed the man who defeated her. The figure tells of Hercules picking the three golden apples of the Hesperides (guarded by Aegle, Arethusa, and Hesperia and their guard-dragon Ladon). The

LE
TABLEAV
DES RICHES
INVENTIONS
Couuertes du voile des feintes
Amoureuſes, qui ſont re-
preſentees dans le
SONGE DE POLIPHILE
Desvoilées des ombres du Songe,
& ſubtilement expoſees
PAR BEROALDE
A. PARIS.
Chez MATTHIEV GVILLEMOT, au Palais,
en la gallerie des priſonniers.
Auec priuilege du Roy.
1600.

FIGURE 33 ■ Engraved title page from Francesco Colonna's *Le Tableau Des Riches Inventions* (Paris, 1600 [after 1610]) (courtesy of The Beinecke Rare Book and Manuscript Library, Yale University). The figure depicts the rising of the Phoenix beginning in chaos.

TRIPVS AVREVS,
Hoc eſt,
TRES TRACTATVS
CHYMICI SELECTISSIMI,
Nempe

I. BASILII VALENTINI, BENEDICTINI ORDInis monachi, Germani, PRACTICA vna cum 12. clauibus & appendice, ex Germanico;

II. THOMÆ NORTONI, ANGLI PHILOSOPHI CREDE MIHI ſeu ORDINALE, ante annos 140. ab authore ſcriptum, nunc ex Anglicano manuſcripto in Latinum translatum, phraſi cuiuſque authoris vt & ſententia retenta;

III. CREMERI CVIVSDAM ABBATIS WESTmonaſterienſis Angli Teſtamentum, hactenus nondum publicatum, nunc in diuerſarum nationum gratiam editi, & figuris cupro affabre inciſis ornati operâ & ſtudio

MICHAELIS MAIERI Phil. & Med. D. Com. P. &c.

FRANCOFVRTI
Ex Chalcographia Pauli Iacobi, impenſis LVCÆ IENNIS.
Anno M.DC. XVIII.

FIGURE 34 ■ Engraved title page from Michael Maier's *Tripus Aureus* ("Golden Tripod"), published in Frankfurt in 1618. The Big Three of Alchemy are Basil Valentine, Thomas Norton, and John Cremer (courtesy of The Beinecke Rare Book and Manuscript Library, Yale University).

ATALANTA
FUGIENS,
hoc eſt,
EMBLEMATA
NOVA
DE SECRETIS NATURÆ
CHYMICA,
Accommodata partim oculis & intellectui, figuris cupro inciſis, adjectisque ſententiis, Epigrammatis & notis, partim auribus & recreationi animi plus minus 50 Fugis Muſicalibus trium Vocum, quarum duæ ad unam ſimplicem melodiam diſtichis canendis peraptam, correſpondeant, non abſq; ſingulari jucunditate videnda, legenda, meditanda, intelligenda, dijudicanda, canenda & audienda:

Authore

MICHAELE MAJERO Imperial. Conſiſtorii Comite, Med.D. Eq. ex. &c.

OPPENHEIMII
Ex typographia HIERONYMI GALLERI,
Sumptibus JOH. THEODORI de BRY,
M DC XVIII.

FIGURE 35 ■ Engraved title page from Michael Maier's *Atalanta Fugiens* ("Atalanta Fleeing") published in Frankfurt in 1618. The engraving recounts the mythology of Atalanta, the fastest mortal, who challenged suitors to a foot race—losers were put to death. But Hippomenes used three golden apples (from Venus) to distract her and win the race and her hand (see discussion in text) (courtesy of The Beinecke Rare Book and Manuscript Library, Yale University).

apples are presented by Venus to Hippomenes who drops them at opportune moments to distract Atalanta and thus wins the race and her hand. "Plucking the apples of the Hesperian ground" (see above) is a metaphor for achieving the Stone. Unfortunately, the couple later profanes the temple of Cybele (the act of profaning is shown in the lower right—clearly they are "fast company" albeit slightly incautious) and are changed into lions.

What *is* going on in Figure 36 (from *Atalanta Fugiens*)? And wouldn't this be a dandy question for a chemistry exam? In Greek mythology, the goddess Athena is born from the head of her father Zeus (she has no mother). One version has the Greek god of fire (Hephaestus, or Vulcan in Roman lore) splitting the head of Zeus. The god Apollo, born of Zeus and Leto, is also depicted. The incestuous *conjunctio,* presided over by Cupid, unites the male principle (sulfur) with the female principle (mercury). Lynn Abraham mentions the myth of Jupiter (Zeus) wherein he is transformed into an eagle, transporting Ganymede to heaven, and converted into a shower (distillation) of gold.[11]

On the other hand, perhaps this is merely an advertisement for taking a name brand of aspirin just prior to the chemistry final exam. I guess I'd have to give at least half credit for that answer.

The first public performance of alchemical music (examples of Maier's fugues) appears to have occurred at the Royal Institution of Great Britain on November 22, 1935. Student members of the St. Andrews University Choir ("The Chymic Choir") and the Music Department faculty "conspired" to give voice to the admirable scholarship of their colleague—Professor John Read.[12]

1. I. MacPhail, *Alchemy and the Occult*, Yale University Library, New Haven, CT, 1968, pp. 189–191.
2. C.G. Jung, *Psychology and Alchemy*, 2nd ed., translated by R.F.C. Hull, Princeton University Press, Princeton, NJ, 1968, p. 38.
3. L. Abraham, *A Dictionary of Alchemical Imagery*, Cambridge University Press, Cambridge, 1998, pp. 150–151, 205.
4. I. MacPhail, op. cit., pp. xv–xxxii (essay by A. Jaffe).
5. N. Schwartz-Salant, *Encountering Jung on Alchemy*, Princeton University Press, Princeton, NJ, 1995.
6. J. Read, *Prelude To Chemistry*, MacMillan, New York, 1937, pp. 228–236.
7. J. Read, op. cit., pp. 169–182.
8. J. Read, op. cit., pp. 183–211.
9. J. Read, op. cit., pp. 236–254.
10. J. Read, op. cit., pp. 281–289.
11. Abraham, op. cit., p. 110.
12. J. Read, op. cit., pp. xxiii–xxiv.

FIGURE 36 ■ Birth of Athena from the head of her father Zeus (no mother here) and a chemical *conjunctio* with her brother Apollo. The *conjunctio*, or chemical marriage, weds Sol and Luna or Sophic Sulfur and Sophic Mercury (courtesy of The Beinecke Rare Book and Manuscript Library, Yale University).

THE WORDLESS BOOK

One of the most beautiful books of the seventeenth century was titled *Mutus Liber* (wordless book), published in France in 1677 and authored by "Altus" a pseudonym representing the "Classic Elder" of alchemy. It was printed at the instigation of one Jacob Saulat and includes 13 folio-sized figures with only slight text in the title figure, which depict The Great Work.[1] In 1702, the *Mutus Liber* became more widely known due to its inclusion at the end of the first volume of Manget's *Bibliotheca Chemica Curiosa* and had 15 figures.[1] The figures are totally allegorical and there is no firm interpretation of them. It is interesting that the pictures depict a man and a woman (possibly husband and wife) apparently working as co-equals. This was a rather novel aspect of the book since women played virtually no significant role in chemistry for a long period. There was, however, an ancient Alexandrian woman called Mary Prophetissa,[2–4] sometimes equated to Miriam, the sister of Moses, who discovered hydrochloric acid and developed an early still called a *kerotakis* as well as the water bath, used for gentle heating. The water bath has survived into modern times and is termed the *Bain Marie*. Moreover, Mary Prophetissa is said to have originated the process of fusing lead–copper alloy with sulfur to make a blackish material.[4] Such black materials were often the starting points for transmutation and represent allegorical death preceding resurrection. This is one of the origins of the term "Black Arts" for alchemical practices.

The six figures shown here are selected from a 1914 Paris reissue of the *Mutus Liber*.[5] The title page (Fig. 37) shows a picture believed to depict Jacob and the Ladder to Heaven and is totally spiritual.[1] His head rests on a rock some say represents the Philosopher's Stone. The translation is as follows:[4]

> The Wordless Book, in which nevertheless the whole of Hermetic Philosophy is set forth in heiroglyphic figures, sacred to God the merciful, thrice best and greatest, and dedicated to the sons of art only, the name of the author being Altus.

The last three lines are biblical references in reverse: *Genesis* 28:11, 12; *Genesis* 27:28, 39; *Deuteronomy* 33:18, 28.

Figure 38 is the second plate in the *Mutus Liber* and depicts the sun above two angels holding a vessel containing Sol and Luna at the sides of Neptune, who is considered to represent a watery or liquid substance needed in the Great Work as the two alchemists kneel at their furnace. The upper part represents the spiritual dimension of the work.[1] The lower section is the earthly part. In the furnace, the bottom section is flame, the middle funnel is a sand or waterbath

→

FIGURE 37 ■ Engraved title page from *Mutus Liber* ("Silent Book"). This figure is from a 1914 Paris reproduction of the *Mutus Liber* in Manget's 1702 *Bibliotheca Chemica Curiosa* which reprinted the plates of the original 1677 book. The image depicts Jacob and the Ladder to Heaven.[5]

MUTUS LIBER, IN QUO TAMEN
tota Philosophia hermetica, figuris hieroglyphicis
depingitur, ter optimo maximo Deo misericordi
consecratus, solisque filiis artis dedicatus,
authore cuius nomen est Altus.
21. 11. 82. Neg:
93 82. 72. Neg:
82. 81 33. Fued.

FIGURE 38 ▪ The second plate (of 15 in Manget's printing) in *Mutus Liber* that depicts, in its spiritual upper section, the sun above two angels holding Sol and Luna in the presence of Neptune, representing the watery substance needed in the Great Work. In the earthly lower section, the male and female alchemists place the Philosophical Egg in the athanor where it is gently heated with a sand or water bath.[5]

FIGURE 39 ■ The fourth plate in *Mutus Liber* that depicts the collection of dew (a kind of *Prima Materia*).[5]

FIGURE 40 ■ The fifth plate in *Mutus Liber* depicts the two alchemists preparing the dew for distillation. The distillate is divided into four bottles and then heated (apparently for 40 days). The residue is spooned into a bottle and given to an old man (Saturn?).[5]

FIGURE 41 ■ In plate 14 in Manget's printing of *Mutus Liber* the man, the child and the woman are trimming the wicks and filling their lamps with oil. Equal parts of Lunar Tincture and Solar Tincture are ground together to provide Sophic Mercury.[1,5]

FIGURE 42 ■ The work is finished (plate 15) and the alchemists proclaim: "Given Eyes To See, Thou Seest."[5]

for controlled heating of the Philosophical Egg.[1] (Any chemist who has tried a new reaction will appreciate the prayerful aspect of this picture.) Figure 39 is the fourth picture in *Mutus Liber* and shows the collection of dew in sheets spread in the pasture under the influence of the sun in Aries and the moon in Taurus (springtime). This illustrates the astrological dimension of the *opus*. Dew is considered to be a type of *Prima Materia* and the two alchemists wring it into a large collection plate.

In the book's next figure (Fig. 40), the man and woman prepare the dew for distillation in an alembic. The man subsequently takes the distillate and pours it into four vessels that are heated, apparently for 40 days. The woman removes the residue from the distillation vessel and spoons it into a bottle that she gives to an old man, holding a child and bearing the mark of Luna. Some interpret the old man as Saturn.

In Figure 41 (Plate 14 in Manget's work) we see three furnaces and the man, child, and woman trimming the wicks on their furnance lamps. Equal parts of Lunar Tincture and Solar Tincture are ground together to make Sophic Mercury. The two alchemists seal their lips and the words read: "Pray, Read, Read, Read, Read again, Labor and Discover."[1] In Figure 42, the last picture in *Mutus Liber* is like the first (Figure 37) and is totally spiritual. The Ladder is no longer needed, a body, possibly Hercules (son of Zeus), lies at the bottom under the influence of Sol and Luna, the Zeus figure is being crowned with laurel wreaths by angels, and the two enlightened alchemists exclaim in unison:

> Given eyes To see, thou seest.[2]

The Great Work is finished.

1. A. McLean, *A Commentary On The Mutus Liber*, Phanes, Grand Rapids, MI, 1991.
2. C.A. Burland, *The Art of The Alchemists*, MacMillan, New York, 1967, pp. 188–198. This book shows all 15 figures in a reasonably large format.
3. *Secrets of the Alchemists*, Time-Life Books, Alexandria, VA, 1990, pp. 70–77. This book depicts all 15 figures (in gold tint, no less!), significantly reduced in size, but with nice textual discussion.
4. J. Read, *Prelude To Chemistry*, MacMillan, New York, 1937, pp. 155–159.
5. *Mutus Liber—Le Livre d'Images sans Paroles, ou toutes les opérations de la Philosophie hermetique sont décribes et représentés. Reédite d'après l'original et precedé d'une Hypotypose explicative par* MAGOPHUN., Librairie Critique, Emile Nourry, Paris, 1914.

SECTION III
IATROCHEMISTRY AND SPAGYRICALL PREPARATIONS

PARACELSUS

Theophrastus Bombast Von Hohenheim (1493–1541), who called himself Paracelsus, applied chemistry to effect medical cures and fathered a field called *iatrochemistry*. His break with the ancient medical doctrines of Galen was total and his tone intolerant and bombastic. He is recognized as having introduced experiment and observation into medical treatment.

Rather than search for Paracelsan quotes, we borrow from the novel by Evan S. Connell, *The Alchymist's Journal*[1] in order to gain insight into his mind and style:

> I have said that all metals labor with disease, except gold which enjoys perfect health by the grace of elixir vitae. I have taught Oporinus how this metal is sweet and exhibits such goodly luster that multitudes would look toward gold instead of the generous sun overhead. In fixity or permanence this substance cannot be exceeded and therefore it must gleam incorruptibly, being derived from an imperial correspondence of primary constituents which makes it capable of magnifying every subject, of vivifying lepers, of augmenting the heart. Conceived by our gracious Lord, it is a powerful medicament. False gold, which is a simulacrum boasting no remedial virtue, assaults internal organs and therefore it should be abjured, since the alchymic physician repudiates meretricious matter. We must not keep true gold beyond its measure but distribute what we hold, allegorically reminding each man of an earthly choice he is obliged to make between damnation and bliss.
>
> Pseudo-Alchymists that labor against quicksilver, sea salt, and sulfur dream of hermetic gold through transformation, yet they fail to grasp the natural course of development since what they employ are literal readings of receipts. Accordingly they bring baskets of gilded pebbles to sell, or drops of silver in cloudy alembics—futile panaceas meant for a charnel house. This is false magistry.
>
> Should it be God's will to instruct an alchymist at his art He will dispense understanding at the appropriate season. But if by this wisdom He concludes that any man was unfit or should He decide that irrevocable mischief would ensue, then that sanction is withheld.

The first novelized quotation indicates the imperfections in baser metals that are converted to gold (perfection) using the elixir vitae (or the Philosopher's Stone). True gold can be used as a medication. The second quotation indicates the hopeless quest of false alchemists, sometimes called "puffers" after their furnace bellows, whose goal is solely gold making without an eye toward the unity of alchemy with nature. The last is perhaps most interesting: failure to duplicate an alchemical recipe is due to God's denial of the secret to the unworthy seeker rather than shortcomings in the original formula or method. It is, by the way, never clear how the Stone or the Elixir brings about its transformations.

We may obtain some feeling for the medicine of the period by visiting some of the cures attributed to Paracelsus in a book published in London in 1652 titled *THREE EXACT PIECES of LEONARD PHIORAVANT, Knight and Doctor in PHYSICK, viz. His RATIONAL SECRETS, and CHIRURGERY, Reviewed and Revived, Together with a Book of Excellent EXPERIMENTS and SECRETS, Collected out of the Practices of Severall Expert Men in both Faculties, Whereunto is Annexed PARACELSUS his One hundred and fourteen EXPERIMENTS: With certain Excellent Works of B.G. a Portu Aquitano, Also Isaac Hollandus his SECRETS concerning his Vegetall and Animal Work, With Quercetanus his Spagyrick Antidotary for GUN-SHOT.* (Nice to know what's in a book before you buy it):

- A certain woman was long sick of the Passion of the heart, which she called *Cardiaca,* who was cured by taking twice our *Mercuriall* vomit, which caused her to cast out a worm, commonly called *Theniam,* that was four cubits long.
- A boy of fifteen years old, falling down a stone staires, had his arme and leg benummed and voide of moving, whose neck with the hinder part of the head, and all the back bone I annointed with this unguent: a) Of the fat of a Fox; b) Oyle of the earthwormes; c) *Oleum Philosophorum.* I mixed them together, and annointed therewith, and in short space no wound nor swelling appeared in him so hurt.
- One that spit bloud, I cured by giving him one scruple of *Laudanum Precipitatum,*[2] in the water of Plantaine, and outwardly I applied a linnen cloth to his brest, dipped in the decoction of the bark of the roots of Henbane.
- One had two Pushes, as it were warts upon the yard, which he got by dealing with an unclean woman, so that for six moneths he was forsaken of all Physitians as uncureable, the which I cured by giving him *Essentia Mercurialis,* and then mixed the oyle of Vitriol with *Aqua Sophia,* and laid it on warm with a suppository four daies.
- A boy of eighteen years old had a tooth drawn, and three months after a certain black bladder appeared in the place of the tooth, the which I daily annointed with the oyle of Vitriol, and so the bladder was taken away, and the new tooth appeared.
- A fat drunken Taverner was in danger of his life by a surfet, who was restored to his health by letting of bloud.
- One who was troubled with paines in the stomack through weaknesse, who took *Oleum salis* in his drink, and caused him to have many seeges or stooles, and so was restored to his health, as we have written on our book called *Parastenasticon.*
- A man that was troubled with the head-ach, I purged by the nostrills, casting in the juice of *Ciclaminus* with a siringe.
- A woman being almost dead of the Collick, I cured with the red oyle of Vitriol, drunk in Anniseed water, and a while after that potion, she voided a worm and was cured.
- To cause nurses to have abundance of milk, I have taken the fresh branches or tops of fennell, and boyled them in water or wine, and given it to drinke at dinner or supper, and at all times, for it greatly augmenteth the milk.

- A man being vehemently troubled a years space with pains in the head, I cured onely by opening of the skull, and in the same manner I cured the trembling of the brain, taking therewithall, *Oleum salis* in water of Basil.
- A Prince in Germany that was troubled with the Frenzie, by reason of a Sharp Fever, whom I cured by giving him five grains of *Laudanum nostrum*,[2] which expelled the Fever, and caused him to sleep six houres afterward.

1. Evan S. Connell, *The Alchymist's Journal*, North Point, San Francisco, 1991.
2. J.R. Partington, *A History of Chemistry*, MacMillan, London, 1961, Vol. 2, p. 150, notes that opium had been employed by the Arabs in their medicine well before Paracelsus. But he also raises doubts over whether Paracelsus' laudanum ever had any opium. If not, then the above cures suggest effective placebos.

THE DREAM TEAM OF ALCHEMY

Figure 43 is from the 1611 edition of *Basilica Chymica*, by Oswald Croll, which was printed in subsequent editions for 100 years. It is credited for passing the knowledge of Paracelsus and his followers into the seventeenth century.

The book's beautiful frontispiece depicts the *Alchemical Dream Team*:

Hermes Trismegistus, Egyptians
Geber, Arabs
Morienes, Romans
Roger Bacon, English
Ramon Lull, Spanish
Paracelsus, Germans

It's a Dream Team in another sense as well. There is no evidence that a Hermes Trismegistus ever existed. The name of the reputed father of alchemy, Hermes-The-Thrice-Great, is a bit suspicious. In any case alchemy came to be called the "hermetic art." When we hermetically seal something, we protect it from air much as some alchemical experiments were sealed in glass and buried literally for years.

FIGURE 43 ■ Title page of the *Basilica Chymica* (Frankfurt, 1611) by Oswald Croll, perhaps the major early source of Paracelsan chemical lore.

DISTILLATION BY FIRE, HOT WATER, SAND, OR STEAMED BOAR DUNG

Conrad Gesner (1516–1565) was born in Zurich into "the very poorest circumstances."[1,2] His early brilliance was noted by his father who sent him to his uncle, who sold medicinal herb extracts, for further education. In that setting, Gesner developed a lifelong interest in plants and the medicines derived from them. His teachers sponsored Gesner's later education, despite his foolishness, at the age of 19, in marrying a bride with no dowry. He compiled a Greek–Latin dictionary and was appointed Professor of Greek in Lausanne Academy by the age of 21. This allowed Gesner to accumulate money, and he attended medical school for one year achieving the Doctorate in Medicine at the age of 25. The remainder of his life was spent as a physician in Zurich and a Lecturer in Aristotelian physics at the Collegium Carolinum. Gesner died of plague at the age of 49.

Figure 44 is from Gesner's *The Practice of the New and Old Physicke . . .* (102-word title!) published in London in 1599. The first edition of this book

The ſecond Booke of Diſtillations,
containing ſundrie excellent ſecret
remedies of Diſtilled
Waters.

FIGURE 44 ■ The title page of Book Two of Conrad Gesner's *The Practice of the New and Old Physicke, Wherein is Contained the Most Excellent Secrets of Phisicke and Philosophie, divided into foure Bookes. In the Which are The Best Approved Remedies for the Diseases as well Inward as Outward of al the Parts of Man's Body, etc.* (London, 1599). Now *that's* a title!

(a)

(b)

(c)

FIGURE 45 ■ Distillation apparatus from Gesner's treatise of 1599: (a), left: A furnace employing a water bath is termed the *Bain Marie* (*Balneum Marie* or bath of Marie); (b) heating samples in closed cucurbits using sand heated by the sun preferably in July or August; (c) A heating bath of boar dung freshly steamed. I suggest calling this the "Bane of Marie" and further advise outdoor use only.

¶ The fourth Booke of Dyſtillations, *containing many ſingular ſecret* Remedies.

FIGURE 46 ▪ The title page of Book Four of Gesner's 1599 treatise. The winged pet dragon represents Sophic Mercury; the Tree of Life flowers in cucurbits that produce Birds of Hermes signaling success of the Great Work.

(*The Treasure of Evonymous* . . . –"evonymous" means anonymous) appeared in 1552.[1] Figure 44 is the title page of the second book of four in this volume. The sun and the moon represent the male (Sophic Sulfur) and female (Sophic Mercury) principles. In Figure 45(a) we see the *Bain Marie* (or *Balneum Marie*; *bain* is French and *balneum* is Latin for "bath"; Marie refers to the third-century alchemist Mary Prophetissa). It is a furnace using a water bath to achieve a gentle and controlled distillation. Similar results can be achieved with the simpler apparatus on the right in this figure. A cucurbit (or retort) is fitted with an alembic (or limbeck) on top, having a beak to condense the vapors into a collecting retort.

Figure 45(b) depicts the heating of distillates in sealed cucurbits in a sand bath heated by the sun (Gesner advises July and August as the best times for this work, which may take periods as long as 40 days). Another technique for gentle distillation is to place cucurbits topped with alembics into a box of continuously steamed boar dung [Fig. 45(c)]. I suggest "Bane of Marie" as the name for this apparatus. The operation is probably best done outdoors.

The title page for Book Four (Fig. 46) is full of wonderful symbols. The sun and moon witness the growth of the Philosopher's Tree (or Tree of Life), representing the growth of The Great Work.[3] The pet dragon eating (eating what?!) from *her* bowl is winged and probably represents Sophic Mercury. The cucurbit, when sealed, can be considered to be a Philosopher's Egg.[3] (In this figure, we are one short of a dozen eggs.) A Bird of Hermes[3] ascends from each egg, symbolizing completion of The Great Work.

Figures 47 to 49 are from *The Art of Distillation* by John French (1653). The first [Fig. 47(a)] represents a steam-distillation apparatus. Figure 47(b) depicts a *Bain Marie* made using a brass kettle and cover and heated in the center by a stack oven. Figure 48(a) illustrates the use of sunlight for heating glass crystals or an iron (or marble) mortar as the heat source for distillation. The heavy-duty furnace in Figure 48(b) promises distillation of large quantities of spirits and oils from minerals, vegetables, bones, and horns in only 1 hour instead of the usual 24 ("time is money" even in 1653). Figure 49(a) depicts the distillation of spirit of salt (hydrochloric acid). Figure 49(b) depicts a still for volatile substances including condensers (one of these water-cooled) at the end: state-of-the-art, maintenance contract available for additional purchase.

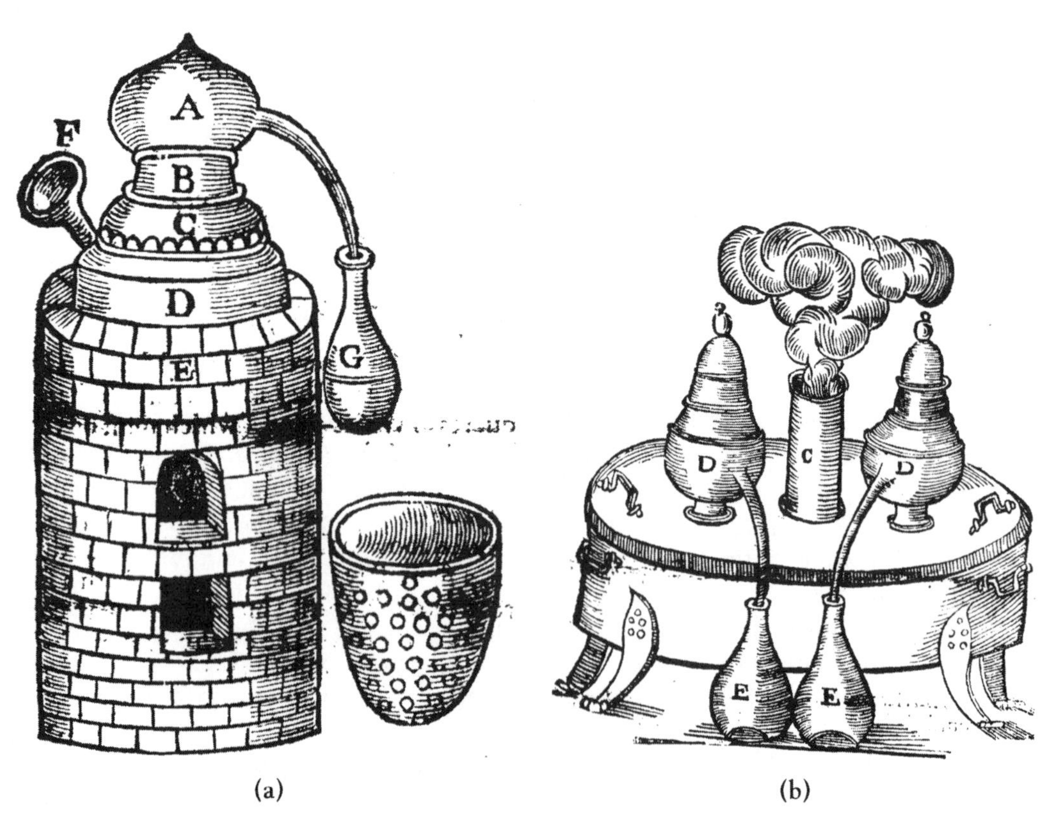

FIGURE 47 ■ Apparatus from *The Art of Distillation* by John French (London, 1653; first edition, 1651): (a) Apparatus for steam distillation; (b) A *Balneum Marie*.

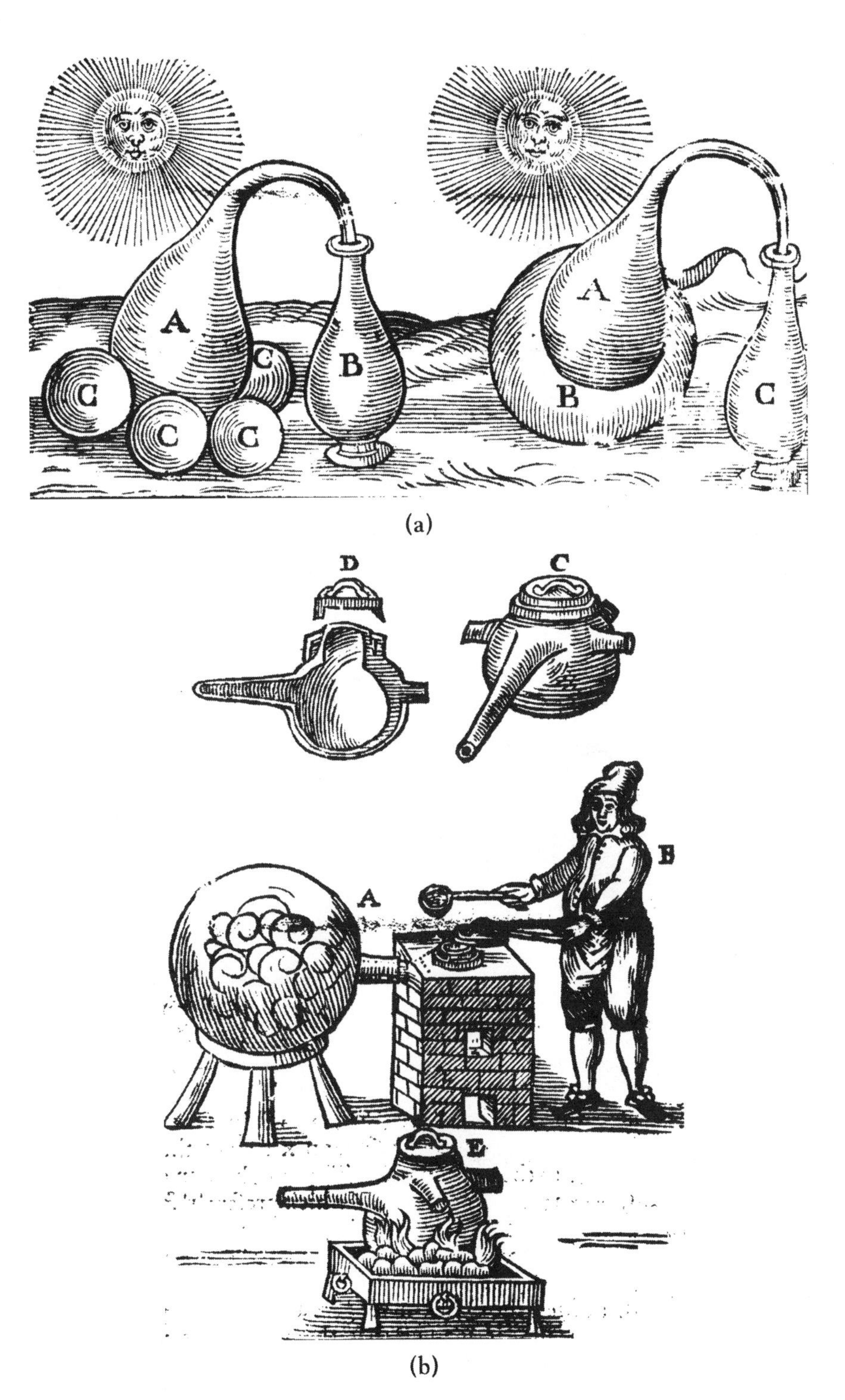

FIGURE 48 ▪ From French's *The Art of Distillation*: (a), left, glass crystals heated by the sun as heat source; right, iron or marble mortar as the heat source; (b) heavy-duty furnace for distillation from large quantities of bones, horns, minerals, and vegetables.

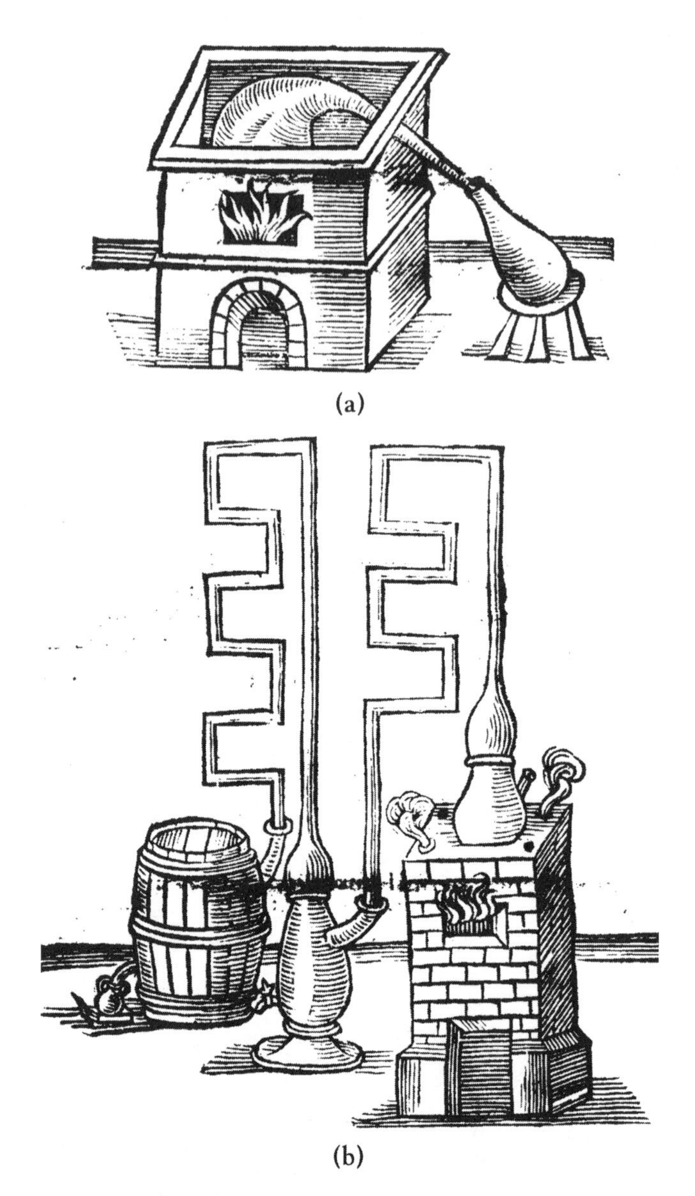

FIGURE 49 ■ From French's *The Art of Distillation*: (a) Apparatus for distilling spirit of salt (hydrochloric acid), a very biting "serpent" indeed; (b) State-of-the-art still with water-cooled condenser for distilling volatile liquids. Similar apparatus are still found in the hills of Kentucky and West Virginia.

1. J. Ferguson, *Bibliotheca Chemica*, Derck Verschboyle, London, 1954, Vol. 1, pp. 315–316.
2. *Encyclopedia Brittanica*, Vol. 5, Encyclopedia Brittanica, Chicago, 1986, p. 225.
3. L. Abraham, *A Dictionary of Alchemical Imagery*, Cambridge University Press, Cambridge, 1998.

SECTION IV
CHEMISTRY BEGINS TO EMERGE AS A SCIENCE

THE FIRST TEN-POUND CHEMISTRY TEXT

The first systematic textbook of chemistry was the *Alchemia*, published in Frankfort in 1597 by Andreas Libavius (ca. 1540–1616).[1] The title page of the beautiful enlarged and illustrated second edition, the *Alchymia* (1606, Frankfurt), is shown in Figure 50. My copy of this book is bound in ornate, Italian-tooled vellum, measures about 9 inches by 13½ inches and weighs about 10 pounds. Libavius had a classical education and, in addition to obtaining the M.D. and serving as a physician, was Professor of History and Poetry at the University of Jena. In the manner of Paracelsus, Libavius employed metallic remedies including potable gold (gold dissolved in *aqua regia*) as well as calomel. However, his opinion of Paracelsus was stated thusly: "Paracelsus, as in many other matters he is stupid and uncertain, so also here writes like a madman."[1] While a believer in alchemy, Libavius performed much practical chemistry and noted that lead gains 8–10% in weight upon calcination.[1]

Alchymia describes the construction of a hypothetical chemical "house" (*Domus chymici*) (Fig. 51) with detailed floor plans. The *Domus chymici* was to have a main laboratory, storeroom for chemicals, preparation room, a room for laboratory assistants, a room for crystallizing and freezing, a room for sand and water baths, a fuel room, a museum, gardens, walks, and . . . a wine cellar.[1,2] The book goes on to describe fume hoods, furnaces, glassware, luting material, mortars, forceps, chemical preparations, and everything else needed to be "state of the art" during the time of Shakespeare.

But Libavius means to cover all bases in the textbook market and concludes with an amply illustrated section on the *Lapidum Philosophorum* (Philosopher's Stone). Figures 52 and 53 are described by John Read as representing the Vase of Hermes heated at the bottom.[2] In Figure 52, we see a serpent, representing Sophic Mercury, eating its tail—a representation of coagulation and fixity.[2] The eagle has multicolored feathers representing color changes during fermentation. The black crow represents putrefaction. The maiden represents the moon or silver; the lion represents the sun or gold. The king and queen similarly represent male and female, sulfur and mercury.[2] Some highlights of Figure 53 include the base representing earthly foundation; two Atlases supporting the vessel; a four-headed dragon representing four stages of fire; the Green Lion representing mercurial liquid, the first matter of the stone; a three-headed silver eagle, a black crow representing putrefaction, the winged serpent biting its own tail again, a pretty nasty white swan between two globes, and a number of male–female, earth–moon, or sulfur–mercury images. Pretty obvious when you know what to look for (or have John Read's *Prelude to Chemistry* at your side).

1. J.R. Partington, *A History of Chemistry*, MacMillan, London, 1961, Vol. 2, pp. 244–267.
2. J. Read, *Prelude To Chemistry*, MacMillan, New York, 1937, pp. 212–221.

FIGURE 50 ■ The title page of the second edition (1606) of Libavius' *Alchymia*—the first chemistry textbook (the first edition, published in 1597, was smaller and not illustrated and said by Partington to be rarer than Newton's *Principia* which itself "hammered down" at almost $400,000 at a 1998 book auction).

FIGURE 51 ▪ The *Domus chymici* ("house of Chemistry") in Libavius' *Alchymia* was never built. I suspect that zoning laws would have kept it out of a respectable residential neighborhood.

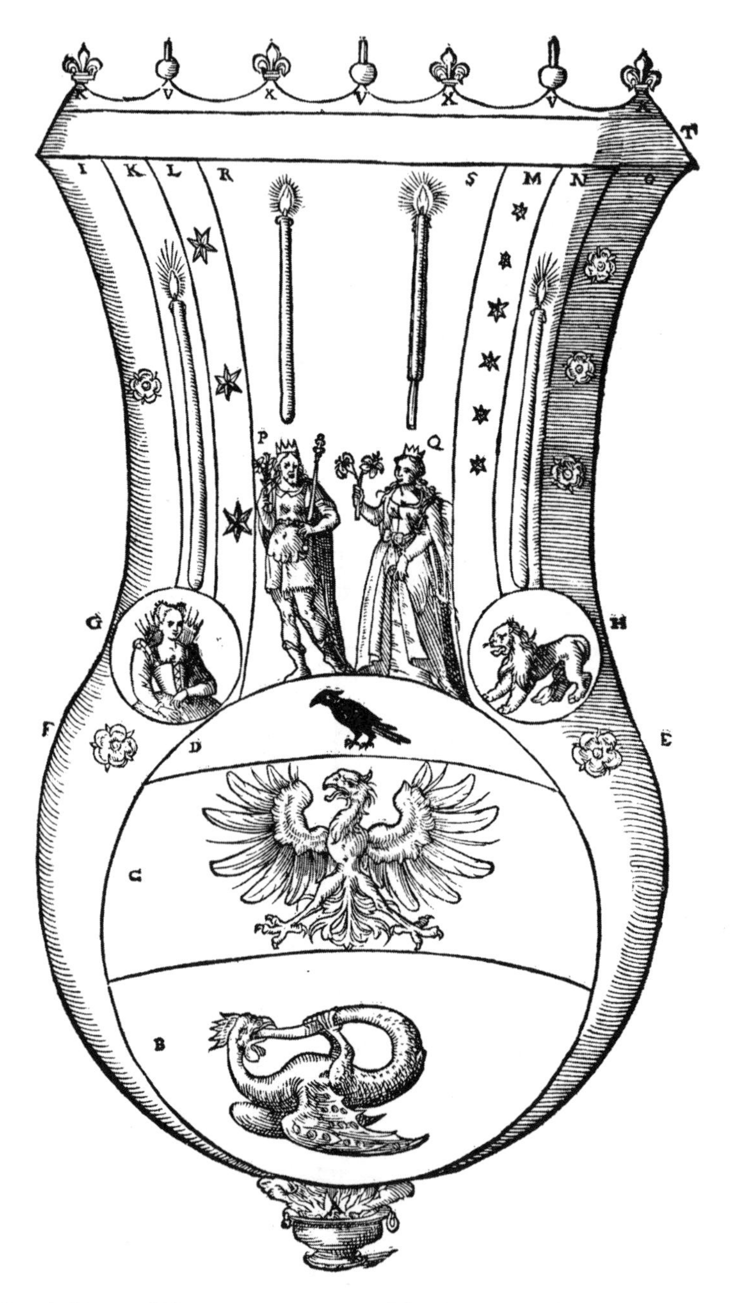

FIGURE 52 ■ A Vase of Hermes representing The Great Work in the section on The Philosopher's Stone in Libavius' 1606 *Alchymia.* Nice to see this schematic after all of the rational description of furnaces, flasks, lutes, forceps, and chemical preparations of the age.

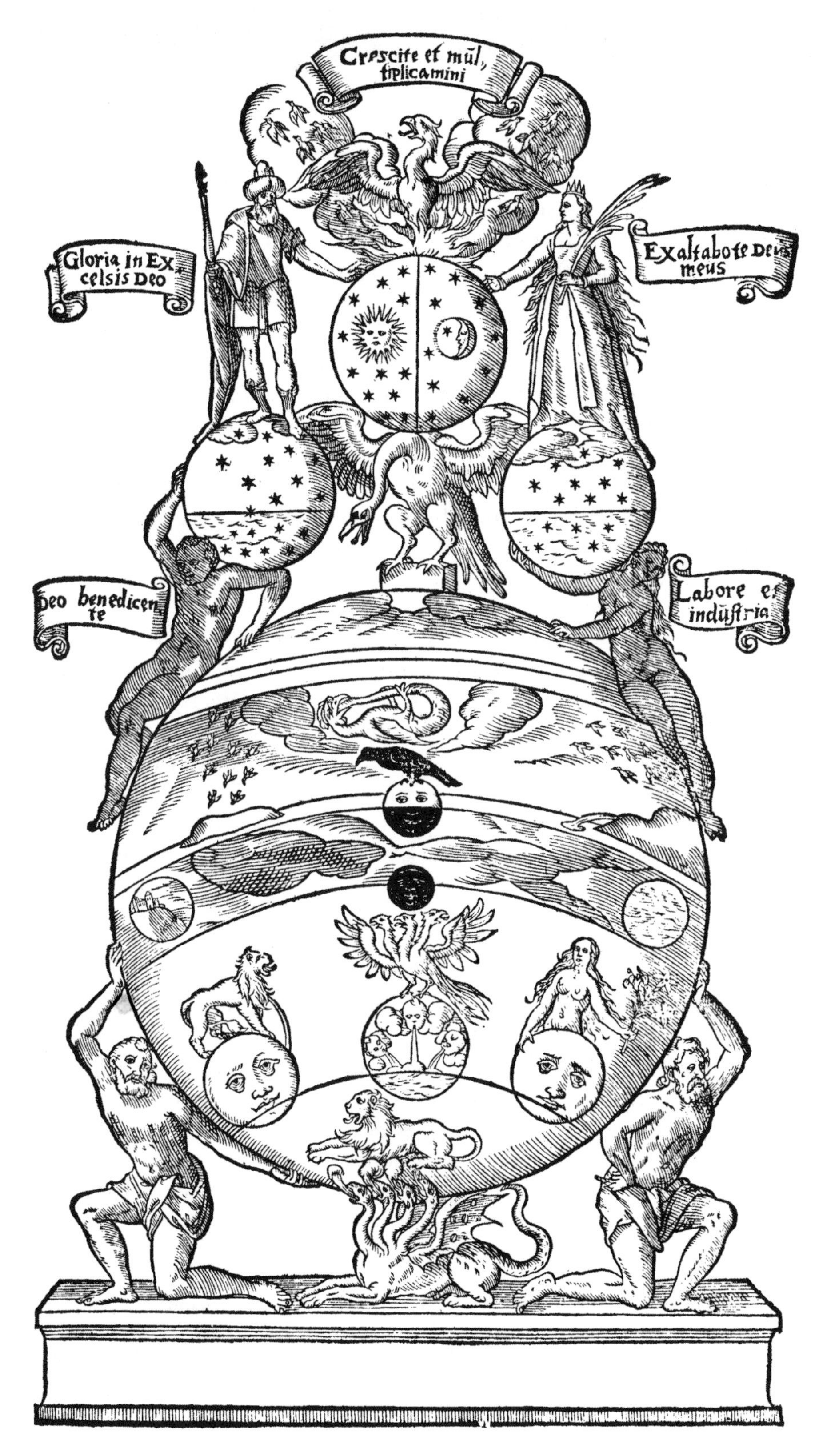

FIGURE 53 ■ Another Vase of Hermes from Libavius' *Alchymia*. What did Libavius' final exam look like?

A TREE GROWS IN BRUSSELS

How ironic that Johann Baptist Van Helmont (1577–1644) refers to "Dame Nature" as the "Proto-Chymist,"[1] for if ever there was a human protochemist it was he. His writings navigate the borders between science, pseudoscience, and superstition. Van Helmont was born in Brussels but travelled extensively. The picture of Van Helmont (Fig. 54, left) is from the *Ortus Medicinae*, compiled by his son the alchemist and polymath Franz Mercurius (Fig. 54, right) and first published in 1648.

At a time when measurement and experiment were just beginning to define science, Van Helmont performed his famous "tree experiment." He believed that there were only two true fundamental elements, water and air, and that trees were composed of the element water. To test this hypothesis, he weighed 200 pounds of dried earth, moistened it with distilled water and added the stem of a willow tree weighing 5 pounds. After five years of judicious watering he determined that the tree weighed 169 pounds, the soil, when separated and dried, still weighed 200 pounds and, thus, the extra 164 pounds could only come from addition of the element water.[2]

These conclusions were, of course, totally erroneous. We now know that the mass of the tree is comprised of cellulose and water. Cellulose is derived from photosynthesis (only discovered some 140 years later) involving carbon dioxide and water. And again, how ironic that the person who coined the term *gas* (from *chaos*) and effectively discovered carbon dioxide did not understand its role in his "tree experiment."

The law of conservation of matter is typically associated with the father of modern chemistry, Antoine Laurent Lavoiser, who worked in the late eighteenth century. Van Helmont's tree experiment demonstrates that this law was a tenable hypothesis over 120 years earlier. And about 150 years after the death of Lavoisier, it evoked near-religious awe in Betty Smith's novel, *A Tree Grows in Brooklyn:*[3]

"Francie came away from her first chemistry lecture in a glow. In one hour she found out that everything was made up of atoms which were in continual motion. She grasped the idea that nothing was ever lost or destroyed. Even if something was burned up or left to rot away, it did not disappear from the face of the earth; it changed into something else—gases, liquids and powders. Everything, decided Francie after that first lecture, was vibrant with life and there was no death in chemistry. She was puzzled as to why learned people didn't adopt chemistry as a religion."

1. J.B. Van Helmont, *A Ternary of Paradoxes* (translated by Walter Charleton), London, 1650, p. 7.
2. H.M. Leicester and H.S. Klickstein, *A Sourcebook in Chemistry 1400–1900*, McGraw-Hill, New York, 1952, pp. 23–27.
3. B. Smith, *A Tree Grows in Brooklyn*, Harper & Brothers, New York, 1943, p. 389. (I thank Professor Susan Gardner for making me aware of this passage.)

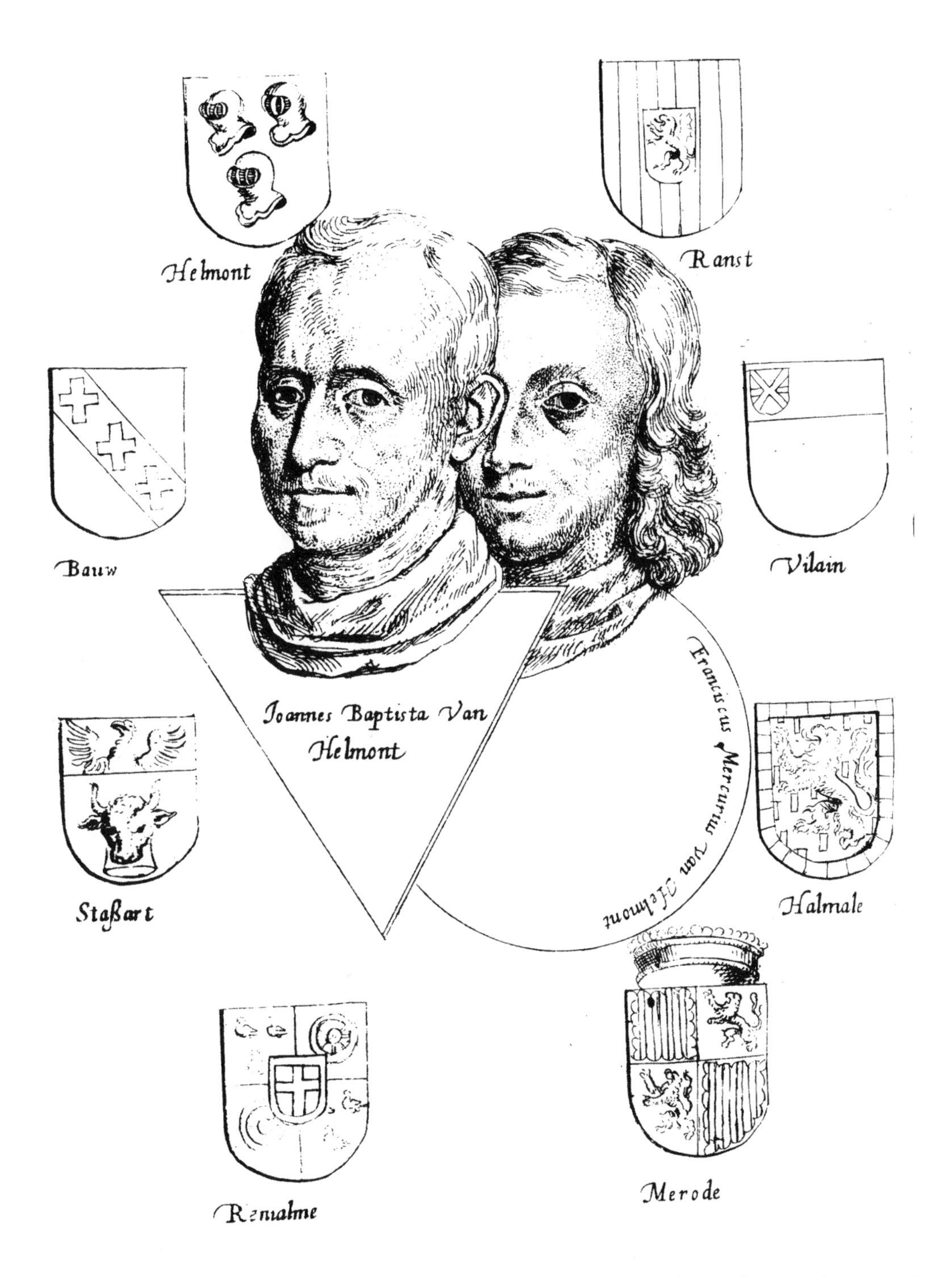

FIGURE 54 ▪ Frontispiece from Johanne Baptist Von Helmont's *Ortus Medicinae* (Amsterdam, 1648) published by his alchemist son Franciscus Mercurius (right).

CURING WOUNDS BY TREATING THE SWORD WITH POWDER OF SYMPATHY

Let us explore Van Helmont's beliefs a bit further. He was a fervent believer in the Powder of Sympathy. Rather than defining it, let us let Van Helmont describe its application[1]:

> . . . Mr. James Howel . . . interceding betwixt two Brothers of the sword, received a dangerous wound through the Arm: By the violent pain whereof, and other grievous accidents concommitant, he was suddenly dejected into extreem Debility and Danger. That in this forlorn plight, despairing to finde ease or benefit, by the fruitless continuance of Chirurgery, and fearing the speedy invasion of Gangraen, he consulted Sir K.D.,[2] who having procured a Garter cruentate, wherewith the hurt was first bound up, inspersed thereon, without the privacy of Master Howel, a convenient quantity of Roman Vitriol. That the powder no sooner touched upon the blood, in the Garter, then the patient cryed out, that he felt an intolerable shooting, and penetrative torment, in his Arm: which soon vanished upon the remove of all Emplasters and other Topical Applications, enjoyned by Sir K.D. That thence-forward, for three days, all former symptoms departed, the part recovered its pristine lively Colour, and manifest incarnation and consolidation ensued: but then Sir K.D. to compleat his experiment, dipt the garter in a fawcet of Vinegar, and placed it upon glowing coals; soon whereupon the Patient relapsed into an extream Agony, and all former evils instantly recurred. And finally, that having obtained this plenary satisfaction, of the Sympathy maintained between blood extravenated, and that conserved in the veins . . . he took again the Garter out from the Vinegar, gently dryed it, and freshly dressed it with the Powder, whereupon the Sanation proceeded with such admirable success, that within few days, there remained only a handsom Cicatrice, to witness there was once a wound.

In other words, treat the dressings that once covered the wound and are covered with blood with the Powder of Sympathy and the cure will be communicated to the blood still in the body. This could be, incidentally, regarded as a conceptual advance beyond Paracelsus' earlier doctrine wherein sprinkling Powder on the sword which caused the wound would heal the wound. Van Helmont argued that it is not the sword, but the blood on the sword, that communicates with the wound. The sympathy concept was the basis for the "wounded dog theory" tested by the Royal Navy in 1687.[3] A dog would be wounded and sent off to sea while its bandage remained in London. At noon, London time, powder of sympathy was sprinkled on the bandage, the dog was supposed to immediately cry and comparison with on-ship time was to provide longitude.

1. J.B. Van Helmont, A *Ternary of Paradoxes* (translated by Walter Charleton), London, 1650, Prologue.
2. Sir K.D. was Sir Kenelm Digby, scientist, physician, privateer, and gifted scoundrel whose Powder of Sympathy (a copper sulfate) was considered the best. He certainly was not faint of heart or capable of much sympathy himself.
3. D. Sova, *Longitude*, Penguin Books, New York, 1995. I am grateful to Professor Thomas W. Mattingly for bringing this to my attention.

DO ANONYMOUS PASSERSBY DEFECATE AT YOUR DOORSTEP? A SOLUTION

Here is another practical use of the Sympathy concept[1]:

> Hath any one with his excrements defiled the threshold of thy door, and thou intendest to prohibit that nastiness for the future, do but lay a red-hot iron upon the excrement, and the immodest sloven shall, in a very short space, grow scabby on his buttocks, the fire torrifying the excrement, and by dorsal magnetism driving the acrimony of the burning, into his impudent anus.

Think about this. First, it is not *only* blood that can "communicate" over long distances. Second, if only the blood of the wounded or the anus of the perpetrator is affected, then the cure and the punishment are "DNA-fingerprinted" —a major advance over late-twentieth-century medical care and forensic science.

1. J.B. Van Helmont, *A Ternary of Paradoxes* (translated by Walter Charleton), London, 1650, p. 13.

A HOUSE IS NOT A HOME WITHOUT A BATH TUB AND A STILL

Johann Rudolph Glauber (1604–1670) is widely considered the father of industrial chemistry and chemical engineering. Although he certainly believed in transmutation, Glauber made numerous important contributions to chemistry. He was the first to describe crystalline sodium sulfate (Na_2SO_4), commonly termed Glauber's salt, and its seemingly amazing medicinal properties[1]:

> Externally adhibited, it cleanseth all fresh wounds, and open Ulcers and healeth them; neither doth it corrode, or excite pain, as other salts are wont to do. Within the body it exerciseth admirable virtues, especially being associated with such things whose virtues it increaseth, and which it conductith to those places to which it is necessary they should arrive . . .

He called sodium sulfate *Sal Mirabile* (Wonder Salt).

The three panels shown in Figure 55 are from *The Philosophical Furnaces*, a work reprinted in the beautiful 1689 folio *The Works of The Highly Experienced and Famous Chemist* . . . . Stills (used for making wine, beer, and medicines) and bathtubs of the period were usually made of copper and thus were extremely expensive and required special furnaces for each application. While not a problem for the wealthy, ownership of these household appliances was often out of reach for the less well off.

FIGURE 55 ■ The men at the bottom are not students who failed Libavius' final exam (see Fig. 53). These figures are from the folio-sized book by Johann Rudolph Glauber, *The Works of the Highly Experienced and Famous Chymist . . . Containing, Great Variety of Choice Secrets in Medicine and Alchymy . . .* (London, 1689). Some consider Glauber to be the first chemical engineer. The figures at the bottom show a common-man's bathtub and sauna designed by Glauber (see text).

Glauber designed a copper globe, about the size of a human head, along with its own furnace that could be moved and plumbed into inexpensive wooden stills and baths. (Plumbing seals were made with "Oxe-bladders" or with starch and paper.) Figure I (Fig. 55) shows the furnace (A) and copper globe (B) and their attachment to distilling vessel C, itself attached to refrigeratory D (with "worm"-twisted copper tubing for condensation), which feeds to receiver E. Figure II depicts a balneum (bath apparatus) with a cover having holes for glasses containing samples for gentle, controlled heating. Figure III shows a wood bath tub as well as a wooden box for a dry bath (to provoke sweat with volatile spirits). The same furnace (A) and copper globe (B) could be used with each appliance.

Although Glauber notes that heat is supplied more slowly than would be the case for a copper appliance with its own customized furnace, the savings are worthwhile (except for the wealthy for whom time is *always* money). As for those foolish enough not to avail themselves of this innovation, Glauber says:

> Let him therefore keep to his copper vessels, who cannot understand me, for it concernes not me.

1. H.M. Leicester and H.S. Klickstein, *A Source Book in Chemistry 1400–1900*, McGraw-Hill, New York, 1952, pp. 30–33.

BOYLE VERSUS ARISTOTLE AND PARACELSUS

Robert Boyle (1627–1691), born to an aristocratic English family, is considered by many to be the true father of chemistry. His most famous book, *The Sceptical Chymist* (1661) (Fig. 56), put an end to the Aristotlean concept of the Four Elements (air, earth, fire, water) and started the descent of the Paracelsian concept of the Three Principles (mercury, sulfur, salt). He demonstrated that air is necessary to support life and to transmit sound. His studies of air as a gas (pneumatic studies) were vital to the discoveries in the late eighteenth and early nineteenth century by Lavoisier, Dalton, Avogadro, and others that led to the atomic theory—the fundamental paradigm of chemistry.

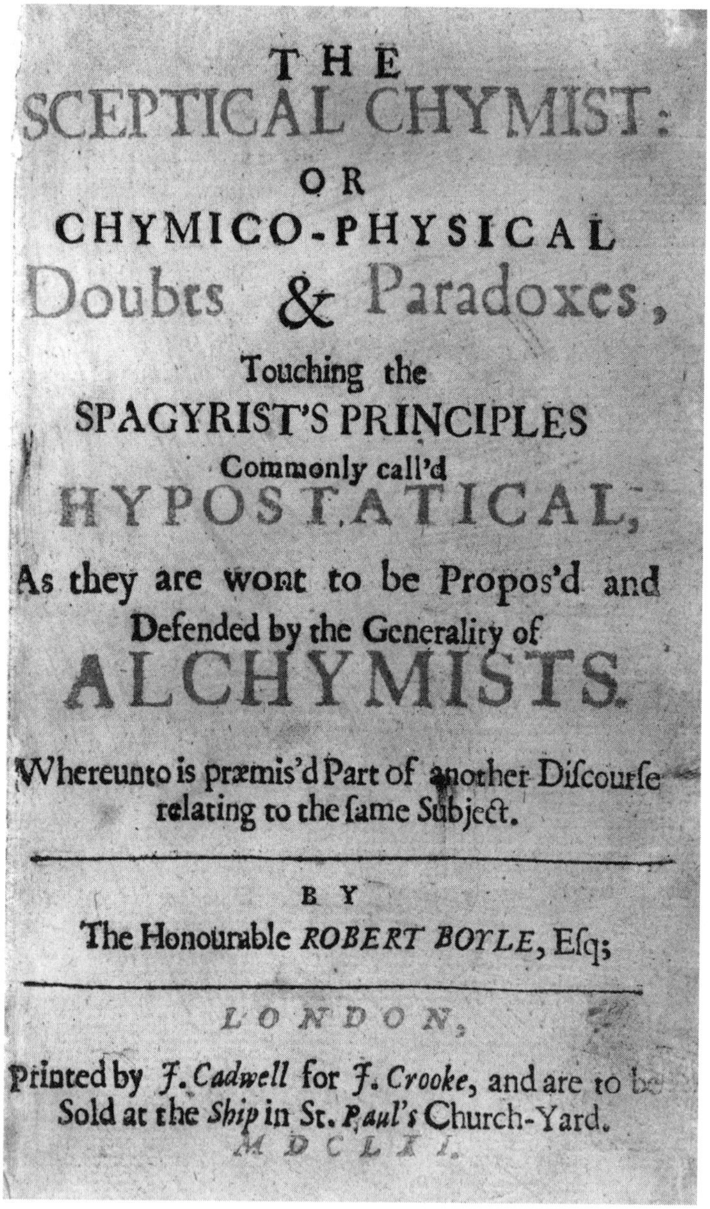

THE
SCEPTICAL CHYMIST:
OR
CHYMICO-PHYSICAL
Doubts & Paradoxes,
Touching the
SPAGYRIST'S PRINCIPLES
Commonly call'd
HYPOSTATICAL,
As they are wont to be Propos'd and
Defended by the Generality of
ALCHYMISTS.
Whereunto is præmis'd Part of another Discourse
relating to the same Subject.

BY
The Honourable *ROBERT BOYLE*, Esq;

LONDON,
Printed by *J. Cadwell* for *J. Crooke*, and are to be
Sold at the *Ship* in St. *Paul's* Church-Yard.
MDCLXI.

FIGURE 56 ■ Title page of Robert Boyle's *The Sceptical Chymist* (London, 1661), which dismissed the four elements of the ancients and helped define the meaning of the term *element* (courtesy Edgar Fahs Smith Collection, Rare Book and Manuscript Library, University of Pennsylvania).

THE ATMOSPHERE IS MASSIVE

What is air? Paraphrasing David Abram[1]: We are immersed in the invisible air, but we barely even perceive it. We sense its effects—it is needed to support life—but not its substance. Perception also rides upon windy drafts, which, in early times, might have been regarded as ethereal breaths of nature.

Why learn the gas laws in chemistry? We have known since the early nineteenth century that the gaseous state is where molecules roam as freely as individuals. This permits understanding of their physical and chemical behavior at the simplest levels. We also learned that two equal-sized balloons of hydrogen gas react totally and precisely with one equal-sized balloon of oxygen gas to produce water identical in mass to the two gases combined.

Galileo (1564–1642) was the first to attempt to determine the density of air (around 1638).[2] He forced air into a narrow-necked bottle, weighed the closed bottle, allowed the excess air to escape, and weighed the closed bottle again. (Galileo, who discovered the moons of Jupiter, spent the last eight years of his life under house arrest for teaching the Copernican view of the solar system.) Evangelista Torricelli (1608–1647) invented the barometer around 1643. At sea level, the atmosphere will support a column of mercury precisely 760 mm (roughly 30 inches) in height. Since mercury is 13.6 times denser than water, this would correspond to a column of water almost 34 ft high. This is the reason why an old-fashioned farm-type pump cannot raise water that is 34 ft deep or more. Figure 57 is a wonderfully stylized diagram of a barometer in the book *Traitez de l'Equilibre des Liqueurs et de la Pesanteur de la Masse de l'air . . .* published in 1663 by Blaise Pascal (1623–1662). In 1648, Pascal sent his brother-in-law Perier to measure the air pressure on the top of a mountain[2] and confirmed that the pressure was lower than that at sea level—clearly the atmosphere has mass even though we do not routinely perceive it. [The modern unit of air pressure, defined as force per unit area (1 newton per square meter) is the pascal (Pa): 760 mm = 101,325 Pa]. The inventor of the first computer, Pascal was a religious philosopher who entered a state of grace late in his life: "He can only be found by the ways taught in the Gospel. Greatness of the human soul. 'Righteous Father, the world has not known thee, but I have known thee.' Joy, Joy, Joy, tears of joy."

Otto von Guericke (1602–1686) invented the first vacuum pump around 1654.[2] During that year he conducted one of the greatest scientific demonstrations of all time. Figure 58 depicts the scene in Regensburg, Germany. In the presence of Emperor Ferdinand III, von Guericke used his pump to evacuate the air from a sphere assembled from two copper hemispheres. Although the sphere was only 14 inches in diameter, two teams of horses could not pull the hemispheres apart. A 760 mm column of mercury with a base of one square inch weighs about 14.7 pounds. Thus, atmospheric pressure is about 14.7 pounds per square inch. Since the surface area of the evacuated sphere was about 616 square inches, the total force on it was equivalent to a weight of 9000 pounds (4.5 tons). The total surface area of an adult human is much larger than that of the copper sphere, and thus the weight of the atmosphere upon us is much greater than a mere 4.5 tons. Fortunately, we are not hermetically sealed. Our internal

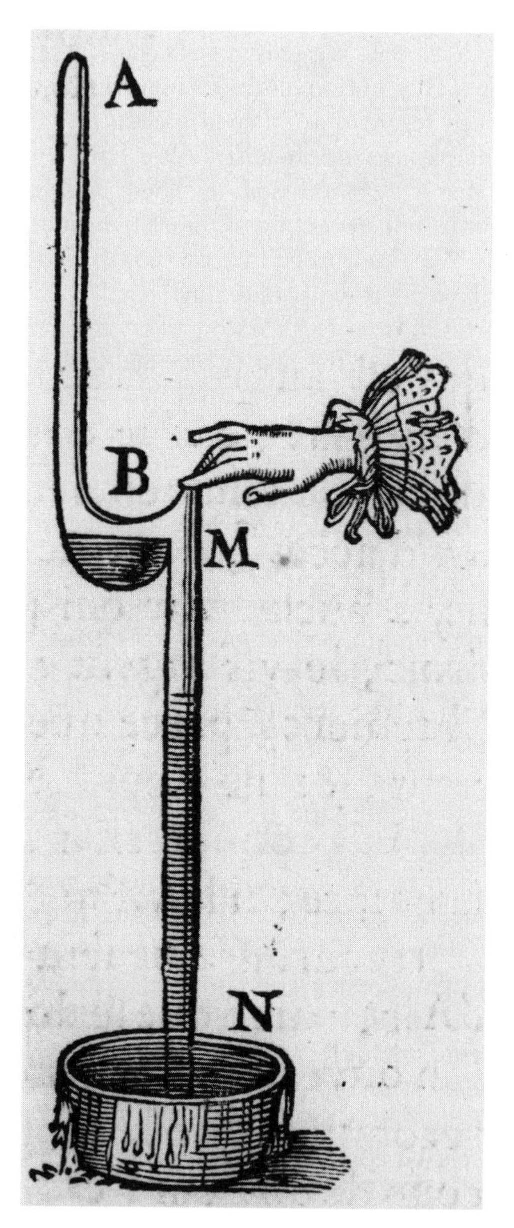

FIGURE 57 ▪ A figure from Blaise Pascal's *Traitez de L'Equilibre des Liqueurs, et de La Pesanteur de La Masse de L'Air* (Paris, 1663) depicting a highly stylized barometer. He sent his brother-in-law Perier to measure the atmospheric pressure on a mountain top (courtesy Edgar Fahs Smith Collection, Rare Book and Manuscript Library, University of Pennsylvania).

FIGURE 58 ▪ One of the greatest science demonstrations of all time: When von Guericke used his vacuum pump to remove the air from a sphere only 14 inches in diameter, teams of horses could not overcome the 9,000-pound (4.5-ton) force of atmospheric pressure pushing the hemispheres together [from Von Guericke's *Experimenta Nova* (1672)] (courtesy Edgar Fahs Smith Collection, Rare Book and Manuscript Library, University of Pennsylvania).

pressure equalizes the outside pressure and so we are blithely unaware of this matching of huge forces within and without our bodies.

In Figure 59, we see the Boylean vacuum pump, built by Boyle's youthful assistant Robert Hooke (1635–1703) in 1655. The large glass globe is sealed at the top with a brass rim and brass key. A stopcock (SN) connects the globe to brass cylinder P, which has a piston in it sealed with leather and run by a rack-and-pinion mechanism worked by hand crank. Plug R fits tightly into a hole in the cylinder. A vacuum is pumped as follows: With stopcock SN open and plug R in place, the piston is drawn down, removing air from the globe. The stopcock is closed, plug R removed, and the airtight piston raised to force out the collected air. The process is repeated.[2] In his 1665 book *Micrographia* Hooke first used the word *cell* to describe the honeycomb structure of cork visible by microscope.

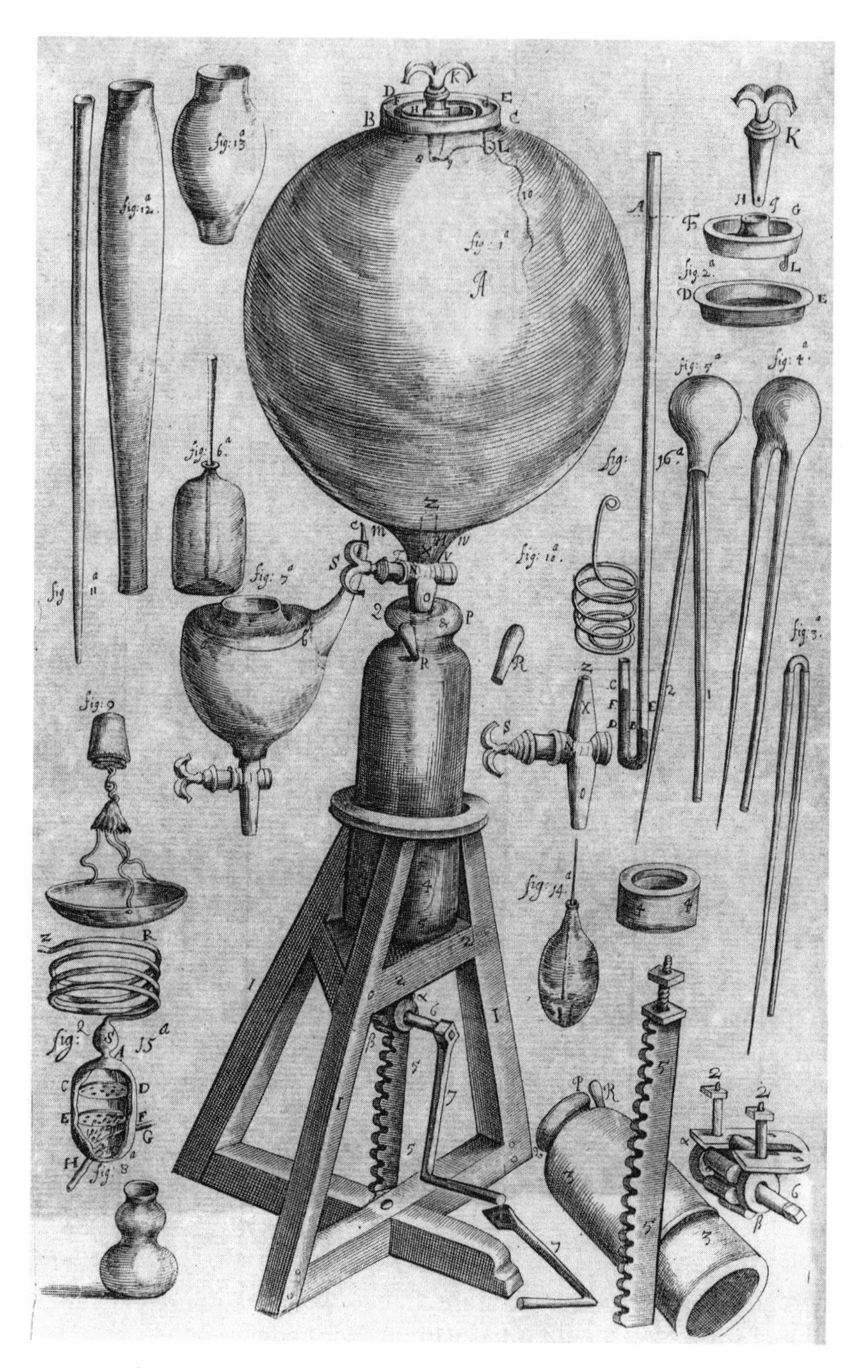

FIGURE 59 ■ The Boylean vacuum pump, built by Robert Boyle's assistant Robert Hooke (from *New Experiments Physico-Mechanical, Touching The Spring of the Air*, 2nd ed., London, 1662).

1. D. Abram, *The Spell of the Sensuous*, Pantheon, New York, 1996, p. 260. I am grateful to Professor Susan Gardner for introducing me to this book and suggesting some of the themes of the present essay.
2. J.R. Partington, *A History of Chemistry*, MacMillan, London, 1961, Vol. 2, pp. 512–519.
3. J. Steinmann, *Pascal* (translated by M. Turnell), Harcourt, Brace & World, New York, 1965, p. 80.

BOYLE'S LAW

The second edition of Boyle's first book, *New Experiments Physico-Mechanical Touching the Air*, was published in 1662 and contained a section titled "A Defense of Mr. Boyle's Explications of his Physico-mechanical Experiments, against Franciscus Linus." In this section, he disclosed the relationship between the pressure and the volume of a gas that we now call Boyle's Law—the first Ideal

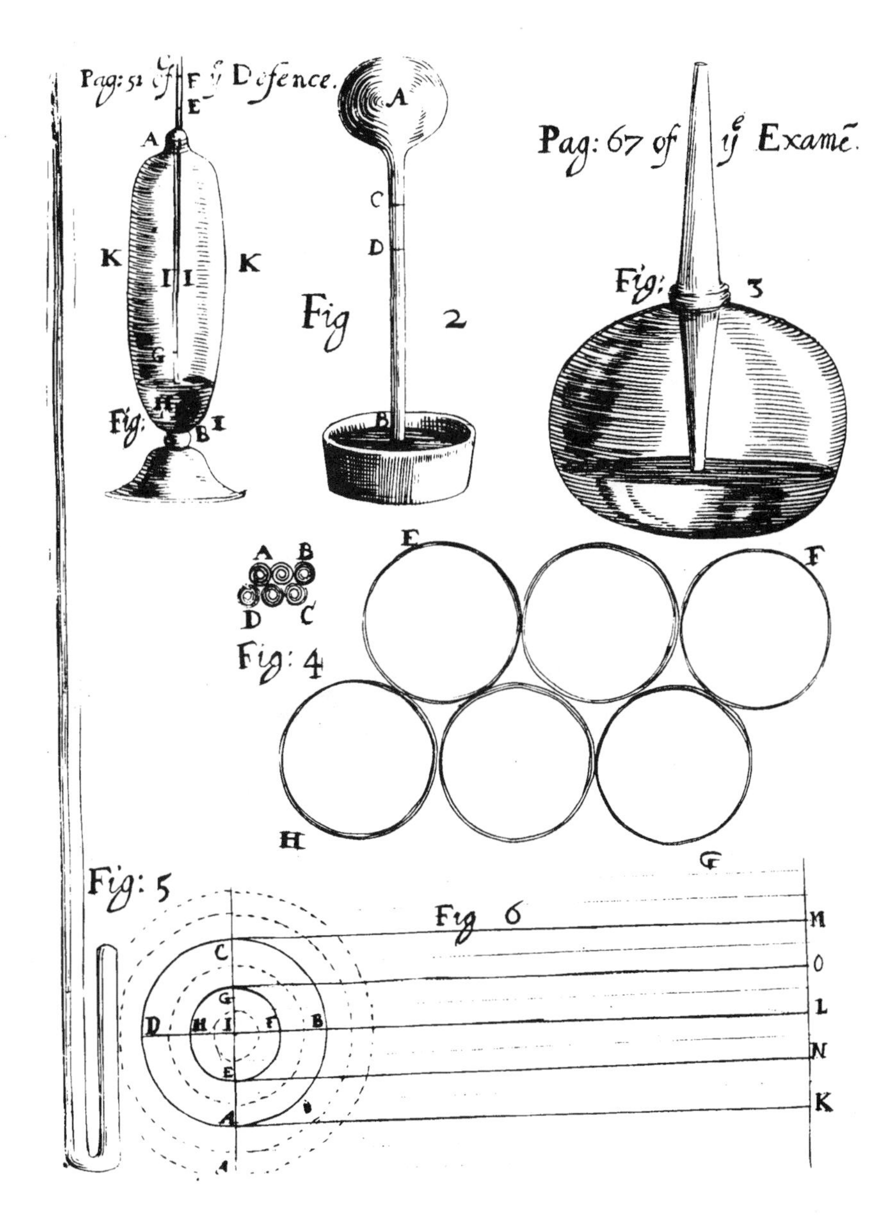

FIGURE 60 ■ In *Fig.* 5 we see Robert Boyle's famous J tube used to demonstrate that $PV = k$ (Boyle's Law). Air is trapped by mercury in the small arm of the J tube. As more mercury is added, the volume of the air decreases. (From *New Experiments Physico-Mechanical . . .*, 1662).

Gas Law. Why must all high school chemistry students learn this simple relationship? In part, because Boyle's Law and the other gas laws helped to establish the reality of atoms and molecules over 150 years later.

In the plate shown here (Fig. 60), *Fig.* 5 depicts the J tube Boyle designed to test the pressure–volume relationship of the only gas he knew—air. The experimental data described here are taken directly from Boyle's book. On the day he performed the experiments, the air pressure measured with a barometer was $29^2/_{16}$ inches of mercury (the pressure of atmospheric air supports a column of $29^2/_{16}$ inches against a vacuum). Boyle poured mercury into the open end of the J tube so as to trap a parcel of air, and he carefully adjusted the amount of mercury so as to have equal heights of mercury in both arms. This means that the pressure on the trapped air sample is $29^2/_{16}$ inches. (Since the two arms of the tube have the same cross-sectional area, the volume is directly related to height, in inches, which Boyle used as his measure of relative volume.) If enough mercury is added to compress the air "volume" to 9 inches ($^3/_4$ of the original volume), the total pressure is $39^4/_{16}$ inches ($29^2/_{16}$ + $10^2/_{16}$) or about $^4/_3$ of the original pressure. If sufficient additional mercury is added to compress the height of the trapped air to 6 inches from its original 12 inches, this air packet is supporting $29^{11}/_{16}$ inches of mercury in addition to the atmospheric $29^2/_{16}$ inches for a total of $58^{13}/_{16}$ inches: double the pressure, halve the volume. When enough mercury has been added to compress the air to 3 inches (one-fourth of original volume), the total pressure on the trapped air packet is $88^7/_{16}$ + $29^2/_{16}$ or $117^9/_{16}$ inches or four times the original pressure.

Thus, the form of Boyle's Law is:

$$PV = \text{constant} \quad \text{or} \quad P_1V_1 = P_2V_2 = P_3V_3 = P_4V_4 = \cdots$$

WHO WOULD *WANT* AN ANTI-ELIXIR?

A strange narrative indeed! Although *The Sceptical Chymist* rid chemistry of the Aristotlean Elements, Boyle was a believer in the possibility of transmutation (as was fellow member of the Royal Society Isaac Newton).

This pamphlet (Fig. 61) is considered to be the rarest of Boyle's works. Of the first (anonymous) edition published in 1678 and this second, attributed edition of 1739, Duveen[1] accounted for only four known copies combined, although Ihde[2] suggests possibly four copies of each edition. In it Boyle narrates a series of one-time-only reverse transmutation experiments he witnessed in which the transmuting agent was a miniscule amount of solid substance. The claim tested was that the substance could transform gold into a baser metal. Why would anybody be interested in such an "anti-elixir"? Using very modern chemical logic, Boyle reasoned that if one learns how to transmute gold into a baser metal, then one would also gain the knowledge to perform the reverse operation.

The experiments narrated in this pamphlet gave tantalizing but inconclusive evidence for the chemical degradation of gold into a lesser metal, perhaps

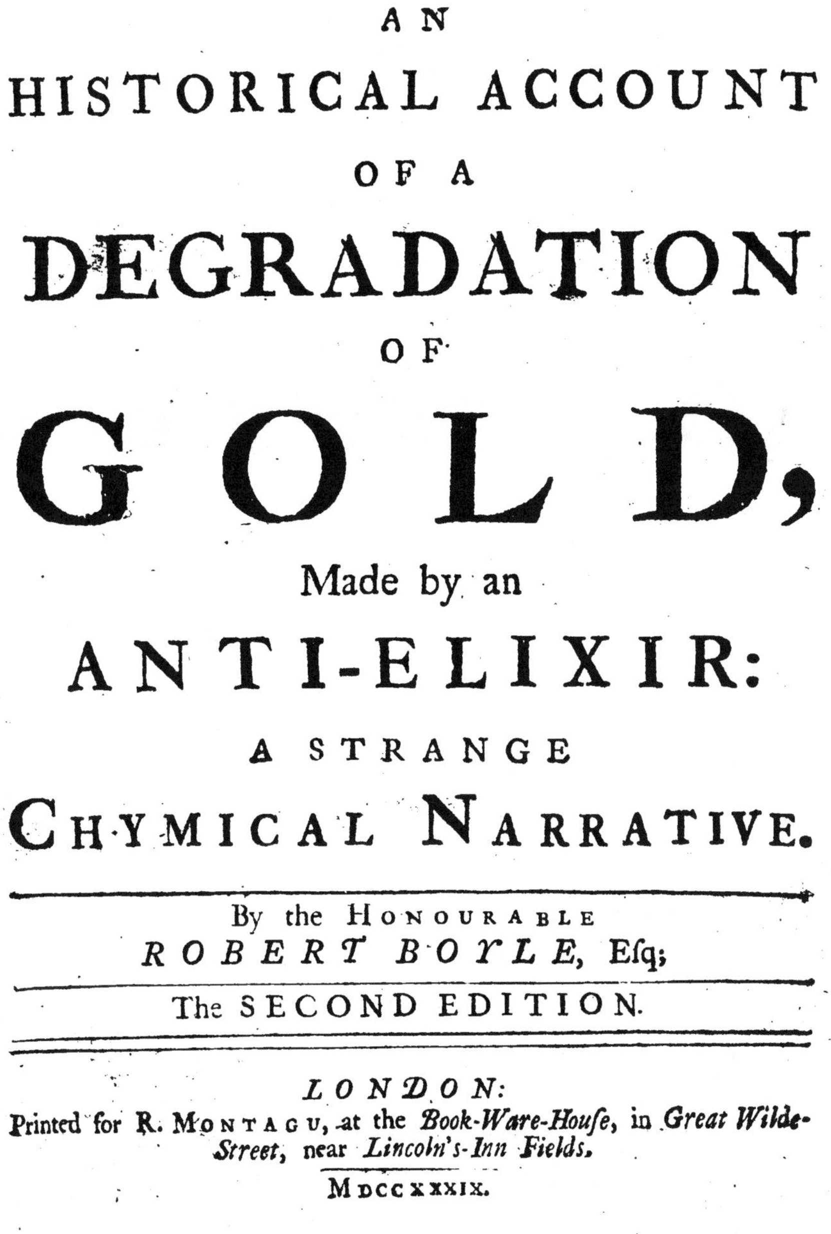

AN
HISTORICAL ACCOUNT
OF A
DEGRADATION
OF
GOLD,
Made by an
ANTI-ELIXIR:
A STRANGE
CHYMICAL NARRATIVE.

By the HONOURABLE
ROBERT BOYLE, Esq;

The SECOND EDITION.

LONDON:
Printed for R. MONTAGU, at the *Book-Ware-House*, in *Great Wilde-Street*, near *Lincoln's-Inn Fields*.
MDCCXXXIX.

FIGURE 61 ■ Title page from the second edition of Boyle's rarest work—his witnessing of a "reverse transmutation." The first edition, published in 1678, was anonymous.

even a salt, but the world's known supply of anti-elixir was consumed—apparently never to be rediscovered. Ihde[2] speculated over whether the experiment was ever done at all, done incompetently, or was possibly a joke by Boyle. His firmer conclusions were that the experiment was, in all likelihood, actually carried out at Boyle's customarily high level of competence and that Boyle had no sense of humor, especially in regard to experimentation. Ihde's tentative conclusion: some sleight of hand by one of Boyle's laboratory assistants to give the chief his desired conclusion and help him recover from an earlier embarrassment at the hands of that young upstart Isaac Newton.[2] Newton was credulous about alchemy and this plays a part in a novel[3] in which the "Aetherial Spirit" is embodied in the 9 lives of a Golden Cat born every 81st generation to parents (Feline Sol and Luna) whose *conjunctio* produces the quintessential cat.

1. D. Duveen, *Bibliotheca Alchemica Et Chemica*, HES, Dordrecht, 1986, p. 97.
2. A. Ihde, *Chymia*, No. 9, 47–57, 1964.
3. S. G. King, *The Wild Road*, Ballantyne, New York, 1997, pp 328–329. I thank Ms. Susan Greenberg for bringing this to my attention.

THE TRIUMPHAL CHARIOT OF ANTIMONY

The plate shown in Figure 62 is from Nicholas Le Fevre's *A Compleat Body of Chymistry* (second English edition, 1670), one of the important texts of the seventeenth century. It depicts a chemist calcining (forming the calx or oxide) of metallic antimony using sunlight.

Antimony was one of the nine elements known to the ancients.[1] It was found as the ore stibnite (Sb_2S_3), and this black sulfide was used by women as an eye cosmetic in biblical times. An early means for obtaining the metal was to roast the ore on charcoal heated to incandescence. Later methods involved

FIGURE 62 ■ Calcining antimony using a heating glass in Le Fèvre's 1670 edition of *A Compleat Body of Chymistry*.

heating stibnite with tartar and nitre or with iron. The resulting "lead" was used to fashion a Chaldean vase of pure antimony around 4000 B.C.[1]

Early chemical books show an amazing fascination with antimony far beyond our modern interest. Why? One reason was its preferred use for releasing gold from metallic impurities. Antimony has a fairly low affinity for sulfur (higher than gold, lower than silver—see Geoffroy's Table of Affinities [Figures 76 and 77]—pure antimony or Regulus of Antimony is represented by a three-pointed crown). Thus, its common ore will release sulfur to baser metals forming "scum" easily scooped from molten gold. It can separate silver from gold since silver captures sulfur from stibnite and the resulting liquid slag of silver sulfide and antimony sulfide is separable from gold antimonide. This last is burned to free the volatile antimony oxide, leaving pure gold.[2]

The wolf depicted in the First Key of Basil Valentine [see Fig. 17(a)], represents antimony (sometimes called *lupus metallorum* or wolf of the metals by the alchemists). Another famous seventeenth-century picture depicts the wolf devouring a dead human (impure gold) with subsequent burning of the wolf (loss of volatile antimony oxide) to release the King (gold).[3]

Now if metals could so effectively be purged of their impurities by antimony, should it not also be an effective human medicine—a purge (or emetic) to remove illness? Paracelsus first described antimony as a purge and set off a violent philosophical debate among physicians. The classical Galenical view was the use of a medicine with properties *contrary* to the disease. The Paracelsans argued for cure by *similitude* (i.e., fight poison with poison). The question of whether antimony was a medicine or a poison raged over centuries but was apparently settled by the cure of Louis XIV with *vin emetique* (emetic wine—yum) in 1658.[4] Le Fevre was much taken with medicinal antimony and particularly with its purification and fixation (as the calx) by the sun.[4] He too noted the increase in weight upon calcination. The book *Triumphal Chariot of Antimony*, first published in 1604 and attributed to the legendary Benedictine Monk, Basil Valentine, used this flashy, Hollywood-like title to strike a blow for antimony in this long and passionate debate. For a modern encore, we eagerly await the movie version starring Charlton Heston as the chariot-driver.

It is worthwhile recognizing that modern anticancer agents "poison" normal cells, but are greater poisons to cancerous cells that multiply much more rapidly. Thus, the Paracelsian view is vindicated in this case but not in neutralizing stomach acid.

1. N.N. Greenwood and A. Earnshaw, *Chemistry of the Elements*, Pergamon, Oxford, 1984, p. 637.
2. F. Ferchl and A. Sussenguth, *A Pictorial History of Chemistry*, William Heinemann, London, 1939, p. 61.
3. J. Read, *Prelude To Chemistry*, MacMillan, New York, 1937, pp. 200–202; 240–241 [see Plate 47 in this book, which is taken from the book by Michael Maier (1687) titled *Secretioris Naturae Scrutinium Chymicum*].
4. A.G. Debus, *The French Paracelsans*, Cambridge University Press, Cambridge, 1991, pp. 21–30, 95–99

A SALTY CONVERSATION

We have previously spoken of Johann Baptist Van Helmont in light of the Powder of Sympathy and his famous Tree Experiment. Although he was a disciple of Paracelsus and a believer in metallic medicines, he did not accept Paracelsus' *tria prima* nor the four elements of the ancients. Van Helmont believed in two elements, Water and Air, with only the first comprising matter. He was an independent thinker and drew the attention of the Spanish Inquisition (Spain occupied the Low Countries during parts of the sixteenth and seventeenth centuries). He spent the final 20 years of his life under house arrest.[1] Following his death in 1644, his son Franciscus Mercurius Van Helmont published his complete medical writings as *Ortus Medicinae* (1648). His work included recognition of the role of acid in digestion, the role of bile in digestion, and the role of acid in inflammation and the production of pus.[2] Van Helmont and Sylvius[1,3] (Francois Dubois, Franciscus de la Boe, 1614–1672) represented the golden age of iatrochemistry. Sylvius rejected the *Archeus* (see the next essay), which Van Helmont had merely modified. He recognized that although bile (for example, dog bile) tasted acidic (!), it was really alkaline. Aware that acidic substances and alkaline substances produced effervescence and/or heat upon mixing, Sylvius envisioned warfare between acid and alkali in living beings.[1] Sylvius' student Otto Tachenius promoted the acid–alkali theory of his master but added the unifying concept of the salt—the union of acid and alkali. This greatly improved the classification beyond the taste test, but it was Robert Boyle who discovered the quantitative test. In his *Reflections upon the Hypothesis of Alcali and Acidium* (1675) Boyle defined acids as bodies that turn syrup of violets red and alkalies as bodies that turn this indicator green.[1]

A Dialogue Between Alkali and Acid (Fig. 63) published by physician Thomas Emes (2nd ed., 1699; 1st ed., 1698) is a wonderful example of invective directed against another physician, John Colbatch, who believed that the causes of diseases were alkaline and the cures acidic.[4] Emes ends his 59-word title thus: *Being a Specimen of the Immodest Self-Applause, Shameful Contempt, and abuse of all Physician gross Mistakes and great Ignorance of the Pretender John Colbatch.* Now why won't my publisher let me use such a nifty title? And so the book begins:

Alkali: Well met Mr. Acid, whither are you hurrying so fast, to Some Heroe run through the Lungs, or the Heart?

Acid: I should hardly have to tell you Mr. Alkali, but that I am engag'd to oppose you where-ever we meet, you Principle of Death and Corruption, I am always provoked by you, you have done so much mischief in the World: And now to your farther reproach, I have a fresh instance of your badness, by a Messenger from my Lord Lazington, whom you have plagued with a fit of the Gout, and that a desperate one if I come not in time to his assistance, none can help him but I, and he thinks it 7 Years ere I come to him.

Alkali: You are very sharp Mr. Acid . . .

Stop! Enough! I hate that pun. Acid in the role of Scarlet Avenger? I suspect

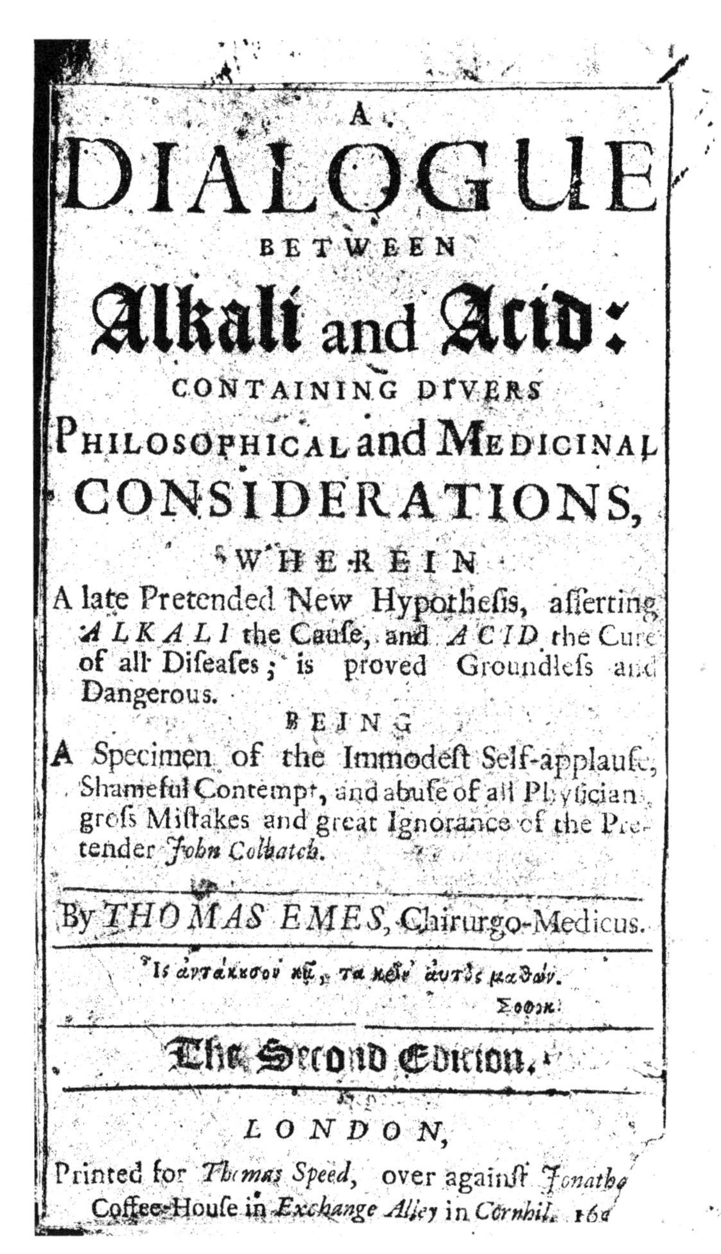

A

DIALOGUE

BETWEEN

Alkali and Acid:

CONTAINING DIVERS

PHILOSOPHICAL and MEDICINAL

CONSIDERATIONS,

WHEREIN

A late Pretended New Hypothesis, asserting *ALKALI* the Cause, and *ACID* the Cure of all Diseases; is proved Groundless and Dangerous.

BEING

A Specimen of the Immodest Self-applause, Shameful Contempt, and abuse of all Physicians, gross Mistakes and great Ignorance of the Pretender *John Colbatch*.

By *THOMAS EMES*, Chirurgo-Medicus.

The Second Edition.

LONDON,

Printed for *Thomas Speed*, over against *Jonatha*[n's] Coffee-House in *Exchange Alley* in *Cornhil*. 16[illegible]

FIGURE 63 ■ Title page of the second edition (1699) of physician Thomas Emes' *Dialogue Between Alkali and Acid*. The question is: Which one is the cause of disease and which the cure?

that when I start to hear test tubes of acid and alkali speaking to each other, it may be the moment to retire from chemistry and open that bookstore I've dreamed about.

1. W.H. Brock, *Norton History of Chemistry*, W.W. Norton, New York, 1993, pp. 41–63.
2. J.R. Partington, *A History of Chemistry*, MacMillan, London, 1961, Vol. 2, pp. 209–216.
3. J.R. Partington, op. cit., pp. 281–290.
4. J.R. Partington, op. cit., p. 290.

THE ALCHEMIST IN THE PIT OF MY STOMACH

Galen's views had dominated medicine for 1400 years. He believed that a balance of four bodily humours (phlegm, black bile, yellow bile, and blood) was required for sound health. Paracelsus was as bombastic as his family name (Theophrastus Bombast von Hohenheim) and joyfully trashed the Galenists and everybody else he disagreed with, leaving scores of enemies wherever he traveled.

Paracelsus believed that the purpose of alchemy was to create new medicines rather than gold.[1] He relied upon synthetic inorganic (metallic) compounds as medicines rather than the extracts of herbs used since ancient times. As noted earlier, he believed in similitude and used poisons to kill poison—only, however, after sweetening or dulcifying them. Thus, liquid mercury metal dissolved in *aqua fortis* (nitric acid) followed by evaporation and calcination produced mercury oxide, which was employed against venereal diseases.[1] Similarly, metallic mercury dissolved in *aqua fortis* could be precipitated by adding salt water to produce solid calomel (Hg_2Cl_2).[1] This was an effective purgative (laxative) that could relieve gastric stress and remove intestinal worms.

Paracelsus' mysticism included a belief that the body is born completely healthy but that during life disease is received from food.[1] In the stomach he felt that there exists a vital force—the *Archeus*—a kind of Alchemist of Nature having a head and hands only. The *Archeus* separates the nutritive part of food from the poisonous part, the latter eliminated, in part, as excrement. Air is similarly digested to produce a nutritive part and a poisonous part ("excrementous air"?[2]). The *Archeus* can become sick if the separation is incomplete, and this results in the individual's illness. In this light, the laxative calomel clearly helps the *Archeus* remain healthy—a sound mind (and a sound *Archeus*) in a healthy body. Figure 64 is artist Rita L. Shumaker's depiction Of the *Archeus*.[3] The artist used tripe to model the human omentum, the inner membrane that connects the stomach with other organs and supporting blood vessels. The androgenous *Archeus* is depicted in the vicinity of the intestines, intimately part of the membranous fabric.

1. J.R. Partington, *A History of Chemistry*, Macmillan, London, 1961, Vol. 2, pp. 115–151.
2. "Thy Worst. I fart at thee" (said by *Subtle*, the second line in the 1610 play, *The Alchemist*, by Ben Jonson).
3. The author thanks Ms. Rita L. Shumaker, a faculty member at the University of North Carolina at Charlotte, for this original drawing and her imaginative evocation of the *Archeus*.

FIGURE 64 ■ Artist Rita L. Shumaker's 1999 rendition of the *Archeus* believed by Paracelsus to inhabit the region near the stomach. It has a head and two hands and separates the nutritive part of food from the excrementous. If the Archeus was overwhelmed with excrementous matter, it would become ill as would the human inhabitant. Take calomel, sayeth Paracelsus, to unburden the Archeus.

A HARVARD-TRAINED ALCHYMIST

Renaissance alchemy conjures up back alleys in Prague and other Old World images. Harvard University is, of course, strictly New World in our minds—a cradle for progressive thought and the home of Nobel Laureates. How delightful that George Starkey (born in Bermuda in 1628, died in the London plague of 1665) provides us with a surprising conjunction of the Old and New Worlds.[1,2] Eirenaeus Philalethes ("A Peaceful Lover of Truth") was the pseudonym provided for his posthumous writings and *Secrets Revealed* [see Fig. 65(a)] is the English translation of his most influential book.

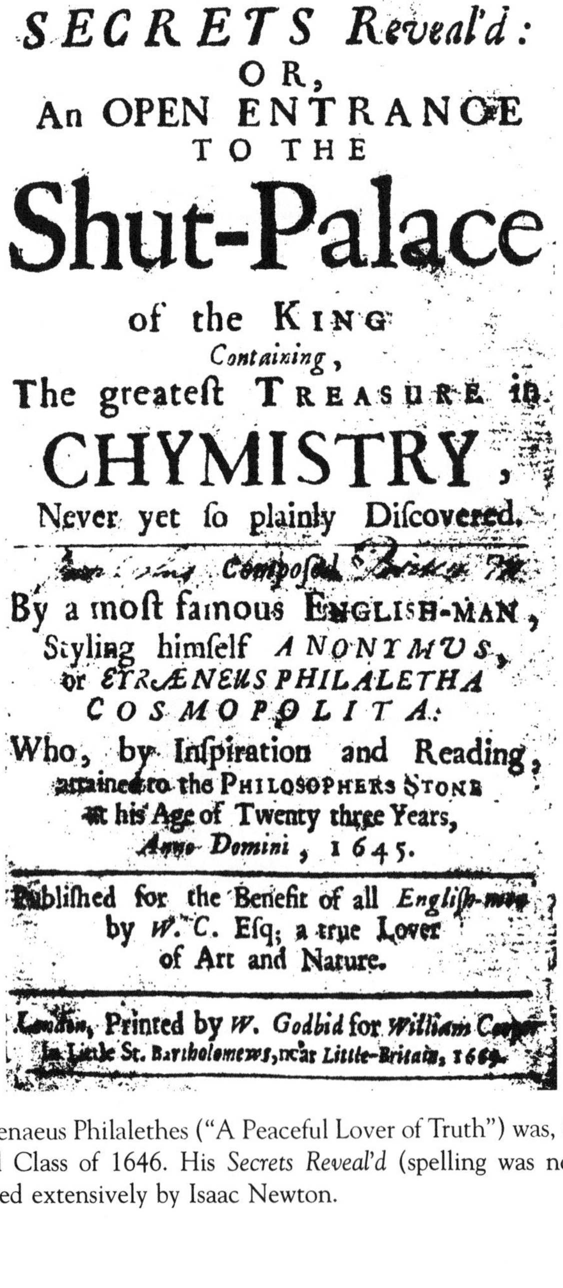

SECRETS Reveal'd:
OR,
An OPEN ENTRANCE
TO THE
Shut-Palace
of the KING
Containing,
The greateſt TREASURE in
CHYMISTRY,
Never yet ſo plainly Diſcovered.
Compoſed
By a moſt famous ENGLISH-MAN,
Styling himſelf ANONYMUS,
or EYRÆNEUS PHILALETHA
COSMOPOLITA:
Who, by Inſpiration and Reading,
attained to the PHILOSOPHERS STONE
at his Age of Twenty three Years,
Anno Domini, 1645.
Publiſhed for the Benefit of all Engliſh-men
by W. C. Eſq; a true Lover
of Art and Nature.
London, Printed by W. Godbid for William Cooper
In Little St. Bartholomews, near Little-Britain, 1669.

FIGURE 65 ■ Eirenaeus Philalethes ("A Peaceful Lover of Truth") was, in reality, George Starkey, Harvard Class of 1646. His *Secrets Reveal'd* (spelling was not emphasized at Harvard) was cited extensively by Isaac Newton.

Starkey graduated from Harvard in 1646, one of a class of four[3] who received their lectures from President Dunster.[2] He shared a dorm room (measuring no more than 7 feet 9 inches by 5 feet 6 inches) with a John Allin.[2] Courses included "Logick," "Physicks," "Ethicks and Politicks," and "Arithmetick and Geometry."[2] The natural philosophy curriculum at Harvard reflected some of the finer points of the great debate between the Aristotleans and Cartesians (matter is continuous; there are no vacant spaces; "nature abhors a vacuum") and those who believed in corpusular matter, including Newton and Boyle. According to Newman,[2] this division was not so clear-cut at Harvard where a late Aristotlean view, which allowed for finite particles, was in currency. In any case, following graduation Starkey rated the natural philosophy curriculum at Harvard as "totally rotten."[2]

On the basis of his examination of Harvard theses from the mid-seventeenth to late-eighteenth centuries, Newman[2] notes the following successfully defended positions:

1687	Is there a stone that makes gold? Yes.
1698, 1761	Is there a universal remedy? Yes in 1698, no in 1761.
1703	Can metals be changed into one another alternately? Yes.
1703, 1708, 1710	Is there a sympathetic powder? Yes.
1771	Can real gold be made by the art of chemistry? Yes.

As Newman notes,[2] "Obviously, Harvard was far from being an uncongenial place for the budding alchemist; as late as 1771, Harvard undergraduates were defending the powers of the philosopher's stone" (and these were not only the "New Age" people).

Moving to England in 1650, Starkey became an important exponent of Van Helmont's approach and worldview. Van Helmont did not support the Galenical view of medication (contraries) or the Paracelsian view (similitude). Instead he believed in cures that produced "healing ideas in the *Archeus*"—the inner architect or life spirit (Figure 64) located in a region between the stomach and spleen.[4] Van Helmont and Starkey shared a belief in the importance of pyrotechny (arts, such as distillation, involving fire) and the utility of practical, experimental work. Starkey had little use for the abstractions of mathematics. He referred to himself as a Philosopher By Fire, in sneering contrast to the safe academicians who eruditely cited published facts. Such fiery rhetoric made him few academic friends. However, he had important correspondence with Robert Boyle, and Newman establishes that Isaac Newton, who seriously studied alchemy, cited Starkey's works far more often than any other alchemist of the period.[2] "Heady stuff" for a young man of modest means from the Colonies.

1. C.C. Gillespie (Editor-In-Chief), *Dictionary of Scientific Biography*, Charles Scribner, New York, 1975, Vol. XII, pp. 616–617.
2. W.R. Newman, *Gehennical Fire: The Lives of George Starkey, an American Alchemist in the Scientific Revolution*, Harvard University Press, Cambridge, MA, 1994.
3. The historical evidence is thus unambiguous: There was no senior basketball club team when Starkey attended Harvard.
4. J.R. Partington, *A History of Chemistry*, MacMillan, London, 1961, Vol. 2, pp. 209–241.

PHLOGISTON: CHEMISTRY'S FIRST COMPREHENSIVE SCIENTIFIC THEORY

The initial concept of phlogiston was due to Johann Joachim Becher (1635–1682) and has clear alchemical roots.[1] For Becher, the important elements were Water and three Earthy Principles. (He regarded Air and Fire to be agents of chemical change rather than elements in the chemical sense). His three Earthy Principles corresponded very roughly to the Paracelsian "salt," "mercury," and "sulfur." This last "sulfur-like" Earthy Principle was termed *Terra Pinguis* (fatty earth) by Becher and was said to be present in combustible matter and released upon combustion. It was this principle that Georg Ernst Stahl (1660–1734) later equated to his phlogiston.

Becher *was* aware, as was Boyle (see effluviums discussion on pp 109–111), that calxes were heavier than the corresponding metals. He too attributed these observations to igneous ("fiery") particles small enough to move through glass and join the metal inside.

Becher was an argumentative man who described himself as follows[2]:

> . . . one to whom neither a gorgeous home, nor security of occupation, nor Fame nor health appeal, for me rather my chemicals amid the smoke, soot and flame of coals blown by the bellows. Stronger than Hercules, I work forever in an Augean stable, blind almost from the furnace glare, my breathing affected by the vapour of mercury. I am another Mithridates, saturated with poison. Deprived of the esteem and company of others, a beggar in things material, in things of the mind I am a Croesus. Yet among all these evils I seem to live so happily that I would die rather than change places with a Persian King.

Clearly, Becher was a truly "hard-core," "gung-ho" chemist. Happily, we modern chemists do not have to recite this pledge as our professional oath.

Figure 66(a) is from the book *Oedipus Chymicus*[3] (1664) and it depicts Oedipus solving the riddle posed by the Sphinx. It is thought to represent the chemist solving the alchemical riddle and is consistent with Becher's firm belief in transmutation. Of course, once Oedipus relieved Thebes of the dreaded Sphinx he was made King but other disasters followed. Perhaps personal disaster would have also afflicted the discoverer of the Philosopher's Stone or the Elixir: King Midas comes to mind. Figure 66(b) is from the 1681 edition[4] of this book. The last edition was published by Stahl in 1738 (see Fig. 1).

1. H.M. Leicester and H.S. Klickstein, *A Source Book In Chemistry 1400–1900*, McGraw-Hill, New York, 1952, pp. 55–58.
2. J.R. Partington, *A History of Chemistry*, MacMillan, London, 1961, Vol. 2, pp. 637–652 (quotation on p. 639).
3. This figure, from the book *Oedipus Chymicus*, is from *Science and Technology: Catalog 5* Jeremy Norman, San Francisco, 1978, p. 13. Courtesy of Jeremy Norman & Company, Inc.
4. I. MacPhail, *Alchemy and the Occult*, Yale University Library, New Haven, 1968, Vol. 2, pp. 472–476.

OEDIPVS
CHIMICUS.

FRANCOFVRTI,
1664.

(a)

FIGURE 66 ■ (a) The title page of *Oedipus Chimicus* (perhaps only the Sphinx knows the riddle of the Stone) (courtesy of Jeremy Norman & Co., from Catalogue 5, 1978). (b) This is the title page of Johann Joachim Becher's *Physicae Subterranae Libri Duo* (Frankfurt, 1681). The first edition, published in 1669, contained Becher's view of matter —the Phlogiston Theory, later modified by Georg Ernst Stahl [see Fig. 1 for the frontis from the final (1738) edition of this book] (courtesy of The Beinecke Rare Book and Manuscript Library, Yale University). *Illustration continued on following page*

DOCT. JOH. JOACHIMI BECH
ERI PISICA SVBTERRANEA
TRIPLEX NATURÆ OFFICINA
GLOBUS TERR-AQU-AERREUS
AER
Vegetabilis
Animalis
AQUA
TERRA
Mineralis

(b)

GUN POWDER, LIGHTNING AND THUNDER, AND NITRO-AERIAL SPIRIT

Gunpowder is a mixture of saltpetre (KNO_3) or nitre ($NaNO_3$), sulfur, and carbon developed possibly as early as 1150 A.D. by the Chinese.[1] Its explosive power is due to the exothermic reaction below in which a large volume of gas (carbon dioxide and nitrogen) is generated suddenly and violently along with a great deal of heat. Gunpowder burns under water or in a vacuum. In modern terms, we see saltpetre as the oxidizer (in place of gaseous oxygen), which converts charcoal to carbon dioxide. Thus, saltpetre and nitre are

$$2\ KNO_3 + 3\ C + S \rightarrow N_2 + 3\ CO_2 + K_2S$$

capable of supporting combustion. Calcination of antimony in air, under a magnifying glass and sunlight (see Fig. 62) formed the same calx (Sb_2O_3 in modern terms) as obtained by dissolving antimony in nitric acid (HNO_3) and heating. Apparently, "nitro-aerial spirit" present in saltpetre and nitric acid is also present in air.

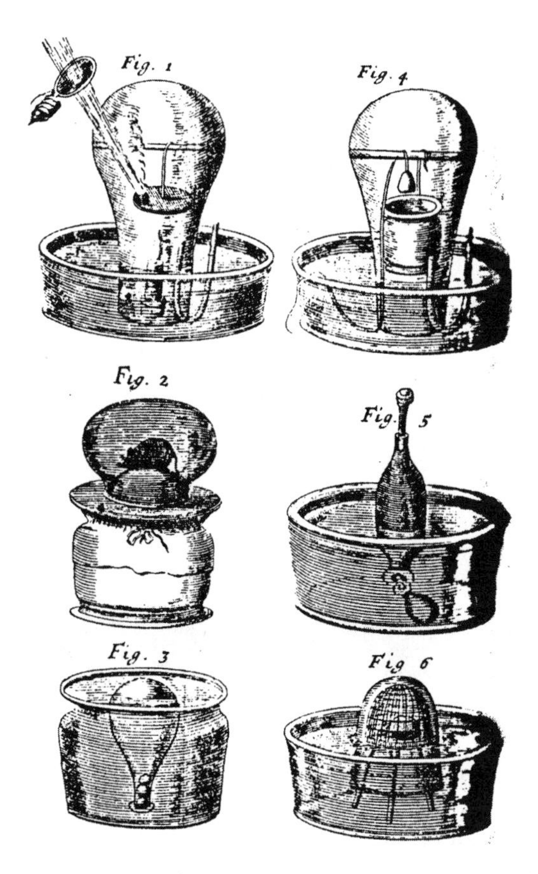

FIGURE 67 ■ This plate is from John Mayow's *Tractus Quinque Medico-Physici* (Oxford, 1674). It shows his experiments in which "nitro-aerial spirit" in saltpetre was transferred to antimony under a heating glass. In effect oxygen was transferred between the two substances (courtesy of distinguished chemist/book collector Dr. Roy G. Neville).

These observations led two of Boyle's assistants, Robert Hooke (1635–1703) and John Mayow (1641–1679), to form theories that appear, at least superficially, to anticipate Lavoisier.[1] Hooke explained lightning and thunder by the violent clash between sulfurous matter in air and nitro-aerial spirit similar to saltpetre. "Sulfurous matter" has a bit of the "smell" of Becher's *terra pinguis* to it.

Mayow took this view a number of steps further.[1] In his 1674 book *Tractatus Quinque Medico-Physici* he reported a brilliantly designed experiment (Fig. 67, *Fig. 1*) in which a candle is burned in a glass bulb inverted over water. In this apparatus there is also a stage with camphor or sulfur on it. When the burning ceased, Mayow cleverly demonstrated a loss in air volume due to depletion of oxygen (the CO_2 generated is water-soluble). He then used a magnifying glass and tried unsuccessfully to ignite the flammable camphor or sulfur in this depleted atmosphere. His experiments further indicated that "nitro-aerial spirit" was needed for calcination and respiration as well as combustion. The mouse in *Fig. 2* in Figure 67 is placed on a moist bladder tightly covered with a cupping glass adhering to the bladder. As the mouse depletes the oxygen supply, the bladder swells. The mouse in the small cage under glass (*Fig* 6) causes water to rise in the bell jar for similar reasons. *Figures 3* to *5* in Figure 67 describe the generation of nitric oxide from iron and nitric acid and its transfer. Mayow also noted, along with other contemporaries, that metal calxes were heavier than the metals.

The idea that addition of a component of air was required for combustion, calcination, and respiration seems to have anticipated Lavoisier by 100 years. Had Mayow heated nitre strongly in his investigations, he might have discovered oxygen[1] and, some might argue, allowed chemists to skip Stahl's phlogiston theory. However, like Van Helmont who coined the term *gas* 30 years earlier for his *spiritus sylvester* that escaped from burning charcoal, Mayow did not really have the expertise to collect gases, and the pneumatic chemistry introduced by Hales was still 50 years in the future.

1. J.R. Partington, *A History of Chemistry*, MacMillan, London, 1961, Vol. 2, pp. 577–614.

THE "MODERN" PHLOGISTON CONCEPT

The Phlogiston concept was chemistry's first truly unifying theory and was developed in its useful form early in the eighteenth century by Georg Ernst Stahl (1660–1734), an irrascible, egotistical, and rather unpleasant chemist and physician. It was said of him that "Stahl seems to have regarded his ideas at least in part due to divine inspiration and the common herd could have no inkling of them."[1] ". . . [H]is lectures were dry and intentionally difficult; few of his students understood them."[1] Stahl attacked adversaries vehemently and while he clearly acknowledged his debt to Becher (Stahl reissued Becher's *Physicae Subterranae*; see Figure 1), he also found much to criticize.

Figure 68 is the title page of Stahl's famous 1723 textbook. It summarizes Stahl's views as early as 1684. Over a half-century later this book was ceremoniously burned by Madame Lavoisier dressed in the outfit of a Priestess (see our later discussion). In the sixteenth century Paracelsus was said to have burned texts of Galen and Avicenna—an earlier act in the theatre of invective.

Phlogiston was postulated to be present in substances that could burn as well as in metals, which were known to form calxes. The concept works like this:

Charcoal (has phlogiston) → Residue + Phlogiston
Metal (has Phlogiston) → Calx + Phlogiston

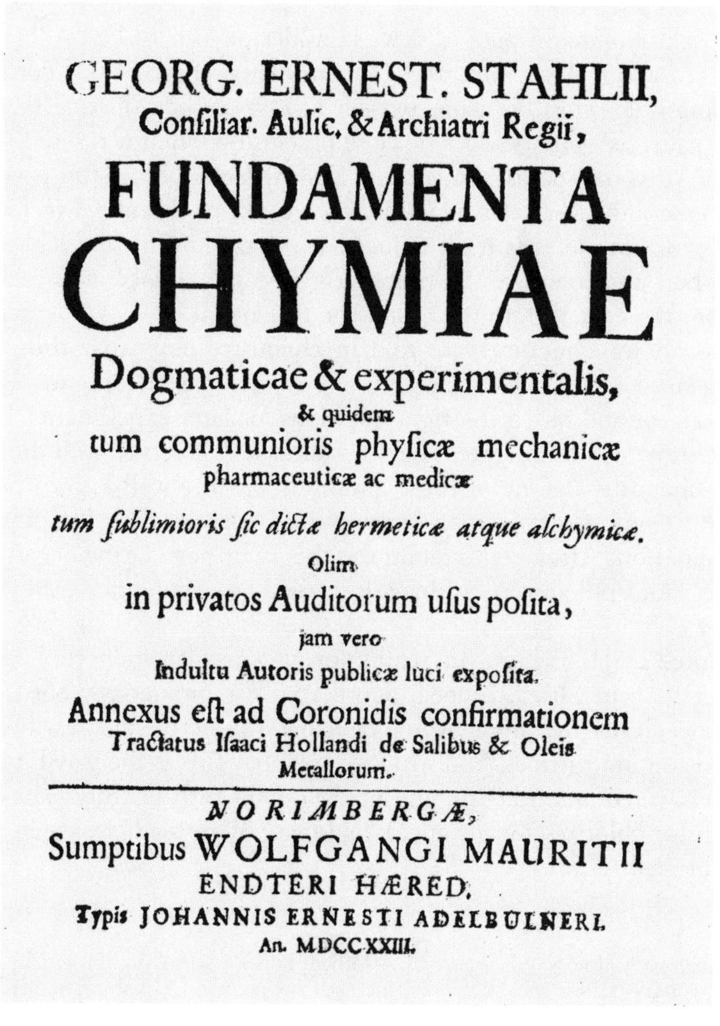

GEORG. ERNEST. STAHLII,
Conſiliar. Aulic. & Archiatri Regii,
FUNDAMENTA
CHYMIAE
Dogmaticae & experimentalis,
& quidem
tum communioris phyſicæ mechanicæ
pharmaceuticæ ac medicæ
tum ſublimioris ſic dictæ hermeticæ atque alchymicæ.
Olim
in privatos Auditorum uſus poſita,
jam vero
Indultu Autoris publicæ luci expoſita.
Annexus eſt ad Coronidis confirmationem
Tractatus Iſaaci Hollandi de Salibus & Oleis
Metallorum.

NORIMBERGÆ,
Sumptibus WOLFGANGI MAURITII
ENDTERI HÆRED.
Typis JOHANNIS ERNESTI ADELBULNERI.
An. MDCCXXIII.

FIGURE 68 ▪ Title page of text by Georg Ernst Stahl who formulated the "modern" phlogiston theory. Madame Lavoisier, dressed as a Priestess, ceremoniously burned this book to mark the publication of Lavoisier's *Traité Élémentaire de Chimie* in 1789.

Aside from relating these two seemingly very different kinds of chemistry, it explained the well-known ability to convert calxes into metals by heating with charcoal:

$$\text{Metal calx} + \text{Charcoal} \rightarrow \text{Metal} + \text{Ashy residue}$$

where charcoal and metal have phlogiston. Similarly, combustion of phosphorus in air formed phosphoric acid; and of sulfur, sulfuric acid. Heating of these acids with charcoal produced elemental phosphorus and sulfur, respectively.

This powerful and conceptually useful theory held sway for about a century. When Priestley discovered oxygen in 1774, he called it *dephlogisticated air* since it supported combustion, which allowed it to attract phlogiston vigorously from substances such as charcoal or iron. Nitrogen was initially called *phlogisticated air* because it did not support combustion and was obviously saturated with phlogiston. When Cavendish discovered hydrogen gas in 1766 and found that its density was less than one-tenth that of air, he thought this flammable gas was phlogiston itself.

As Roald Hoffmann puts it, if we consider oxygen to be "A," then phlogiston works as "not A" or "minus A."[2] Thus, when charcoal (C) burns, carbon does not lose phlogiston but gains oxygen to form carbon dioxide (CO_2). Similarly, iron gains oxygen and does not lose phlogiston when it rusts. If nitric acid (HNO_3) or saltpetre (potassium nitrate, KNO_3) gains phlogiston from a metal such as magnesium (Mg), it is really losing oxygen to the metal to form a calx or oxide (MgO) while it is itself reduced (to potassium nitrite, KNO_2, for example). When charcoal loses its phlogiston to a metal calx, it is really taking oxygen from the calx to form CO_2 and the free metal.

Although we sometimes are told in chemistry texts that the phlogiston concept delayed modern chemistry by 100 years, this theory was a powerful unifying concept and raised the right questions for later experiments. Hoffmann calls phlogiston " . . . an incorrect but fruitful idea that served well the emerging science of chemistry."[2] One of these questions was the well-known problem of the gain in weight of metals upon forming calxes despite their *loss* of phlogiston during calcination. Attempts to retain the theory by postulating negative weight (buoyancy) for phlogiston ultimately failed to convince the scientific community.

As noted by Hoffmann, the realization that oxygen supported combustion would later be generalized. Indeed, fluorine will spontaneously burn metals to form fluorides. If magnesium is heated by flame, this active metal will even burn in nitrogen to form nitrides. You will see later that this is the way Rayleigh and Ramsay discovered argon at the end of the nineteenth century. Thus, fluorine or nitrogen (or chlorine, for instance) could be "A" instead of oxygen under the right conditions.[2]

1. J.R. Partington, *A History of Chemistry*, MacMillan, London, 1961, Vol. 2, p. 655.
2. R. Hoffmann and V. Torrence, *Chemistry Imagined—Reflections On Science*, Smithsonian Institution Press, Washington, D.C., 1993, pp. 82–85.

WHAT ARE "EFFLUVIUMS"?

Effluvium: Now there's a rare word! *Websters's New World Dictionary of the American Language* (College Edition) defines it as follows: 1. A real or supposed outflow of a vapor or stream of invisible particles; aura. 2. A disagreeable or noxious vapor or odor (plural: effluvia).

Boyle believed in a corpuscular theory of matter—something of a forebearer to atomic theory. In this pretty little *Effluviums* book (Fig. 69) he conducts *gedanken* (thinking) experiments to calculate the upper limits to the measurable masses of effluvia. But before we illustrate some of these, let Boyle define the contemporary debate:

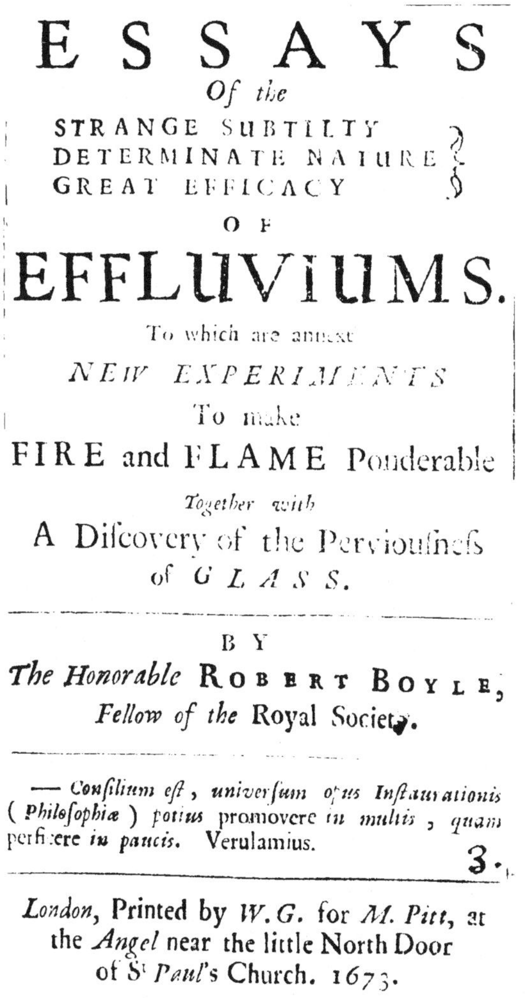

ESSAYS
Of the
STRANGE SUBTILTY
DETERMINATE NATURE
GREAT EFFICACY
OF
EFFLUVIUMS.
To which are annext
NEW EXPERIMENTS
To make
FIRE and FLAME Ponderable
Together with
A Discovery of the Perviousness
of *GLASS*.

BY
The Honorable ROBERT BOYLE,
Fellow of the Royal Society.

— *Consilium est, universum opus Instaurationis (Philosophiæ) potius* promovere *in multis, quam* perficere *in paucis.* Verulamius.

London, Printed by *W. G.* for *M. Pitt*, at the *Angel* near the little North Door of St *Paul's* Church. 1673.

FIGURE 69 ■ Title page of Robert Boyle's wonderful essay in which he estimates the mass of the smallest measurable "effluvia" of silver, gold, silk, and alcohol vapors. In the Denis I. Duveen collection there is a copy of this book autographed by Robert Boyle for Isaac Newton (kind of like the Old Testament autographed by Abraham for Jesus).

> Whether we suppose with the Ancient and Modern ATOMISTS, that all sensible Bodies are made up of Corpuscles, not only insensible, but indivisible, or whether we think with the CARTESIANS, and (as many of that Party teach us) with ARISTOTLE, that Matter, like Quantity, is indefinitely, if not infinitely divisible: It will be consonant enough to either Doctrine, that the EFFLUVIA of Bodies may consist of Particles EXTREMELY SMALL. For if we embrace the OPINION OF ARISTOTLE or DES-CARTES, there is no stop to be put to the subdivision of Matter, into Fragments, still lesser and lesser. And though the EPICUREAN Hypothesis admit not of such an INTERMINATE division of Matter, but will have it stop at certain solid Corpuscles, which for their not being further divisible are called ATOMS ("atomos") yet the Assertors of these do justly think themselves injured, when they are charged with taking the MOTES or small Dust, that fly up and down in the Sun-Beams, for their Atoms, since, according to these Philosophers, one of those little grains of Dust, that is visible only when it plays in the Sun-Beams, may be composed of a multitude of Atoms, and may exceed many thousands of them in bulk.

(Modern) English Translation: Do not think, for a moment, that I am so foolish, as to assume that the effluvia whose masses I will estimate are the same things as my version of atoms. I am estimating an upper limit for the masses of effluvia each of which are composed of many thousands of atoms. In any case, stay tuned and see how effluvia explain my observations of metals and their calxes.

Here are some thought experiments by Boyle:

1. One grain (0.0648 grams or g) of silver has been drawn by a master silversmith into a wire 27 ft long. Boyle had a special ruler subdivided into 200 divisions per inch. Therefore, the wire can be subdivided into $27 \times 12 \times 200 = 64{,}800$ silver "cylinders" each weighing 0.0000010 gram (1.0×10^{-6} g).
2. If it were possible to gild this silver wire, the mass of the gold sheath would be even much less per "cylinder of sheath."
3. "An Ingenious Gentlewoman of my Acquaintance, Wife to a Learned Physician" drew 300 yards of silk gently from the mouth of a silkworm. The silk strand weighed 2.5 grains. The division of the silk gave $300 \times 3 \times 12 \times 200 = 2{,}160{,}000$ silk "cylinders" each of mass 0.000000075 g (7.5×10^{-8} g).
4. Six minute pieces of gold were each beaten into squares with $3^1/_4$ inch sides. The total mass of the six square leaves of gold was $1^1/_4$ grains. Therefore the six square leaves could be subdivided into a total of $6 \times (3.25 \times 200)^2 = 2{,}535{,}000$ squares of gold each weighing 0.000000032 g (3.2×10^{-8} g).

It is most wonderful to note that some 240 years later, the 1926 Nobel Laureate in Physics, Jean Perrin, performed similar calculations in his 1913 book *Les Atomes* (English, 1916): "an upper limit to molecular size."[1] Gold leaf of 0.1 micron (10^{-5} cm) thickness implies that, at a maximum, gold atoms occupy cubes of 10^{-15} cm^3. Using gold's density, this means a mass of 10^{-14} g per gold atom. Since a hydrogen atom is $^1/_{197}$ the mass of a gold atom, its mass can be given an upper limit of 5×10^{-17} g. Actually, we have known for about 100 years that 1 mole of gold has a mass of 197.0 g and is comprised of 6.02×10^{23}

atoms (Avogadro's number). Therefore, an individual gold atom weighs 3.27×10^{-22} g—about 100 trillion times less than Boyle's tangible gold effluvium and 100,000 times less than Perrin's upper limit. However, neither Boyle nor Perrin claimed to be weighing individual atoms.

And why this interest in effluviums? Toward the end of the book Boyle describes his accurate measurement of the *increase* in weight upon heating a metal, such as iron, in air to form a calx such as rust. In 1673 this was an extremely important observation (also noted by Rey, Becher, Stahl, Mayow, and Le Fèvre among others). Boyle's explanation: minute effluviums in the flame (Becher's igneous particles) penetrate the pores of the sealed glass vessel containing metal and air and "adhere" to the metal, thus forming a calx weighing more than the metal. This was a near avoidance of the Phlogiston Theory that was already in its embryonic stage.

It is also worth noting that Van Helmont explained the sympathy concept as well as magnetic phenomena as arising from contact between effluvia (for example, between the blood on a sword and the blood in the wounded person's body; see pp 82–83).

Toward the end of the *Effluviums* book Boyle explicitly raised the possible health issue of the effect of effluviums from fire landing on cooked meat and being consumed. We now know that when the fats from meat drip onto hot coals during charbroiling, they pyrolyze to form carcinogenic polycyclic aromatic hydrocarbons which rise up from the grill and deposit on the surface of the meat. Thus, in this regard, Boyle anticipated the human exposure health specialists by 300 years. It is also worth noting that William Penn corresponded with Boyle from Pennsylvania and during the early 1680s sent him samples of ore and medicinal plants from the New World.[2]

1. J. Perrin, *Atoms*, Constable, London, 1916, pp. 48–52.
2. C. Owens Peare, William Penn—A Biography, Dobson Books, London, 1959, p. 268. (I thank Professor Susan Gardner for calling this to my attention).

BEAUTIFUL SEVENTEENTH-CENTURY CHEMISTRY TEXTS

Following Libavius' *Alchymia*, a series of useful and beautifully illustrated textbooks appeared throughout the seventeenth century.[1–3] Let us begin the Beguin —Jean Beguin's *Tyrocinium Chymicum* ("The Chemical Beginner") was first published in 1610. It went through more than 50 editions before the last in 1669. Figure 70 shows the title page depicting an alchemical Cupid for the 1660 edition. Nicolas Le Fèvre first published his *Traicté de la Chimie* in 1660. The second French edition appeared in 1669, English editions in 1664 and 1670, and the last German edition in 1688. Figure 62 is from the 1670 edition of Le Fèvre. Christophe Glaser first published his *Traité de la Chymie* in 1663. An English edition was published in 1677. German editions (see Fig. 71) were published through 1710. Nicolas Lemery published an incredibly successful text (Lemery was Glaser's student). The first French edition of *Cours de Chimie* was

FIGURE 70 ■ The first edition of Jean Beguin's *Tyrocinium Chymicum* ("The Chemical Beginner") was published in 1610. This 1660 edition used a Chemical Cupid to entice readers to love chemistry.

published in Paris in 1675 (Fig. 72 is from the 1686 Paris edition). The final French edition was published in 1756—an incredible 81-year run! (See discussions of early glassware in figures.) Figure 73 is from Moses Charas' *Royal Pharmacopoea* (London, 1678). Figure 74 is from the text *Pyrosophia* (Utrecht, 1698) by Johann Conrad Barchusen. The figure shows his Utrecht laboratory. Whether the chemist is Barchusen himself is not known.

Figure 75 is from the first English edition of Herman Boerhaave's *Elements of Chemistry* (London, 1735). Boerhaave was a renowned physician and teacher of chemistry.[4] His lectures were so excellent that a pirated edition was published by his students in 1724 (translated into English in 1727). Although he was not a significant primary contributor to chemical science, he was rigorous and skeptical about the phlogiston concept. The first authorized edition of Boerhaave's *Elements* was published in 1732 (Leiden). He signed each copy of the huge tome as verification of its legitimacy. Boerhaave's *Elements* included perhaps the first really comprehensive history of chemistry. Boerhaave was the first great exponent of clinical teaching and he made the medical college at Leiden one of the best in Europe. Following his death, Dr. Samuel Johnson wrote a piece titled "Life of Herman Boerhaave" in the *Gentleman's Magazine* (1739). Johnson's biographer Boswell wrote that Johnson then "dicovered the love of chymistry which never 'forsook him.' "[4] At least twenty years of Johnson's life were spent in his own chymical laboratory.[4]

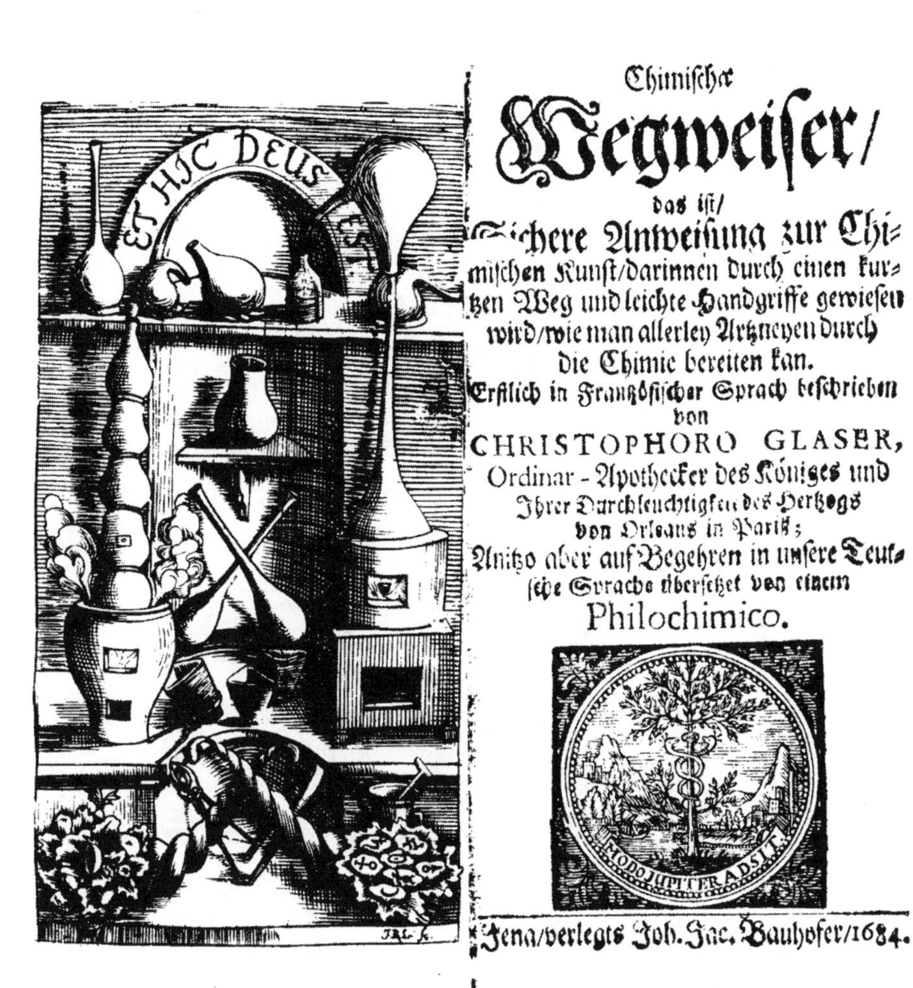

Chimischer
Wegweiser/
das ist/
Sichere Anweisung zur Chi-
mischen Kunst/darinnen durch einen kur-
tzen Weg und leichte Handgriffe gewiesen
wird/wie man allerley Artzneyen durch
die Chimie bereiten kan.
Erstlich in Frantzösischer Sprach beschrieben
von
CHRISTOPHORO GLASER,
Ordinar-Apothecker des Königes und
Ihrer Durchleuchtigkeit des Hertzogs
von Orleans in Pariß;
Anitzo aber auf Begehren in unsere Teut-
sche Sprache übersetzet von einem
Philochimico.

Jena/verlegts Joh. Jac. Bauhofer/1684.

FIGURE 71 ■ The 1684 German edition of Christophe Glaser's 1663 *Traité de la Chymie*.

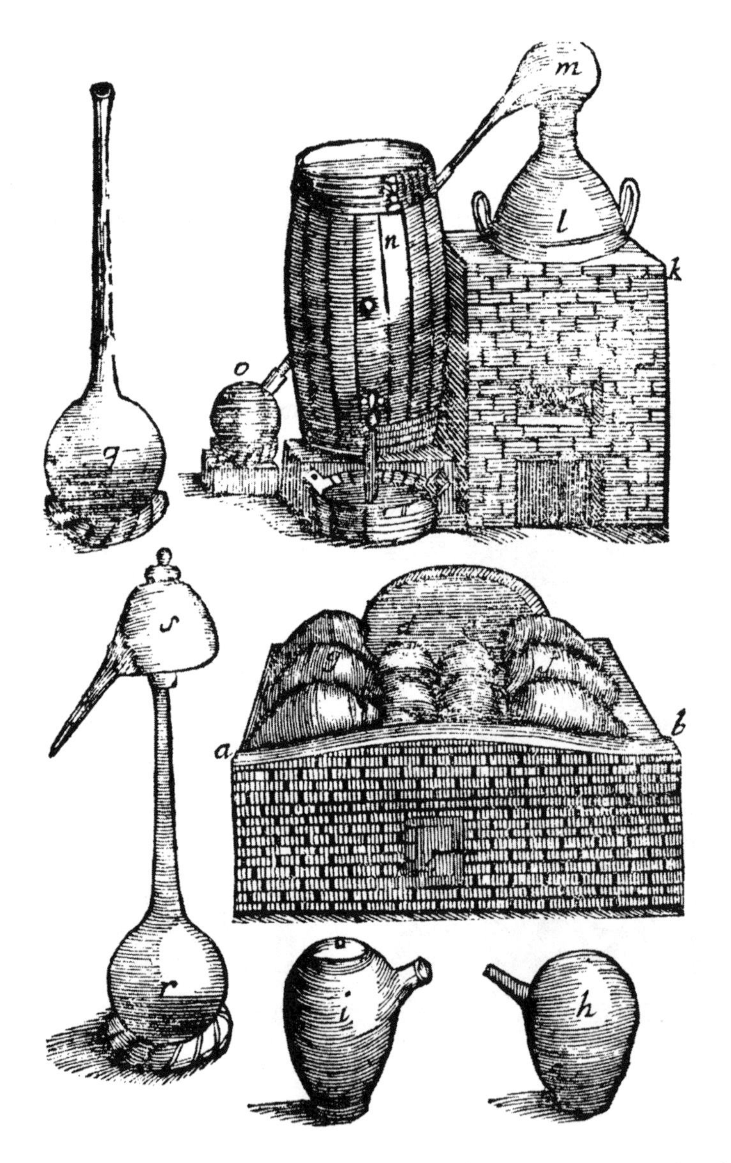

FIGURE 72 ■ Glassware in the 1686 edition of Nicolas Lemery's *Cours de Chimie*. This book was first published in Paris in 1675; the last edition was also published in Paris in the year 1756. If *A Chemical History Tour* is published in numerous editions over 81 years, my book-collecting cash-flow problems will disappear.

FIGURE 73 ■ Seventeenth-century glassware in Moses Charas' *The Royal Pharmacopoea* (London, 1678). Note the double pelican (*KK*) and alembics O and *E*.

FIGURE 74 ■ Is that author Johann Conrad Barchusen weighing the "midnight oil" in his Utrecht laboratory? (*Pyrosophia*, Utrecht, 1698).

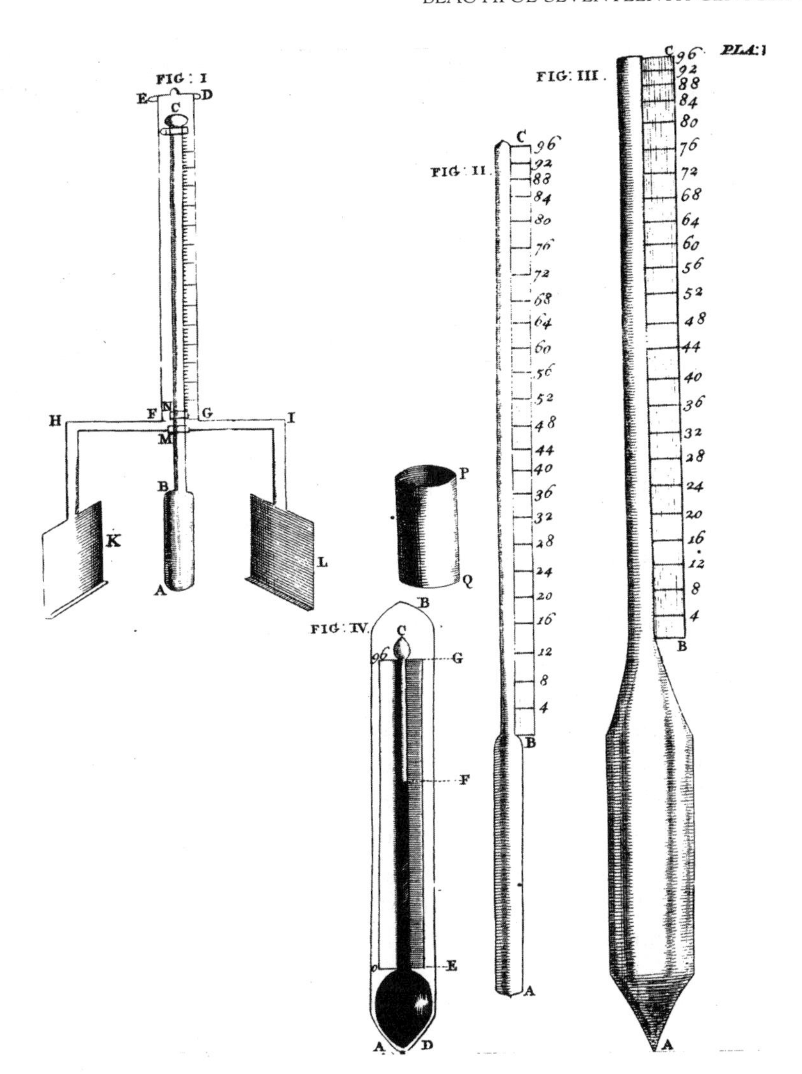

FIGURE 75 ■ These thermometers (see text) are found in the 1735 English edition of Herman Boerhaave's *Elements of Chemistry*. Boerhaave was not a distinguished chemist who made primary discoveries but rather a distinguished teacher of chemistry and medicine who helped introduce clinical teaching into medical school curricula.

In Figure 75 we see (Fig. I) a thermometer designed to be free standing so that the bulb *AB* can sit in the vessel *PQ* into which liquids can be poured or mixed. In Fig. II we see Fahrenheit's first thermometer meant to be filled with alcohol containing red dye. Figure III shows Fahrenheit's second thermometer, this to be filled with mercury. Figure IV shows Fahrenheit's third thermometer, to be used to measure "the Heat of the Human Body." This one can use mercury or alcohol and red dye. It is placed in a hermetically sealed glass chamber. The thermometer is to be used under the arm, upon the breast under one's clothes, or in the mouth . . . whew!

1. J. Ferguson, *Bibliotheca Chemica*, Derek Verschoyle, London, 1954 (reprint of 1906 edition).
2. D. Duveen, *Bibliotheca Alchemica et Chemica*, HES, Utrecht, 1986 (reprint of 1949 edition).
3. J. Read, *Humour and Humanism in Chemistry*, G. Bell, London, 1947, pp. 79–123.
4. J. Read, op. cit., pp. 128–153.

CHEMICAL AFFINITY

Figures 76 and 77 depict two halves of the first logically organized tables of the properties of chemical substances. It was composed by Étienne-François Geoffroy (1672–1731) in 1718.[1] The top horizontal row depicts 16 substances (elements and compounds), classes of substances, and even mixtures in a fairly arbitrary order from left to right. Each column rank orders substances according to *affinity*. Those substances closest to the top have the highest affinities for the substance at the top (the *header*), while those toward the bottom have the lowest affinities.

Let's examine some brief illustrative examples. In Column 16, we see water as the header with alcohol above salt. This means that alcohol has a greater affinity for water than salt. Thus, if you added alcohol (say, ethanol or 200 proof vodka) to salt water (saline) the liquids would mix and the salt would precipitate, forming a filterable solid. Alcohol has displaced salt from water. In contrast, if you took a 50:50 alcohol–water mixture (100 proof vodka), you could not dissolve salt in it since water has greater affinity for alcohol.

Column 1 shows the chemical affinities of substances for acids. Most metals react chemically with acids and release hydrogen gas—they appear to "dissolve" and release "air." However, if we first mix an alkali (base) such as potassium carbonate (K_2CO_3) with the acid and neutralize it, the solution will no longer dissolve metals. If a metal is dissolved in acid and alkali is added, a solid will precipitate (actually the insoluble metal carbonate or hydroxide). Thus, the alkalis have higher affinity for acids than do metallic substances.

But wait a minute, dear readers. Those of you who have had some high-school chemistry realize that solubilities of alcohol and salt in water are physical properties while "solubilities" of metals in acids are chemical properties. You would *not* have received a good grade from me for confusing the two. Clearly, the differences were not yet fully clear to early eighteenth-century scientists.

Human history is writ large in Column 9! Let us look at the affinities of sulfur. Of the metals shown, iron has the highest affinity, with tin and copper (which form the alloy bronze) having lower affinities. Tin and copper ores (commonly sulfides) could be smelted relatively easily, and the Bronze Age thus began around 3000 B.C. Higher temperatures and therefore more modern furnace technology were required to win iron from its sulfides and the Iron Age only began around 1200 B.C.

Notice gold at the bottom of Column 9. This noble metal has little affinity for sulfur and can often be found in nature as shiny nuggets or granules in an uncombined state.

Geoffroy's table, a somewhat arbitrary conglomeration of chemical and physical properties, elements, compounds, classes of substances and mixtures, is,

FIGURE 76 ▪ Top half of Étienne Geoffroy's 1718 Table of Affinities taken from *Recueil de Dissertations Physico-Chimiques Presentés a Differentes Académies* (Jacques Francois Demachy, Paris, 1781).

TABLE DE MR GEOFFROY en 1718.

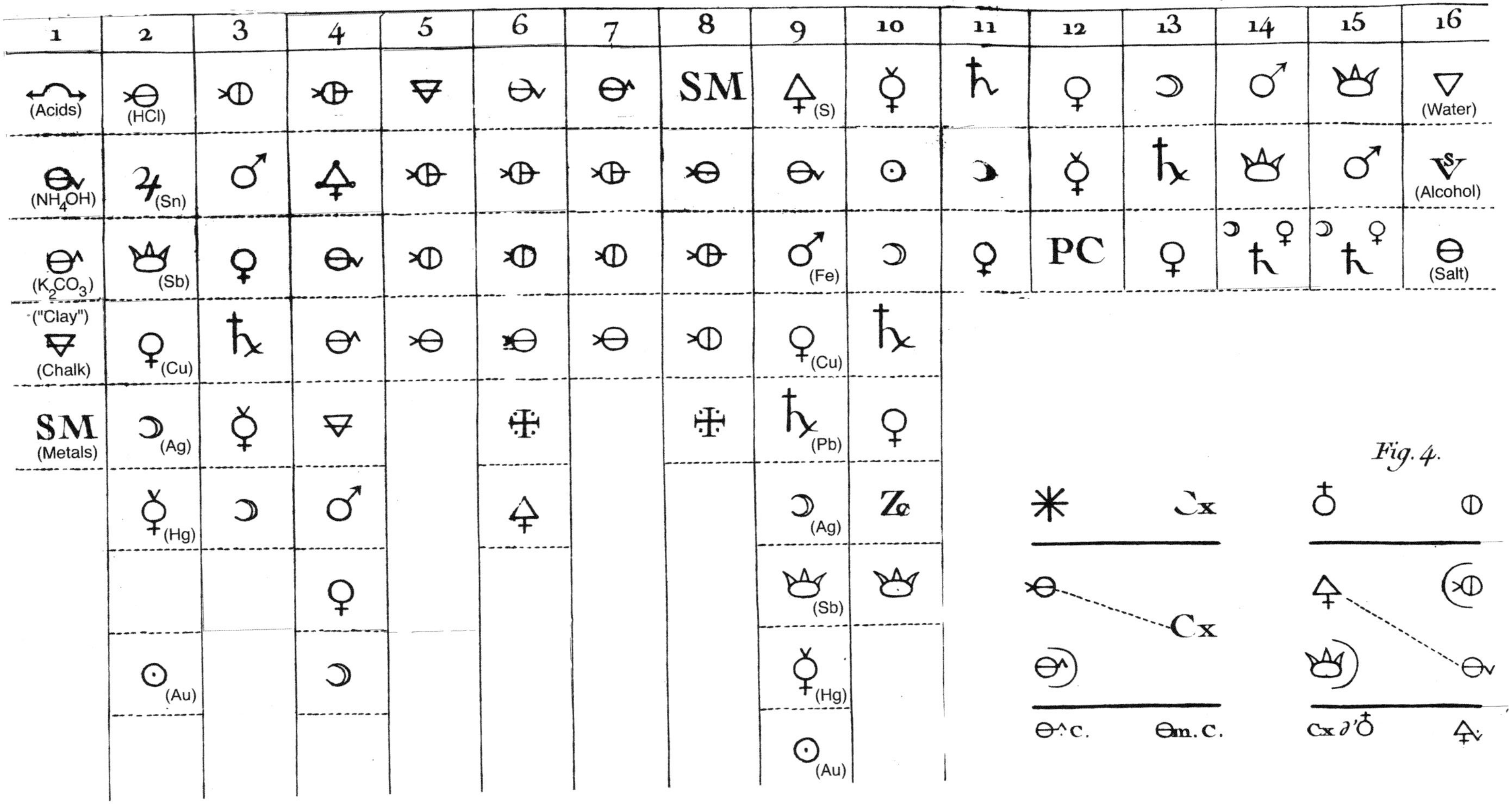

1	2	3	4	5	6	7	8	9	10	11	12	13	14	15	16
⟷ (Acids)	x⊖ (HCl)	x⦶	x⊕	🜃	⊖v	⊖^	SM	🜍 (S)	☿	♄	♀	☽	♂	♁	▽ (Water)
⊖v (NH_4OH)	♃ (Sn)	♂	🜍	x⊕	x⊕	x⊕	x⊖	⊖v	☉	☽	☿	♄	♁	♂	🜈 (Alcohol)
⊖^ (K_2CO_3)	♁ (Sb)	♀	⊖v	x⦶	x⦶	x⦶	x⊕	♂ (Fe)	☽	♀	PC	♀	☽ ♄ ♀	☽ ♄ ♀	⊖ (Salt)
("Clay") 🜃 (Chalk)	♀ (Cu)	♄	⊖^	x⊖	x⊖	x⊖	x⦶	♀ (Cu)	♄						
SM (Metals)	☽ (Ag)	☿	🜃		✠		✠	♄ (Pb)	♀						
	☿ (Hg)	☽	♂		🜍			☽ (Ag)	Zc						
			♀					♁ (Sb)	♁						
	☉ (Au)		☽					☿ (Hg)							
								☉ (Au)							

Fig. 4.

✳ ⊃x

x⊖ ---- Cx

⊖^)

⊖^c. ⊖m.c.

♁ ⦶

🜍 ---- (x⦶

♁) ⊖v

Cx ∂'♁ 🜍:

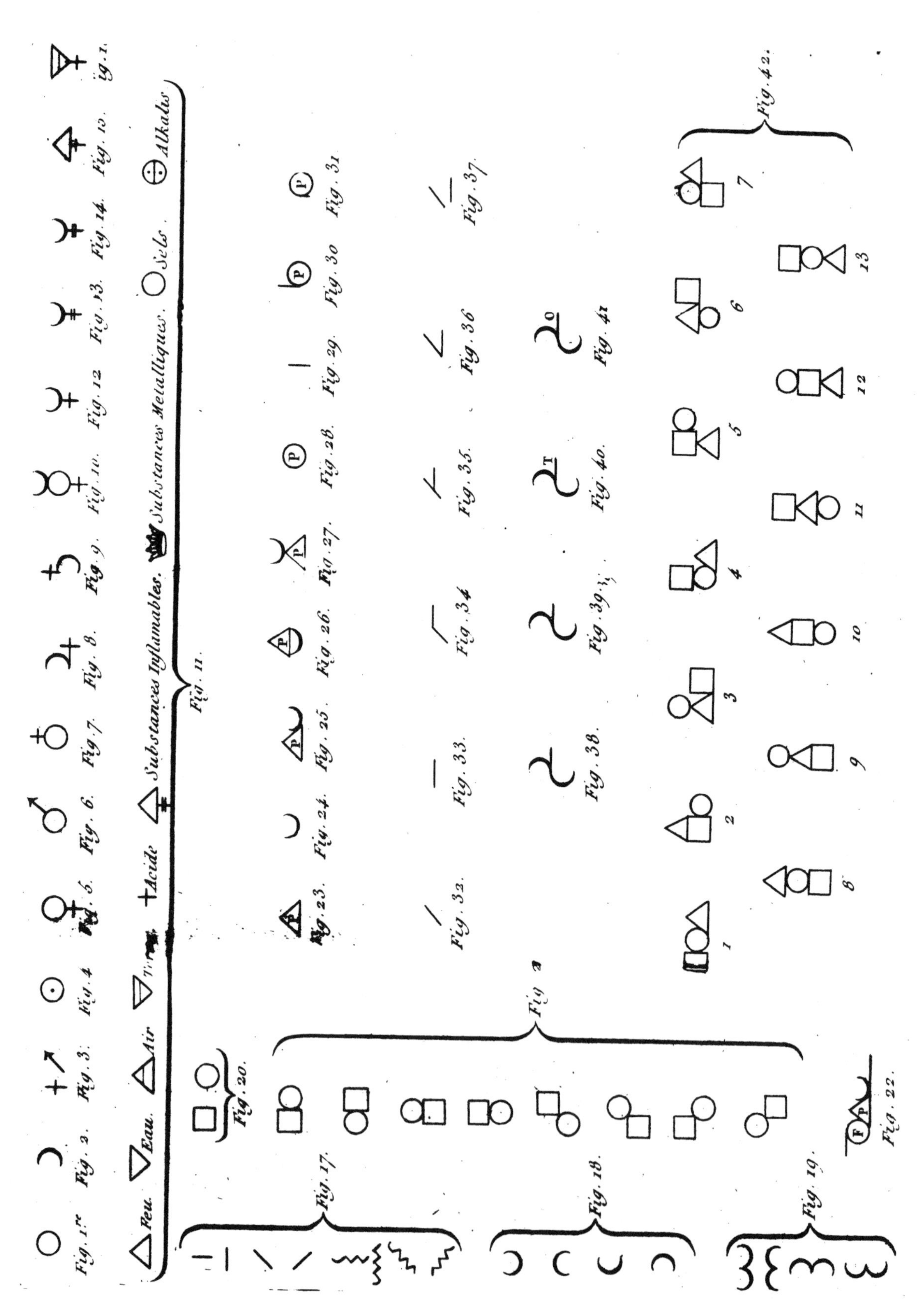

FIGURE 77 ■ Bottom half of Geoffroy's Table of Affinities (see Fig. 76).

nevertheless, a distant relative of the Periodic Table, formulated some 150 years later.

1. This version of Geoffroy's table is from M. De Machy, *Recueil de Dissertationes Physico-Chimiques*, Paris, 1781 (Plate 1). Also, see the discussion in H.M. Leicester and H.S. Klickstein, *A Source Book In Chemistry 1400–1900*, McGraw-Hill, New York, 1952, pp. 67–75.

DOUBLE-BOTTOM CUPELS, HOLLOW STIRRING RODS, AND OTHER FRAUDS

It is interesting to note that although alchemy was essentially dead by the end of the seventeenth century, there remained popular interest well into the eighteenth century. Gullible savants and the just plain greedy were prey for "alchemists," and the venerable *Journal des Savans* continued to publish occasional papers on transmutation. Amazingly, a mainstream scientist, the aforementioned Geoffroy, was moved to publish a paper in *Histoire de l'Académie Royale des Sciences* in 1722 warning against such gullibility.[1] Among the frauds he warned against were the following:[1]

1. Double-bottom cupels
2. Hollow stirring rods
3. Amalgams concealing precious metals
4. Acids containing dissolved gold and silver
5. Filter papers with minute amounts of concealed gold or silver to be recovered upon ashing of the paper

Now, ladies, gentlemen, and children of all ages—watch as I scratch this little black crystal of samarium oxide (SmS). Presto—it changes to gold (colorwise, that is).[2] Who will be the first to purchase some of this "black gold"?

Apparently, even today there are alchemists busily working away in France earnestly trying to discover the Stone of the Philosophers.[3] Eureka! There may yet be a customer to buy the famous *Pont de Brooklyn* from me.

1. A. Debus, in *Hermeticism and the Renaissance*, I. Merkel and A.G. Debus (eds.), Associated University Press, Cranbury, NJ, 1988, pp. 231–250.
2. H. Rossotti, *Diverse Atoms: Profiles of the Chemical Elements*, Oxford University Press, Oxford, 1998, p. 439.
3. A. McLean, *A Commentary on the Mutus Liber*, Phanes, Grand Rapids, MI, 1991, pp. 8–10.

PEAS PRODUCE LOTS OF GAS

Stephan Hales (1677–1761) studied theology in Cambridge and became an active priest but preferred scientific pursuits.[1] He performed important studies on the hydrostatics of fluids in plants (*Vegetable Staticks* . . . , London, 1727) and blood flow (*Statical Essays: containing Haemastaticks* . . . , London, 1733—"a gruesome book"[1]). His studies on "airs" were performed between 1710 and 1727. Hales is considered to be the originator of pneumatic chemistry—the collection and manipulation of isolated gases.[1] The distinguishing characteristic of his apparatus was the separation of the collected gases from their sources.

In Figure 78 (see *Fig. 33*) retort r, holding matter for distillation, is joined to the large long-necked flask *ab* using cement (tobacco pipe clay and bean flour well mixed with some hair) covered by a bladder. The large hole in the bottom

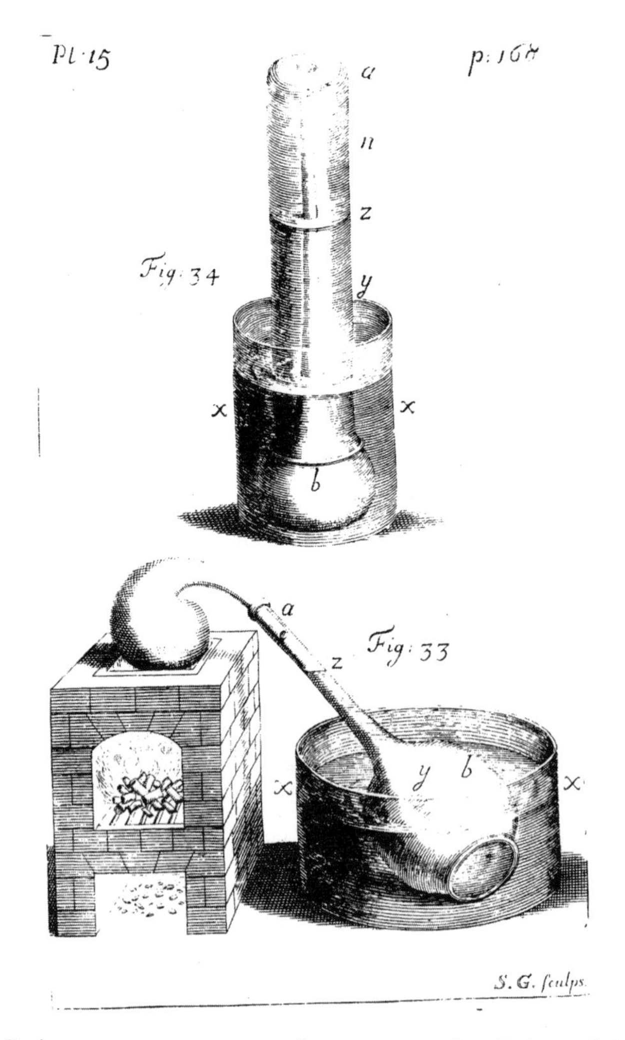

FIGURE 78 ■ Early pneumatic apparatus for measuring "airs" derived from distillation of vegetable matter [from the second edition of Stephan Hales' *Vegetable Staticks* (London, 1731); first edition, 1727].

of *ab* is for insertion of a glass siphon reaching to *z* inside the flask with the other end extending above the surface of the water container (*xx*) holding *ab*. Flask *ab*, while attached to *r*, is first immersed in a large bucket of water up to level *z* as excess air is pushed out through the siphon. Flask *ab* is then immersed into *xx*, which is filled with water. Heating of the vegetable matter in *r* produces "airs" that press the water level down from *z* to a new level *y*, which is carefully marked on the flask. The apparatus is allowed to cool to room temperature, *r* disconnected, and the top of *ab* corked. Inverted flask *ab*, first emptied of water, is filled to *z* and the mass of water determined. It is then filled to *y* and the total mass determined. The difference is the mass of water and therefore the volume of the gas generated. (Sometimes after cooling, there is actually net uptake of gas by the matter remaining in *r*.)

In Figure 79 (see *Fig.* 36) we see a "strong Hungary-water bottle" having mercury at the bottom and otherwise filled with peas soaking in water. An evacuated glass column closed at the top and extending below the mercury pool

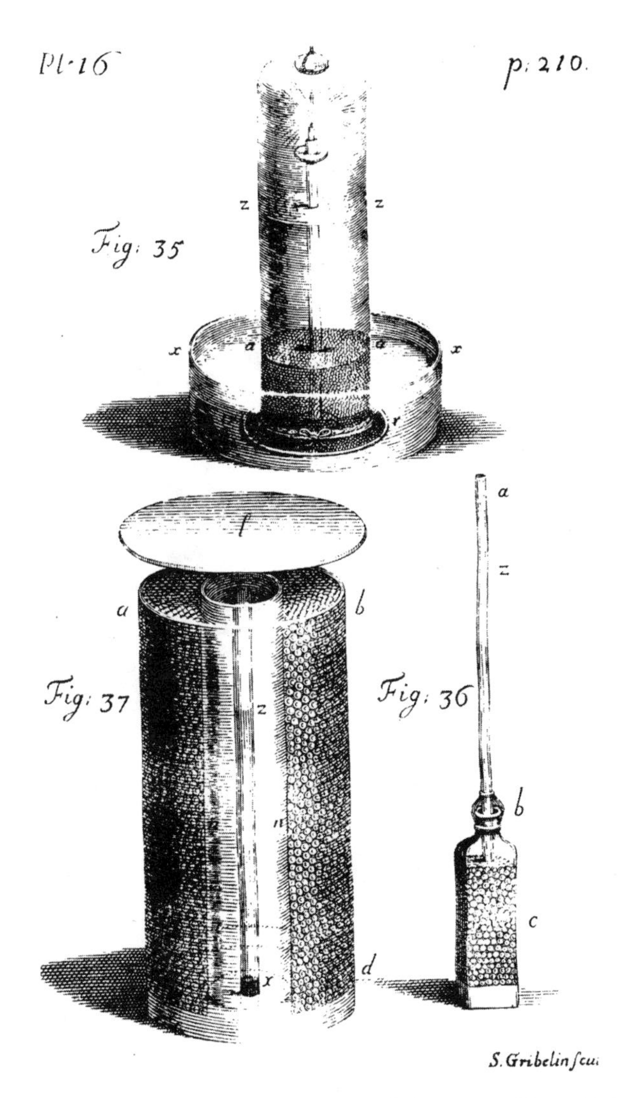

FIGURE 79 ■ Experiments measuring gases derived from peas (from Hales' 1731 edition of *Vegetable Staticks*).

at the bottom is tightly sealed at the top of the bottle. The peas absorb all of the water, and after two or three days, the gas they produce supports a column nearly 80 inches high (about 2.5 atmospheres of pressure). In Figure 79 (see *Fig. 37*) we see a strong iron vessel *abcd* that is 2.5 inches in diameter and 5 inches deep, filled with peas soaking in water over a pool of mercury. In this homely but clever apparatus, a glass tube inside a concentric iron cylinder (for protection) has a drop of honey (*x*) at the bottom. The iron cover, closely fitted and sealed to the vessel with leather, is held closed with a cider press. After a few days, the press is loosened, pressure released, and the cover removed. Although the mercury column has fallen back to zero, a little dab of honey marks the spot (*z*) it arose to. The pressure was, again, about 2.5 atmospheres and corresponded to a force of about 189 pounds against the iron cover.

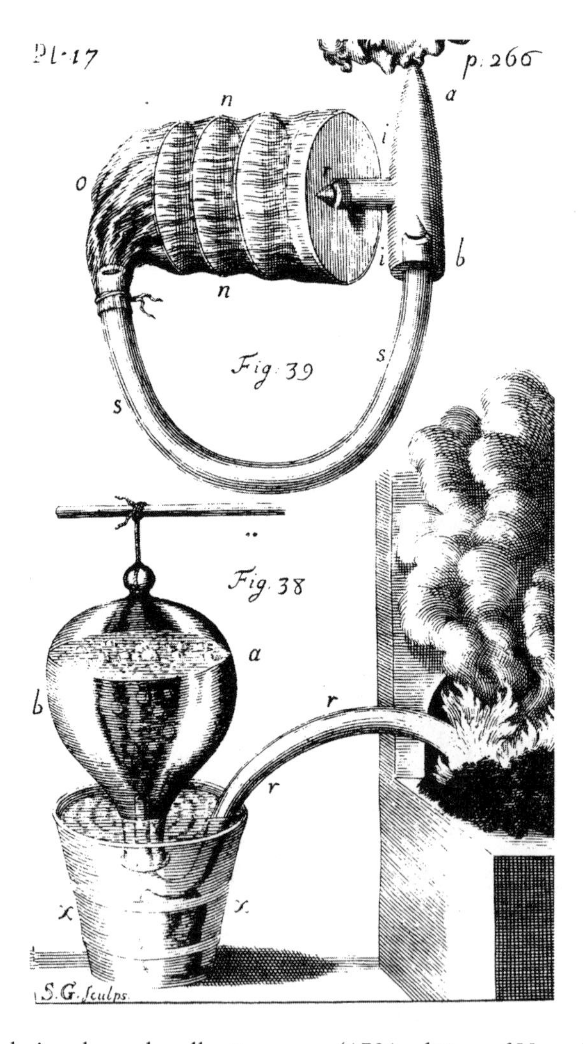

FIGURE 80 ■ Hales' early work collecting gases (1731 edition of *Vegetable Staticks*). The bottom figure shows gases collected from substances decomposed in the barrel of a gun and collected over water. This is the forerunner of the pneumatic troughs used by Scheele, Priestley, and Lavoisier to ignite the chemical revolution. The top figure depicts a bellows for collecting and recycling exhaled air. When the four diaphragms in the bellows were imbued with alkaline potassium carbonate, the breathing cycles would continue over longer periods due to removal of acidic carbon dioxide.

In Figure 80 (see *Fig. 38*) we see the very famous Hales pneumatic apparatus in which various materials are heated in an iron gun barrel. Gases are collected in the inverted, suspended flask, which is initially filled with water. Only water-insoluble gases can be collected in this manner. Water-soluble gases were subsequently collected over mercury or water coated with an oil layer.

Note the fascinating apparatus at the top of Figure 80 (*Fig. 39*). Hales (*his* face?) breathes air from the sealed sieve bag by sucking through wooden soffet *ab*. At the bottom of *ab* (see *ib*) there is a valve that opens upon inhalation. A similar valve at *x*, entering the bag, is closed upon inhalation. The two valves switch roles upon exhalation. Hales found that he could perform inhalation–exhalation cycles for about 1.5 minutes using an empty bag. When the bag contained four flannel diaphragms [dipped in salt of tartar (K_2CO_3) solution and dried—this absorbs CO_2], he could breathe for 5 minutes. If the salt of tartar had been well-calcined (slightly basic due to some loss of CO_2 to form K_2O), he could breathe for 8.5 minutes.

1. J.R. Partington, *A History of Chemistry*, MacMillan, London, 1962, Vol. 2, pp. 112–123.

BLACK'S MAGIC

While Stephan Hales devised techniques of pneumatic chemistry to separate "airs" from their sources, he did not explore their differences in great detail. However, in 1756, in a continuation of his M.D. thesis (1754), Dr. Joseph Black (1728–1799) described the generation of an "air" that had been "fixed" in magnesium alba ($MgCO_3$) and released upon heating.[1] Moreover, he tested this "fixed air" and found that its properties were very different from those of everyday air. For example, it extinguished flames rather than supporting them. The same "fixed air" is also generated when chalk ($CaCO_3$) is dissolved in acid. This "fixed air," when diffused into lime (CaO) water, would turn it cloudy by forming insoluble chalk. It is often said that until 1756 the only gas known was common air and that Black's discovery was the first of a pure gas. Actually, Van Helmont's studies in the seventeenth century involved discoveries of other gases that he recognized as different from common air, usually CO_2 often mixed with others, and he performed some characterizations. For example, Van Helmont knew that the poisonous gas that collects in mines (CO_2 with some CO) extinguishes flames.[2] However, his studies were not readily controllable and generally involved different mixtures of gases depending upon the source.

Black was a gifted teacher and his classic text *Lectures on the Elements of Chemistry* was published posthumously (Edinburgh, 1803; Philadelphia, 1807). He undoubtedly delighted and puzzled audiences by pouring "fixed air" (which is denser than common air) out of a glass to extinguish a candle flame. Black also showed that the same gas was generated by fermentation as well as by respiration since these emissions also turned lime water milky and were therefore CO_2.

Sometime during 1767–1768, Black filled a small balloon with hydrogen gas (newly discovered by Cavendish) and showed that it rose to the ceiling, surprising his audience who suspected that it was secretly raised by a black thread.[1] However, he argued against using hydrogen for manned balloons. In fact, the first hydrogen-filled balloon was flown by Jacques Alexandre Cesar Charles [yes, the ($V = kT$)-Charles-Law Charles] in 1783, the English Channel was crossed in 1785, and military balloons were flown as early as 1796.[1] Of course helium's discovery was about 100 years into the future. The explosion of the zeppelin Hindenburg over Lakehurst, New Jersey in 1937, with the loss of 36 lives, ultimately proved Dr. Black correct.

1. J.R. Partington, *A History of Chemistry*, MacMillan, London, 1962, Vol. 3, pp. 130–143.
2. J.R. Partington, *A History of Chemistry*, MacMillan, London, 1961, Vol. 2, pp. 229–231.

CAVENDISH WEIGHED THE EARTH BUT THOUGHT HE HAD CAPTURED PHLOGISTON IN A BOTTLE

Although we modern chemists go to some lengths to let the public know that we play tennis, like fast cars and stylish clothes, and are down-to-earth social-mixer types, we must admit that our passion for smelly, smoky mixtures will likely get us booted from most respectable country clubs. Henry Cavendish (1731–1810) was definitely an unworldly type. He lived with his father until the latter died in 1783, did not marry, communicated with his housekeeper using daily notes, and dressed in shabby, outdated clothing despite inheriting a fortune when he was 40.[1] The French physicist Jean-Baptiste Biot described him as "the richest of all learned men, and very likely also the most learned of all the rich."[1]

In our modern era when university tenure decisions are sometimes based upon the sheer poundage of publications, it is interesting to note that Cavendish published 18 papers in the *Philosophical Transactions of the Royal Society* (and no books).[2] He left many unpublished works and unstylishly referred to them in his published works.

But what works they were! In his first paper, published in 1766, Cavendish employed the pneumatic studies of Stephen Hales and Joseph Black to isolate hydrogen gas by pouring acids on metals such as zinc, copper, and tin. Indeed, the well-known affinities of these baser (more active) metals (see Geoffroy's Table of Affinities in Figures 76 and 77) for acids were long known to produce calxes. Moreover, the amount of gas collected did not depend on the identity of the acid (hydrochloric or sulfuric) or its amount but only on the quantity of metal. Thus, the metals were believed to lose their phlogiston to the air. The ignitable gas collected, which appeared to escape from the metal, was named "inflammable air" by Cavendish. It was less than one-tenth the density of atmospheric air and for a period Cavendish felt that phlogiston itself had been isolated. Figure 81 is taken from the 1766 work ("Three papers containing experiments on factitious Airs")[2,3] and shows in the panel labeled *Fig. 1* the col-

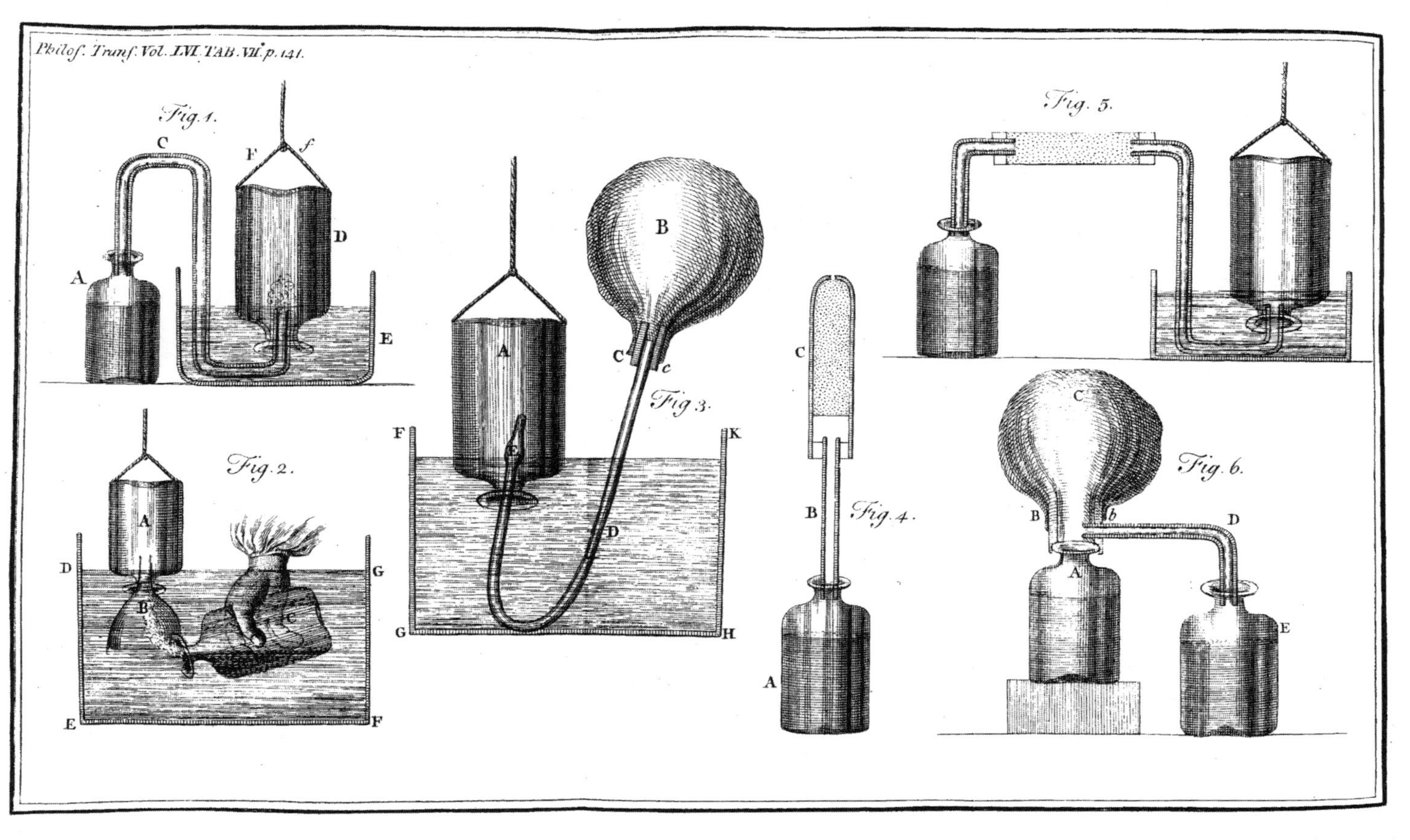

FIGURE 81 ▪ Apparatus used by Henry Cavendish to discover hydrogen and manipulate gases [*Philosophical Transactions of the Royal Society* (*London*), **LVI**: 141, 1766]. He thought that he had isolated phlogiston itself.

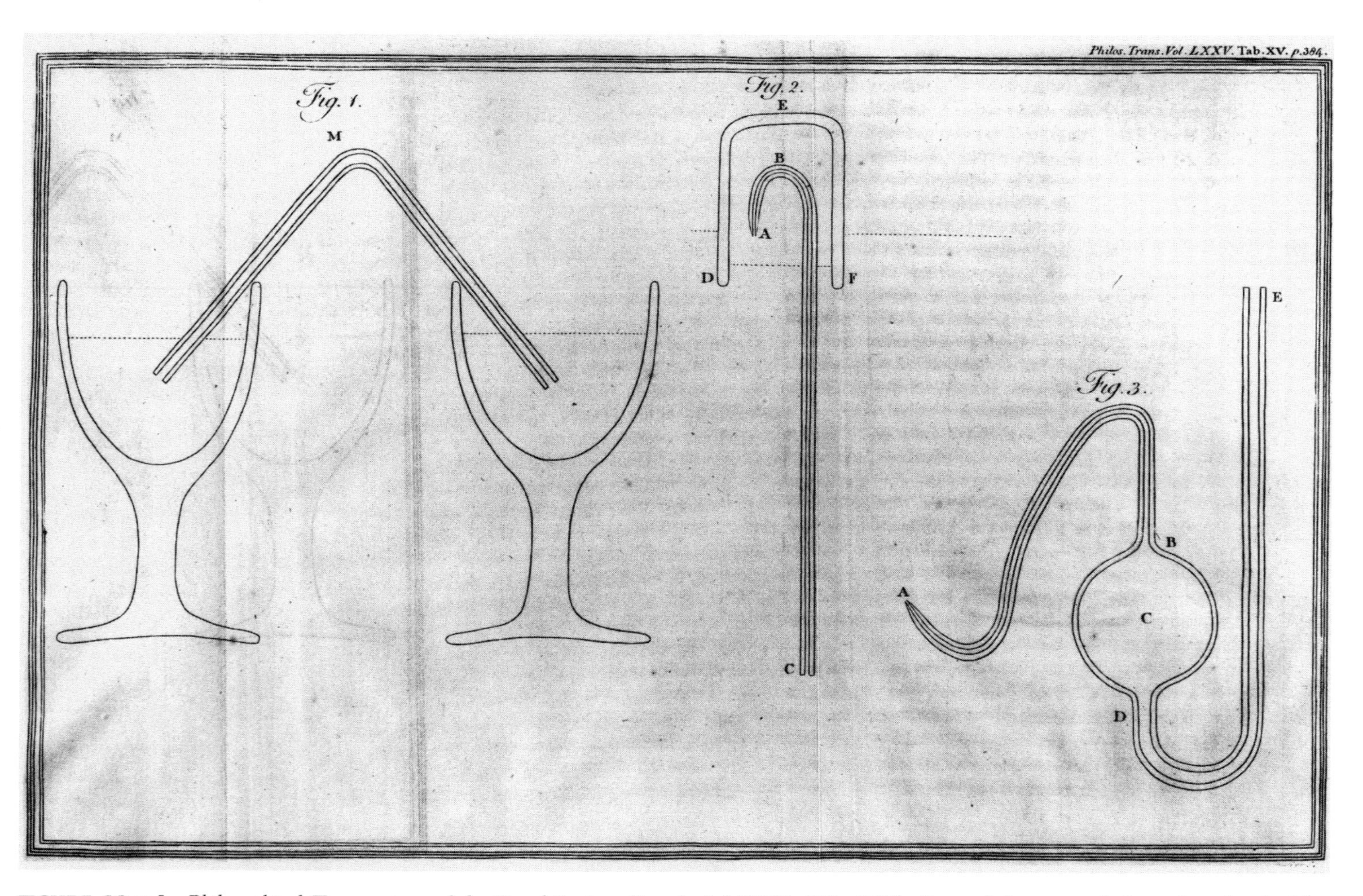

FIGURE 82 ▪ In *Philosophical Transactions of the Royal Society (London)* **LXXV**: 372, 1785, Cavendish reported that after the dephlogisticated air (oxygen) was removed from atmospheric air, the remaining phlogisticated air (nitrogen) could be sparked with oxygen introduced into the vessel and the gas produced yields nitric acid when combined with water. However, total reaction of all of the nitrogen left a tiny bubble of unreactive residue. Cavendish had isolated the rare gases (mostly argon). Over 100 years later Rayleigh and Ramsay, who discovered argon, would express their great admiration for Cavendish's amazingly accurate work.

lection of "inflammable air" over water; gases are transferred through funnels under water (*Fig. 2*). In *Fig. 3* we see transfer of gas into a bladder (a bit of wax is fixed to the end of the pewter siphon tube and then scraped off against the inside of the upper part of the bottle so as to keep water out of the tube). By pushing vessel *A* completely below the surface of the water in trough *FGHK*, all of the gas is pushed into bladder *B*, which is tied tightly around wood collar Cc, itself forming an airtight connection with the siphon tube with the aid of lute (almond powder made into a patty with glue). *Figure 4* shows the gas-generation vessel *A* filled with acid with metal added; glass tube *B* connects to C, which is filled with pearl ash (dry K_2CO_3, for removing aqueous acidic aerosols) and has a small opening at the top. The apparatus in *Fig. 4* allows determination of the weight of hydrogen lost from the top of C. *Figure 5* shows collection of a gas through a drying tube (containing pearl ash) for probably the first time, and *Fig.* 6 is an apparatus used to investigate the water solubility of "fixed air" (carbon dioxide).[2]

In 1784 Cavendish published his work on the composition of water based upon his experiments igniting hydrogen in air. (Primacy for the discovery that, once and for all, water is a compound and not an element was later given to James Watt). At this time he also noted that absorption of all of the oxygen (dephlogisticated air) and nitrogen (phlogisticated air) by chemical reaction left a tiny, but reproducible trace of unreactive gas. The apparatus is shown in Figure 82 ("Experiments on Air").[2] In *Fig. 1* of Figure 82 we see the apparatus used by Cavendish for conducting the experiment. Tube M is initially filled with mercury as are the two glasses. Gases are collected using the j tube in *Fig. 2* from a glass containing nitrogen or oxygen that is inverted in water. Exact gas volumes are cleverly introduced through tip A into tube M. Liquid containing the base and litmus indicator is also similarly introduced into tube M. Mercury serves as the container and electrical conductor for the sparking of known amounts of the two gases in the upper part of tube M. *Figure 3* depicts an apparatus for repeated introduction of quantities of gas into tube M through tip A. In this work, Cavendish anticipated the discovery by Rayleigh and Ramsay of the inert gases (e.g., argon) 110 years later. With great admiration and respect, they quoted him extensively in their own prize-winning report.

In 1798, Cavendish applied Newton's gravitation law to an experiment involving two lead balls and two smaller spheres. In so doing, he accurately determined the mass of the earth.

Let's examine his tenure file: On the one hand, he had only 18 published papers and no books. On the other hand, he discovered hydrogen, was a vital contributor to understanding the composition of water, discovered nitrogen and the composition of the atmosphere, separated the inert gases from atmospheric air, and weighed the planet. His student evaluations indicate that they disapprove his choice of clothing and that they don't "identify with" him. He also has a low profile on campus and seems to avoid committee work. Looks like this will be a difficult tenure decision.

1. *Encyclopedia Brittanica*, 15th ed., Chicago, 1986, Vol. 2, pp. 974–975.
2. J.R. Partington, *A History of Chemistry*, MacMillan, London, 1962, Vol. 3, pp. 302–362.

3. "Factitious air" refers to gases derived from heating or other chemical actions on solids. Thus, hydrogen appears to be "liberated" from an active metal upon addition of acid and is, therefore, a "factitious air."

MAKING SODA POP

Joseph Priestley (1733–1804) began a religious odyssey at an early age and is now recognized as one of the founders of Unitarianism.[1,2] At 19 he entered the Dissenting Academy of Daventry to study for the Nonconformist Ministry, refecting the early influence of his aunt. By the age of 28 he taught languages (including Hebrew), history, law, logic, and anatomy at the highly regarded Dissenting Academy at Warrington. His scientific interests were well under way by this time—he had earlier purchased an air pump and an electrical machine. His scholarship was recognized with an LL.D. from Edinburgh in 1765, and, with Benjamin Franklin's encouragement, Dr. Priestley published his *History of Electricity* in 1767 (he published a similar book on vision, light, and colors in 1772).

Priestley began his studies on "airs" (he disliked Van Helmont's term *gas*) around 1770. His home in Leeds was next to a brewery and Priestley collected "fixed air" (CO_2) directly from the surface of the brewing mixtures and investigated its properties. He also obtained this gas by heating natural mineral waters and recommended it for revitalizing flat beer.[1] His 1772 pamphlet (Fig. 83) was addressed to the Right Honourable John Earl of Sandwich, First Lord Commissioner of the Admiralty. Any modern grantsman will recognize a final report for a Department of Defense contract in it. By pouring dilute oil of vitriol (sulfuric acid) on chalk (calcium carbonate) Priestley generated "fixed air" and impregnated water with it. This articial soda was more readily available and cheaper than the carbonated waters from spas so many of which were, unfortunately, located in the borders of the hated enemy France.

Carbonated water had long been reputed to prevent "the sea scurvy" on long voyages and to slow the putrefaction of water. In addition, it settled upset stomachs and acted as something of a substitute for the fresh vegetables that aid digestion. Priestley thus helped Brittania to "rule the waves." There is nothing like soda pop to help sailors down a ship's store of salt pork. Whether or not the meat was pressed between slices of bread is not clear from this slim pamphlet.

During this period of time, a Portuguese monk named Joao Jacinto de Magalhaens[3] (Magellan for short, a descendant of the famous navigator) was employed in England as a spy for France. He recognized the importance of a potential treatment for scurvy on the high seas and sent a copy of Priestley's pamphlet to France. Clearly, a strategic "soda-pop gap" between England and France was intolerable. The person in France who was requested (i.e., assigned) to study this chemistry? Antoine Laurent Lavoisier. It was the start of his pneumatic researches that ultimately revolutionized chemistry.[3]

1. J.R. Partington, *A History of Chemistry*, MacMillan, London, 1962, Vol. 3, pp. 237–297.

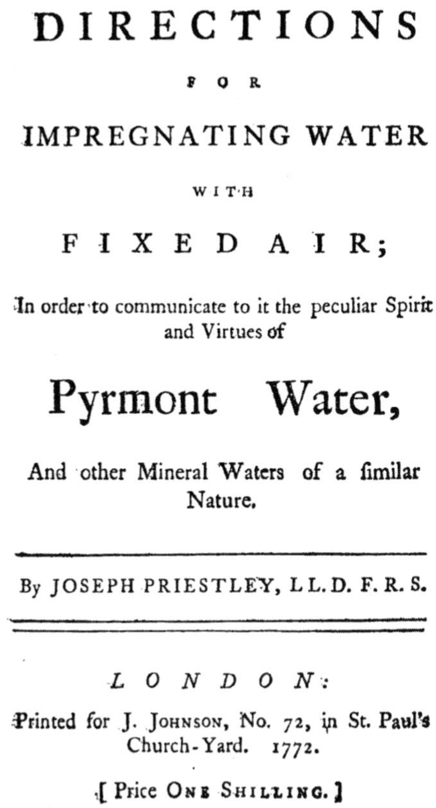

DIRECTIONS

FOR

IMPREGNATING WATER

WITH

FIXED AIR;

In order to communicate to it the peculiar Spirit and Virtues of

Pyrmont Water,

And other Mineral Waters of a ſimilar Nature.

By JOSEPH PRIESTLEY, LL.D. F.R.S.

LONDON:

Printed for J. JOHNSON, No. 72, in St. Paul's Church-Yard. 1772.

[Price ONE SHILLING.]

FIGURE 83 ■ Joseph Priestley made artificial soda by pressurizing water with chemically generated carbon dioxide. The work was vital to the strategic interests of the Royal Navy since carbonated water remained fresh longer than untreated water and was useful for treating upset stomachs. Fearful of a strategic "soda-pop gap," French spies reported this scientific advance and research was ordered. The young researcher? Antoine Laurent Lavoisier, who commenced the chemical revolution.

2. A.J. Ihde, *The Development of Modern Chemistry*, Harper & Row, New York, 1964, pp. 40–50.
3. J.-P. Poirier, *Lavoisier—Chemist, Biologist, Economist*, R. Balinski (translator), University of Pennsylvania Press, Philadelphia, 1996, pp. 51–54.

IF YOU *DO* FIND THE PHILOSOPHER'S STONE: "TAKE CARE TO LOSE IT AGAIN"—BENJAMIN FRANKLIN

Benjamin Franklin (1706–1790) was a brilliant and worldly polymath and a force behind both the American Declaration of Independence and the Constitution. The story of his arrival in Philadelphia at age 17 is well known—walking up Market Street on his first day, munching one "great Puffy Roll" with the other two under each arm and meeting his future wife Deborah Read.[1] According to Franklin, Miss Read " . . . thought I made as I certainly did a most awkward ridiculous Appearance."[1] Starting in the printing trade, he spent two years in England before setting up business in the American Colonies. Money-making ventures starting around 1730 included the printing of *Poor Richard's Almanacs* and the concessions for printing the currencies of Pennsylvania, New Jersey, Delaware, and Maryland. From this period through the 1740s Franklin accumulated wealth, became active in politics, and successfully promoted many ventures including the nascent University of Pennsylvania.[1] During the 1740s Franklin turned his attention increasingly to scientific pursuits.

Contemporary interest in electricity intrigued Franklin. He demonstrated that lightning and electricity were the same by flying a kite into an electrical storm and was lucky to have escaped electrocution. He believed electricity was a fluid that flowed from a body rich in it to a body poor in it. These considerations led to his invention of the lightning rod. The electrical terms *positive* and *negative*, *battery*, and *conductor* were coined by Franklin. His book *Experiments and Observations on Electricity* was first published in 1751 and went through four additional English editions as well as editions in French, German, and Italian (see Fig. 84).

Franklin met Joseph Priestley in London around 1765 and encouraged him to write his book *The History and Present State of Electricity* (1767). One bit of correspondence between the two has Priestley writing to Franklin in 1777 that he "did not quite despair of the philosopher's stone"; Franklin's response was that if he (Priestley) found it, "to take care to lose it again."[2] Franklin spent parts of the late 1770s soliciting and receiving military support from the French. He became a popular figure and a virtual cult hero in France and spent the years immediately following the American Revolution as a diplomat and business agent in France. He was an intimate in Lavoisier's scientific and social circle and Madame Lavoisier painted his portrait—apparently one of his favorites.[3,4] Apparently, Madame Lavoisier painted the portrait, and a copy of it, following the portrait by Duplessis.[5] The painting given to Franklin remains today (1999) in the possession of one of his descendents,[5,6] while Madame Lavoisier's personal copy appears to be unlocated.[5]

1. *Encyclopedia Brittanica*, Encyclopedia Brittanica, Chicago, 1986, Vol. 19, pp. 556–559.
2. J.R. Partington, *A History of Chemistry*, MacMillan, London, 1962, Vol. 3, pp. 241, 245–246.
3. B.B. Fortune and D.J. Warner, *Franklin and His Friends—Portraying the Man of Science in Eighteenth-Century America*, Smithsonian Portrait Gallery and University of Pennsylvania Press, Washington, D.C. and Philadelphia, 1999.
4. D. McKie, *Lavoisier–Scientist. Economist. Social Reformer*, Collier, New York, 1962, p. 68.
5. C. C. Sellars, *Benjamin Franklin in Portraiture*, Yale University Press, New Haven, 1962, pp 273–274.
6. Personal correspondence of the author with Franklin relative.

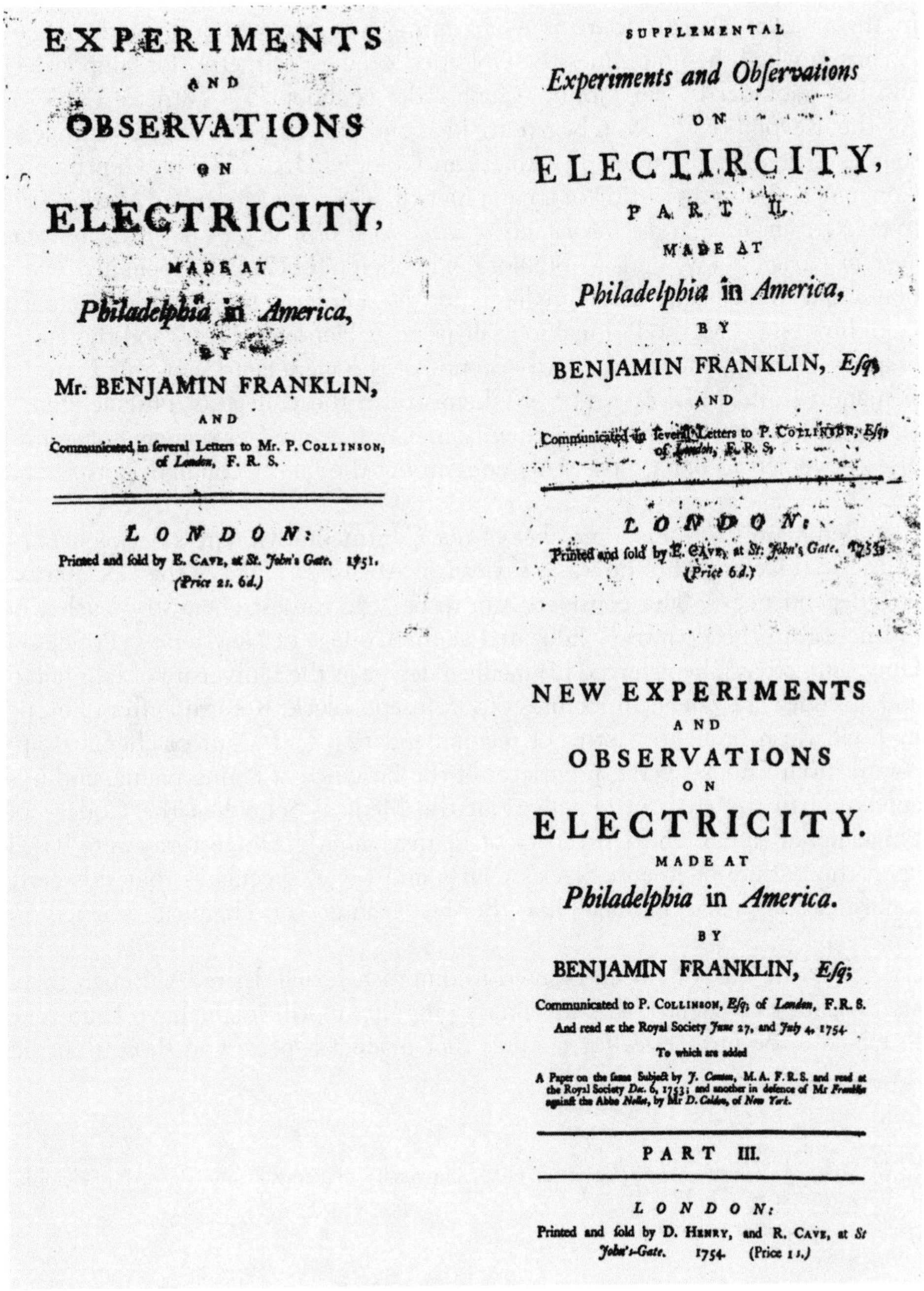

EXPERIMENTS
AND
OBSERVATIONS
ON
ELECTRICITY,
MADE AT
Philadelphia in *America*,
BY
Mr. BENJAMIN FRANKLIN,
AND
Communicated in several Letters to Mr. P. COLLINSON,
of *London*, F. R. S.

LONDON:
Printed and sold by E. CAVE, at St. *John's Gate*. 1751.
(*Price 2s. 6d.*)

SUPPLEMENTAL
Experiments and Observations
ON
ELECTIRCITY,
PART II.
MADE AT
Philadelphia in *America*,
BY
BENJAMIN FRANKLIN, *Esq*;
AND
Communicated in several Letters to P. COLLINSON, *Esq*;
of *London*, F. R. S.

LONDON:
Printed and sold by E. CAVE, at St: *John's Gate*. 1753.
(*Price 6d.*)

NEW EXPERIMENTS
AND
OBSERVATIONS
ON
ELECTRICITY.
MADE AT
Philadelphia in *America*.
BY
BENJAMIN FRANKLIN, *Esq*;
Communicated to P. COLLINSON, *Esq*; of *London*, F.R.S.
And read at the Royal Society *June* 27, and *July* 4, 1754.
To which are added
A Paper on the same Subject by *J. Canton*, M.A. F.R.S. and read at the Royal Society *Dec.* 6, 1753; and another in defence of Mr *Franklin* against the Abbe *Nollet*, by Mr *D. Colden*, of *New York*.

PART III.

LONDON:
Printed and sold by D. HENRY, and R. CAVE, at *St John's-Gate*. 1754. (Price 1s.)

FIGURE 84 ■ Title pages of Benjamin Franklin's works on electricity. Franklin considered electricity to be a fluid and coined the terms "positive" and "negative" to denote electrical charges (courtesy of Jeremy Norman & Co., from Catalogue 5, 1978).

SALTPETRE, ABIGAIL. *PINS*, JOHN

In the musical "1776," there is a charming duet between John and Abigail Adams in which he underlines the Colonies' need for saltpetre (for gunpowder) and her rejoinder is "*pins*, John." Gunpowder is about 75% saltpetre (KNO_3). As the Revolutionary War began to heat up, the British blocked European sources of ingredients from the American Colonies. In 1775, the Continental Congress authorized printing of a pamphlet titled *Several Methods of Making Salt-petre; Recommended to the Inhabitants of the United Colonies, by Their Representatives in Congress* (W. and T. Bradford, Philadelphia, 1775).[1] Hopefully, every household would make the ingredients for gunpowder. The pamphlet included Franklin's essay "Method of making salt-petre at Hanover, 1766" and the larger essay by Dr. Benjamin Rush: "An account of the manufactory of Salt-Petre by Benjamin Rush, M.D. professor of Chemistry in the college of Philadelphia."[1] (During the Civil War, advertisements in Confederate newspapers constantly pressed women to donate the daily contents of the family chamber pots to the cause).

Benjamin Rush was a member of the Continental Congress between 1774 and 1778. Like Franklin, he was a signer (on August 2, 1776) of the Declaration of Independence.[1] Myles considers him to be "the earliest chemistry teacher of distinction in this country."[1] Educated at the College of New Jersey (Princeton University today), he received his medical degree at the University of Edinburgh and attended the chemical lectures of Dr. Joseph Black. Returning from Europe in 1769, Rush brought a letter of recommendation and a gift of chemical apparatus from Thomas Penn, proprietor of the Province of Pennsylvania, and was appointed to the chair of chemistry at the Medical School of the College of Philadelphia (today, the University of Pennsylvania).[1] His lectures were based upon the outline of Joseph Black's course and Myles speculates that this early sophistication made Philadelphia the first center for chemical science in America.

Saltpetre has long been reputed to diminish sexual desire. Although there are no data that support this imaginary property, it still might have been nice to tabulate the birth rates for families that made salt-petre and those that did not.

1. W. Myles, in *Chymia*, H.M. Leicester (ed.), University of Pennsylvania Press, Philadelphia, 1953, Vol. 4, pp. 37–77.

SECTION V
MODERN CHEMISTRY IS BORN

FIRE AIR (OXYGEN): WHO KNEW WHAT AND WHEN DID THEY KNOW IT?

Carl Wilhelm Scheele (1742–1786) was the seventh child of eleven in a Swedish family raised in very modest circumstances. Higher education was never an option for him and at 14 Scheele was apprenticed to the Bauch Apothecary in Gothenburg. He began to learn his craft and read the great chemical texts of Lemery, Kunckel, Boerhaave, Neumann, and Rothe.[1] Moving to Malmo in 1765, his Master Kjellstrom described the young Scheele's reactions as he pored through texts: "that may be; that is wrong; I will try that."[1] He moved to Uppsala and met Torbern Olof Bergman (1735–1784) in 1770. Bergman was Professor of Chemistry and Pharmacy at the University of Uppsala, a member of the Academies of Uppsala, Stockholm, Berlin, Gottingen, Turin, and Paris, a Fellow of the Royal Society in London, and, for a time, Rector of the University.[1] The influential Bergman helped to guide and promote the younger Scheele. Partington[1] notes that Scheele's contributions to chemistry "are astonishing both in number and importance" and quotes the great nineteenth-century chemist Humphrey Davy: "nothing could damp the ardour of his mind or chill the fire of his genius: with very small means he accomplished very great things."

Scheele is now regarded as the uncontestable discoverer of oxygen. His work began with a complaint by Bergman that a sample of saltpetre (KNO_3) purchased in the shop that employed Scheele gave off red vapors upon contact with acid. Scheele quickly established that heating saltpetre produced another salt. Impressed, Bergman suggested that Scheele investigate manganese dioxide (MnO_2).

Scheele believed in the Phlogiston Theory and continued to do so throughout his life, as did Joseph Priestley. He felt that heat was a combination of phlogiston and what he called "fire air." His theoretical basis for this belief is nicely and succinctly described by Ihde.[2] Scheele reasoned that when a substance burned, it lost phlogiston, which combined with air, to some extent, increasing its mass and decreasing its volume. However, he found that the remaining "foul air" ("mephitic air" or nitrogen) was less dense rather than more dense than air. Thus, he reasoned, there was a component of common air he termed "fire air" that combined with phlogiston to produce heat, a kind of ethereal fluid, which escaped through the glass vessel. Scheele then decided to isolate "fire air" from heat by capturing the phlogiston using nitre. [Remember Mayow's experiments published almost 100 years earlier and depicted in Fig. 67—nitre or saltpetre dephlogisticate (burn) charcoal or sulfur to produce the respective "acids."] Scheele's investigation involved heating saltpetre ("fixed" nitric acid) and capturing the "fire air":[2]

$$\text{Heat} + \text{Nitric acid} \rightarrow \text{Fire air} + \text{Red fumes}$$

Here, heat was composed of fire air and phlogiston; the red fumes were nitric acid and phlogiston.

Figure 85 is from Scheele's complete works (*Opuscula Chemica et Physica*, two volumes, Leipzig, 1788 and 1789). The first edition of his book, *Chemische Abhandlung von den Luft und dem Feuer* (Leipzig, 1777) is retailing for over $20,000 these days; the English translation, *Chemical Observations and Experiments on Air and Fire* (London, 1780) is following in its price wake. *Figure 3* (in the lower left of Figure 85) depicts the heating of saltpetre with oil of vitriol (concentrated sulfuric acid). The red fumes are nitrogen dioxide arising from the well-known decomposition of concentrated nitric acid ($2HNO_3 \rightarrow 2NO_2 + H_2O + \frac{1}{2}O_2$). The collection bladder (*Fig. 4*) contained milk of lime (a sus-

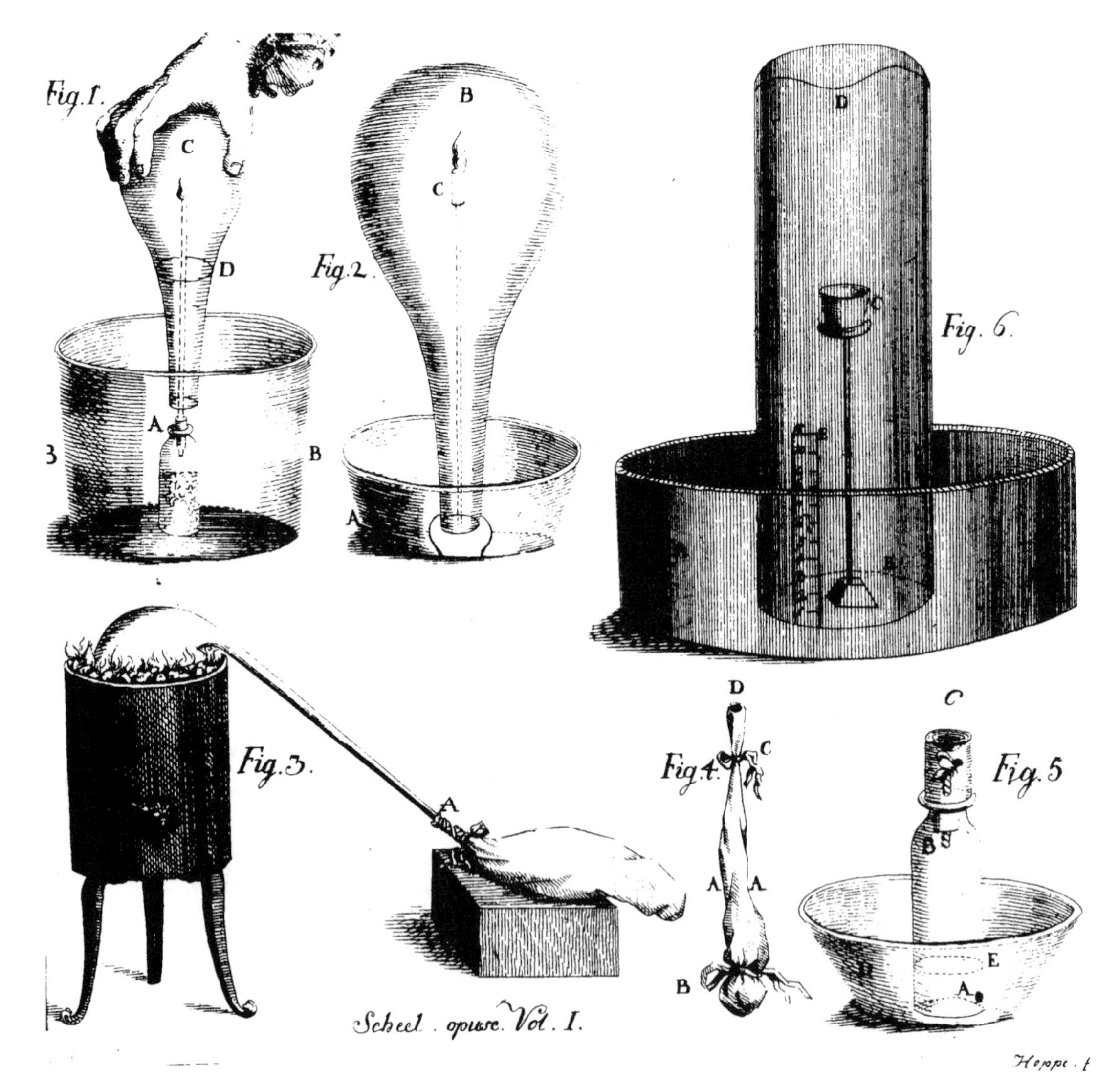

FIGURE 85 ■ Pneumatic experiments in which Carl Wilhelm Scheele was the first to discover oxygen ("fire-air"). These findings were first published in the exceedingly rare 1777 *Chemische Abhandlung von den Luft und dem Feuer* (Leipzig); first English edition (*Chemical Observations and Experiments on Air and Fire*, London, 1780). This figure is derived from *Opuscula Chemica et Physica* (Leipzig, 1788–89).

pension of calcium hydroxide) to collect the acidic fumes. The bladder filled with fire air (oxygen), which was transferred to glass bottles. The new gas supported flames and life. *Figure 1* in Figure 85 depicts the combustion of hydrogen gas (prepared in the lower bottle by combining metal and acid) in common air. When burning stops, water has filled the vessel to *D*—about 20% of the original air volume. *Figure 2* in Figure 85 depicts a candle burning in air. When the burning stops and the vessel is unsealed and immersed open-end down in lime water (CaO), the "fixed air" (CO_2) generated by the candle forms chalk with the solution that then rises into the glass vessel. *Figure 5* depicts bees (not mice, for once) in vessel C—they are shown to produce "fixed air" via the lime-water test.

Poirier[3] notes that on November 16, 1772, and probably as early as 1771, Scheele had heated manganese dioxide (MnO_2) and obtained "fire air." During this early period he also obtained oxygen by heating mercuric oxide, silver carbonate, magnesium nitrate, and saltpetre.[3] On September 30, 1774, he wrote to Lavoisier suggesting preparation and heating of silver carbonate using Lavoisier's powerful burning glass. The letter was received by Lavoisier on October 15, 1774. Poirier notes that it was never answered: "The Swedish historians of science have still not forgiven him for what was much more than simple rudeness. It is difficult to disagree with them."[3] Joseph Priestley announced his truly independent discovery of "dephlogisticated air" (oxygen) in August, 1774 and later in that year Lavoisier effectively rediscovered it. Lavoisier's real discovery was not oxygen but the weight gain of the metals and acids formed by its absorption in chemical combination.[3]

Scheele was apparently already working on his book by November, 1775.[1] Apparently, his first knowledge of Priestley's discovery came in August, 1776. Long delays by the book publisher compounded by a further delay awaiting Bergman's Introduction in the book delayed it to 1777. By this time, the self-effacing Scheele no longer claimed primacy, fearing charges of plagiarism.[1] He continued his incredibly productive chemical investigations, dying at the age of 44 from "a complication of disorders, including rheumatism contracted by work in unfavorable circumstances."[1]

1. J.R. Partington, *A History of Chemistry*, MacMillan, London, 1962, Vol. 3, pp. 205–234; see also pp. 179–199.
2. A.J. Ihde, *The Development of Modern Chemistry*, Harper and Row, New York, 1964, pp. 50–53.
3. J.-P. Poirier, *Lavoisier—Chemist, Biologist, Economist*, R. Balinski (translator), University of Pennsylvania Press, Philadelphia, 1996, pp. 76–83.

NICE TO HIS MICE

Priestley's first original scientific paper (1770) was on charcoal and had a number of errors.[1,2] However his 1772 paper "Observations on Different Kinds of Air"[3] was a "powerhouse" and was the start of his six volumes published between 1774

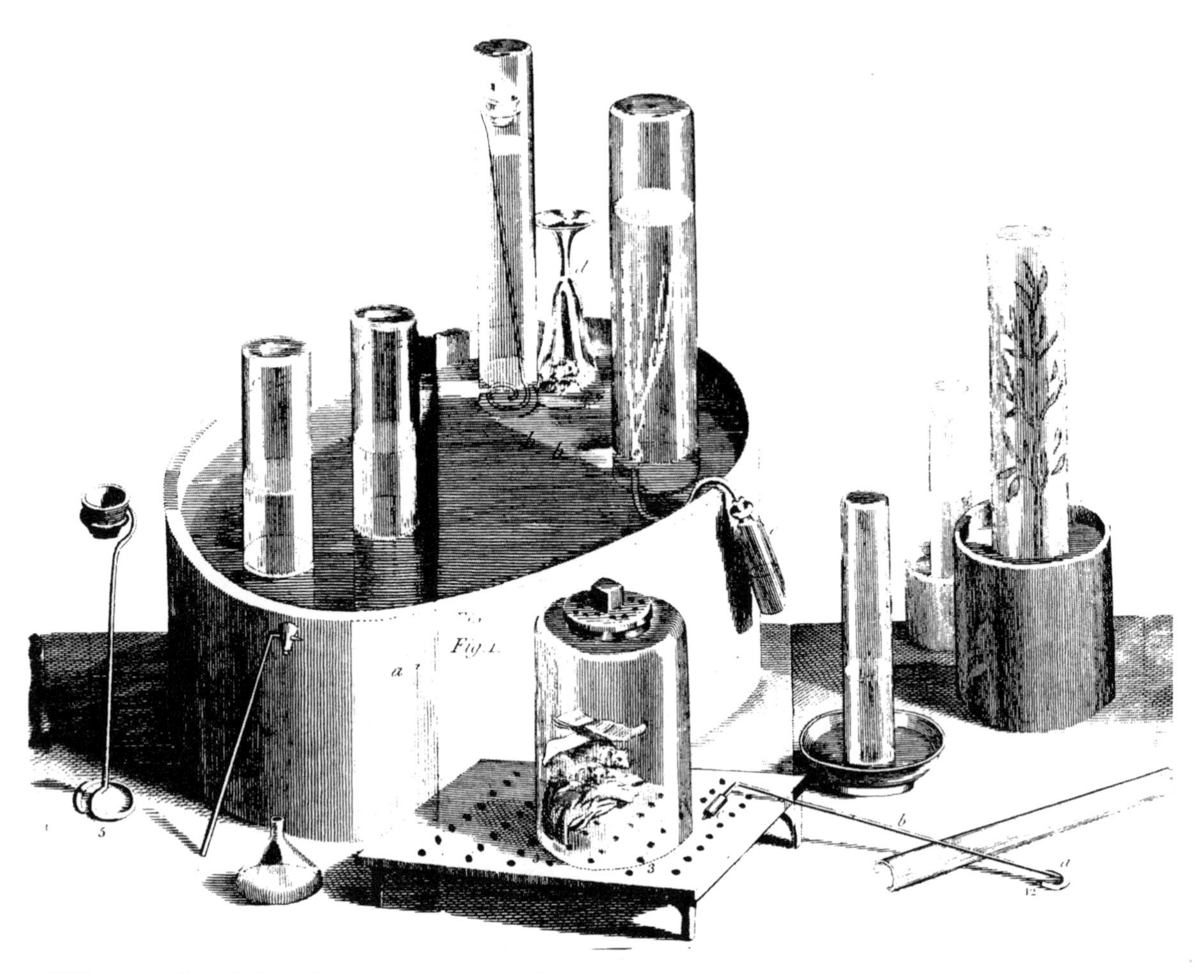

FIGURE 86 ■ Joseph Priestley's pneumatic trough for isolation of "factitious airs" (gases derived from solids). Although Scheele was the first to discover oxygen, Priestley published first (1774). He was gentle to his experimental mice (from the later abridged edition, *Experiments and Observations on Different Kinds of Air*, Birmingham, 1790).

and 1786. Priestley's pneumatic trough[4] (Fig. 86) evolved from Hales' apparatus (Fig. 80) through modifications by William Brownrigg. Priestley capitalized on Cavendish's technique for collecting water-soluble gases such as carbon dioxide over mercury instead of water.[1,2]

In the landmark 1772 paper Priestley describes the isolation and properties of gases first observed by others but not so systematically. He described carbon dioxide ("*fixed air*"—sometimes termed *mephitic air*), nitrogen (the air remaining after a candle had burned out in common air and following CO_2 precipitation in lime water—he termed it "phlogisticated air," often also termed by others "mephitic air"), hydrogen (Cavendish's "inflammable air"—sometimes confused by Priestley with carbon monoxide), hydrogen chloride ("acid air"—later "marine air"), and nitric oxide (NO—"nitrous air").

Nitrous air was generated by exposure of brass, iron, copper, tin, silver, mercury, bismuth, or zinc to nitric acid.[5] Priestley discovered that it reacted instantly with common air to produce a reddish-brown gas (NO_2), which dissolved in water to produce nitric acid. After his own discovery of oxygen in

1774, two to three years after Scheele (Priestley was scrupulously honest and unaware of Scheele's work), Priestley realized that he had discovered a simple and reliable technique for testing the "goodness" of air: "every person of feeling will rejoice with me in the discovery of nitrous air, which supersedes many experiments with the respiration of animals."[6] Although the inverted beer glass in Figure 86 (see *Fig. 1*, part d, and also *Fig. 3*) depicts an experimental mouse, Partington[6] notes that Priestley "always took pains to keep his mice warm and comfortable."

Priestley's discovery of oxygen on August 1, 1774 was made by heating red HgO (*mercurious calcinatus*), itself obtained by heating mercury in air or by reaction of mercury with nitric acid (remember Mayow's work, Fig. 67). A firm believer in Stahl's Phlogiston Theory to the end of his life, Priestley called the amazing new air, which supported combustion and respiration, "dephlogisticated air." The idea is that a burning candle, for example, loses its phlogiston to something "dephlogisticated" that hungrily grabs it. Indeed, Priestley also found that exposure of "nitrous air" (NO) to iron filings produced a new gas, capable of supporting a brilliant flame, which he called "dephlogisticated nitrous air."[1,2] This was actually nitrous oxide (N_2O, or "laughing gas") which had apparently first been made prior to 1756 by Joseph Black, who heated ammonium nitrate and found vapors whose "effect on breathing and sensation was very far from being unpleasant."[7] Other gases explored by Priestley included ammonia (NH_3, "alkaline air"), sulfur dioxide (SO_2, "vitriolic acid air"), and silicon tetrafluoride ("fluor acid air").

The politically liberal Priestley was sympathetic to the aspirations of the American Colonies and was a regular correspondent of Franklin. In a climate of fear and conservative backlash to the American and French Revolutions and on July 14 (Bastille Day) of 1791 a wild Birmingham mob burned Priestley's meeting house to the ground (the family had fled earlier). In his entertaining book *Crucibles* Jaffe seems to have discovered an eighteenth-century video recording or a "fly on the Church wall" that overheard one rioter yell: "Let's shake some powder out of Priestley's wig."[8] Even the more cosmopolitan London was no longer friendly. In 1794 Priestley moved to the United States, modestly declining a Professorship at the University of Pennsylvania and the charge of a Unitarian chapel in New York, for the peace of living and writing in Northumberland, Pennsylvania.

1. J.R. Partington, *A History of Chemistry*, MacMillan, London, 1962, Vol. 3, pp. 237–297.
2. A.J. Ihde, *The Development of Modern Chemistry*, Harper & Row, New York, 1964, pp. 40–50.
3. J. Priestley, *Philosophical Transactions of the Royal Society*, **62**:147–267 (1772).
4. This figure is from the 1790 edition (*Experiments and Observations on Different Kinds of Air and other Branches of Natural Philosophy*, three volumes, Birmingham, 1790) of the six books published between 1774 and 1786. The same figure appeared in Volume 1 of that series.
5. It is important to note that nitric acid is different from an acid such as hydrochloric in that the nitrate part (NO_3^-) is a stronger oxidizing agent than aqueous hydronium ions (H_3O^+). Thus, copper and iron, which have more positive (more favorable) reduction potentials than H_3O^+, are not oxidized readily in HCl to produce H_2 gas. However, they are oxidized by the powerful NO_3^-, which has a very high reduction potential and is therefore readily reduced to NO. Magnesium, which is very easily oxidized (very hard to reduce), will produce H_2 gas in both hydrochloric and nitric acids.
6. J.R. Partington, op. cit., p. 253.
7. J.R. Partington, op. cit., p. 142.
8. B. Jaffe, *Crucibles*, Simon and Schuster, New York, 1930, p. 52.

WHERE IS THE INVECTIVE OF YESTERYEAR?

Invective was employed as an art form in scientific discourse centuries ago. A wonderful example is from the Preface to the 1776 edition of a book called *Phosphori*,[1] written by Benjamin Wilson (1721–1788). Like Priestley, he believed in phlogiston and held that the glow of phosphorescence was visible evidence of phlogiston, the fire trapped in many types of matter.

Wilson[2] endured his family's poverty until not yet 20, worked in poor circumstances, commenced artistic studies in these circumstances, and started to enjoy some success in his 40s, being appointed by the Duke of York to succeed William Hogarth as Sergeant-Painter in 1764. He speculated in stocks and was declared a defaulter on the Stock Exchange in 1766. During the 1740s he also developed an interest in electricity and later engaged in a highly charged public debate with Benjamin Franklin on the shape of lightning rods. (Wilson had painted a portrait of Franklin in 1759.)[3] Franklin argued for a sharp point, and Wilson correctly argued for a rounded point that would not actually attract lightning. He won the debate but his arguments were so excessive that he received the following criticism in the *Philosophical Transactions*:[2]

> But he has been chiefly distinguished as the ostensible person whose perverse conduct in the affair of the conductors of lightning produced such shameful discord and dissensions in the Royal Society, as continued for many years after, to the great detriment of science.

The scorn so evident in the Preface of *Phosphori* is generally missing in scientific discourse. After all, Dr. X may eventually review Dr. Y's research grant proposal. In reading this excerpt one should note that Doctor Priestley was a painfully honest English clergyman and a friend of Franklin (and sympathetic to the American Colonies fight for independence) who had immense standing in the scientific community and had criticized Wilson's experiments:

> Now why may not such a plain philosopher (with the good Doctor's gracious leave) be supposed capable of, at least stumbling upon discoveries, which had escaped the observation of preceding philosophers, even of the highest and most respectable characters? For it is well known, that it is not always men of "vast and comprehensive understandings," that have been favoured by Providence with making discoveries sometimes the greatest, and most useful to the world: but on the contrary (to allude to the words of an eminent writer with whom Dr. Priestley is intimately acquainted), the Great Author of Nature hath frequently chosen "weak things," in the philosophical, as well as the spiritual world, to confound the mighty, and things that are not, to bring nought the things that are.

1. B. Wilson, *Phosphori*, 2nd ed., London, 1776.
2. *Dictionary of Scientific Biography*, Charles Scribner, New York, 1976, Vol. XIV, pp. 418–419.
3. B.B. Fortune and D.J. Warner, *Franklin and His Friends—Portraying the Man of Science in Eighteenth-Century America*, Smithsonian National Portrait Gallery and University of Pennsylvania Press, Washington, D.C. and Philadelphia, 1999, pp. 74–77.

FIGURE 32 ■ For full description, see page xv.

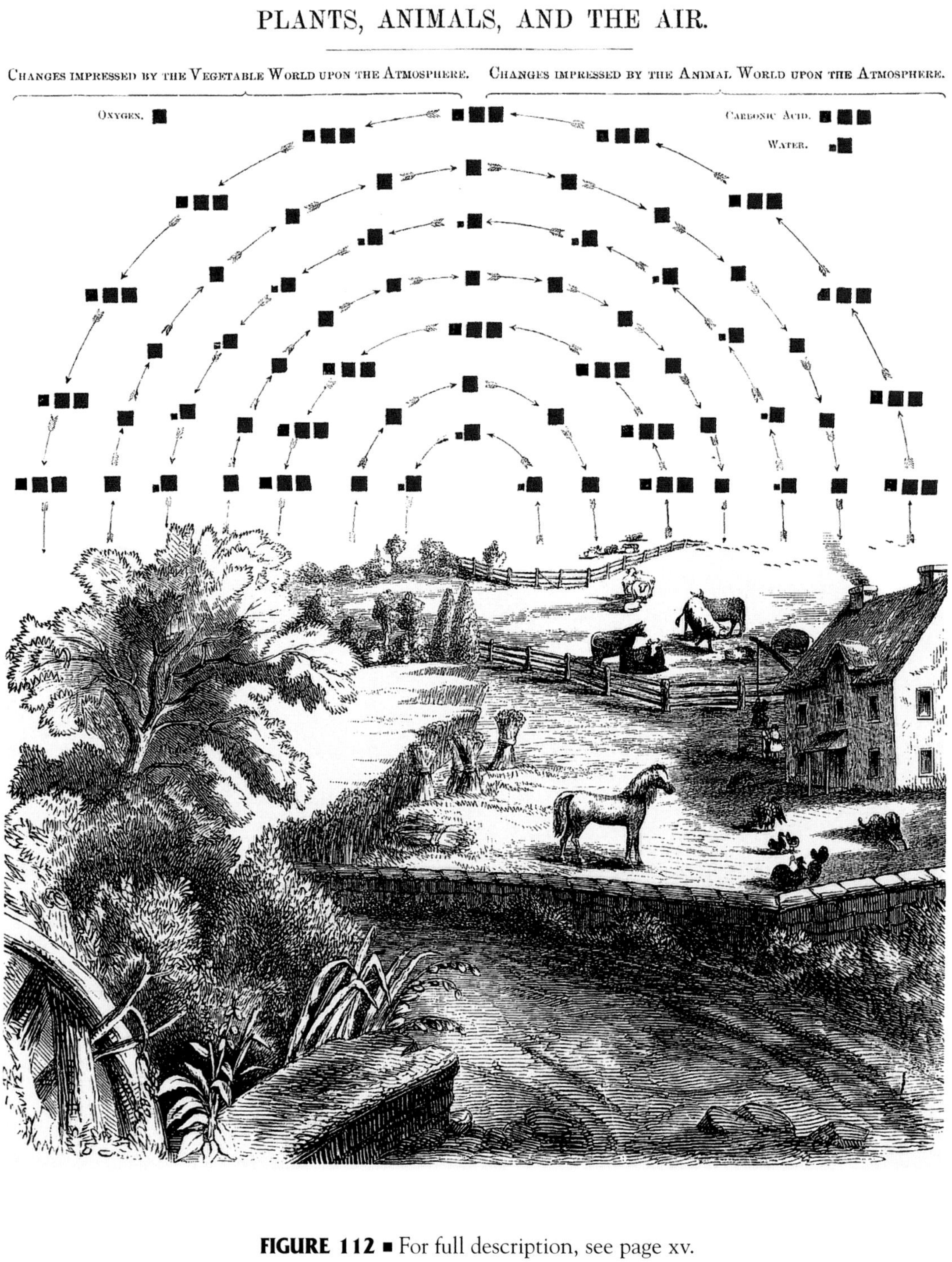

PLANTS, ANIMALS, AND THE AIR.

CHANGES IMPRESSED BY THE VEGETABLE WORLD UPON THE ATMOSPHERE. CHANGES IMPRESSED BY THE ANIMAL WORLD UPON THE ATMOSPHERE.

OXYGEN.

CARBONIC ACID.

WATER.

FIGURE 112 ▪ For full description, see page xv.

HOMOLOGOUS SERIES OF COMPOUNDS.

$C_2 H_2$ is the common difference, that is, each member of the series differs from the one preceding it and the one following it by two equivalents of Carbon and two of Hydrogen.

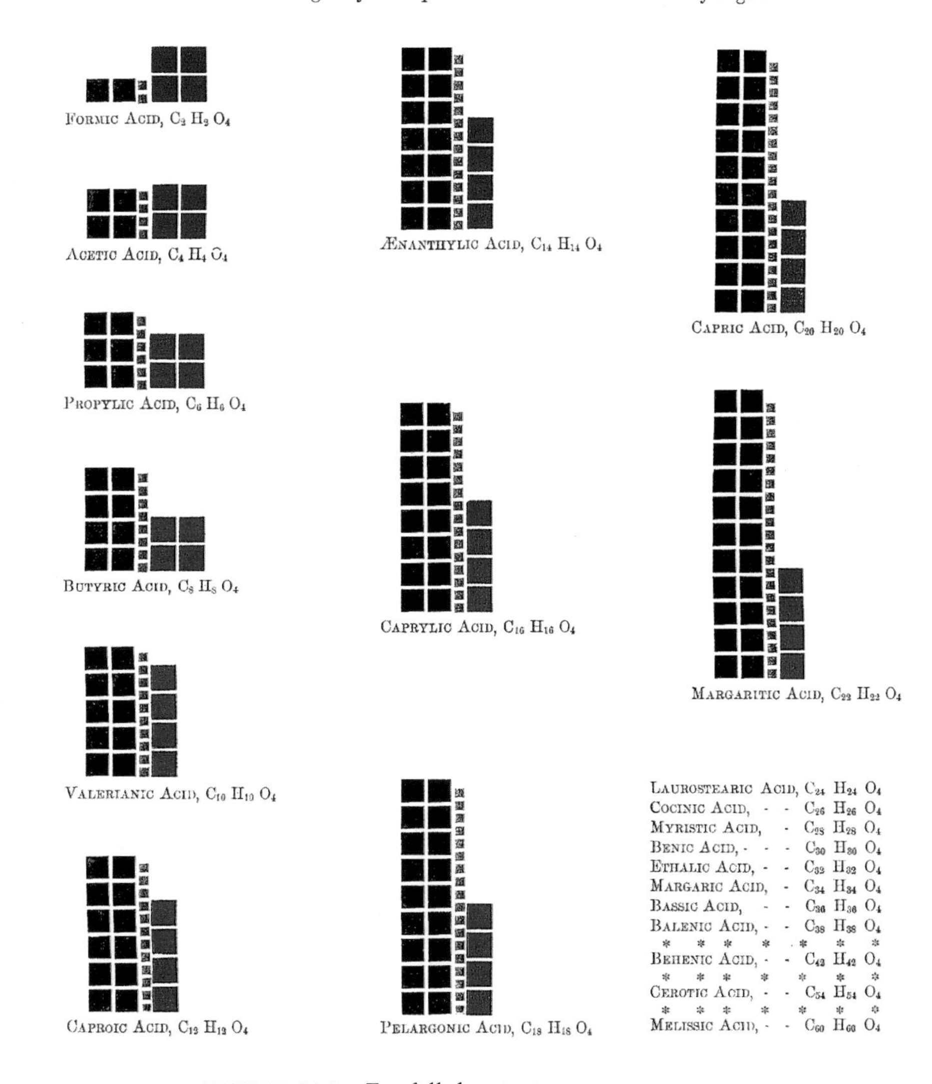

FIGURE 114 ▪ For full description, see page xv.

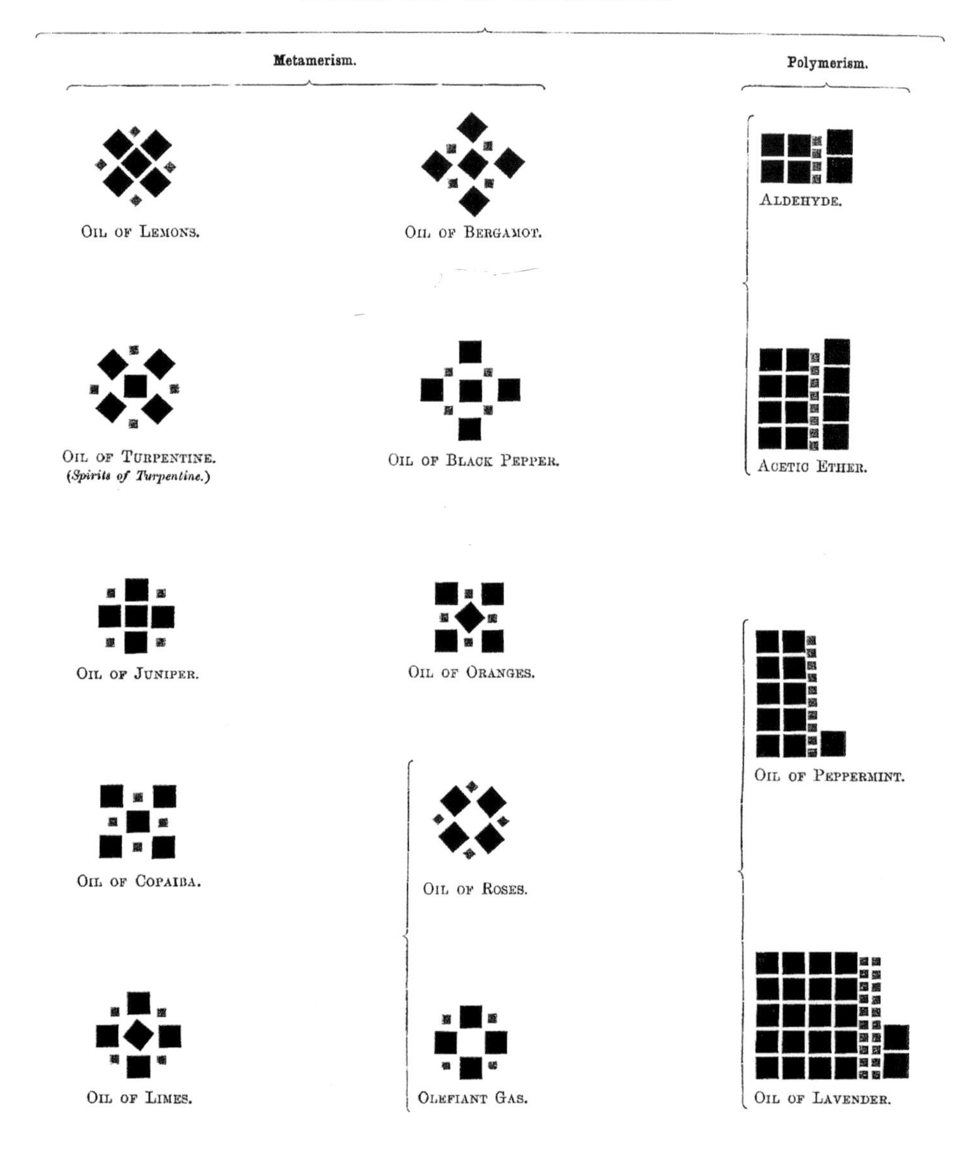

FIGURE 115 ▪ For full description, see page xv.

PLATE VI.

ILLUSTRATION OF THE THEORY OF COMPOUND RADICALS.

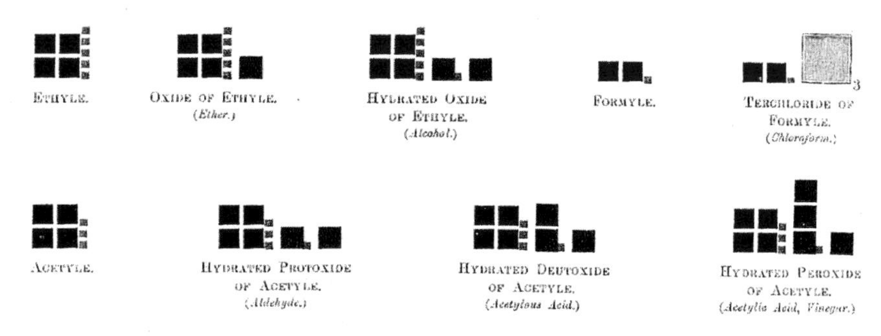

THEORY OF CHEMICAL TYPES—DOCTRINE OF SUBSTITUTION.

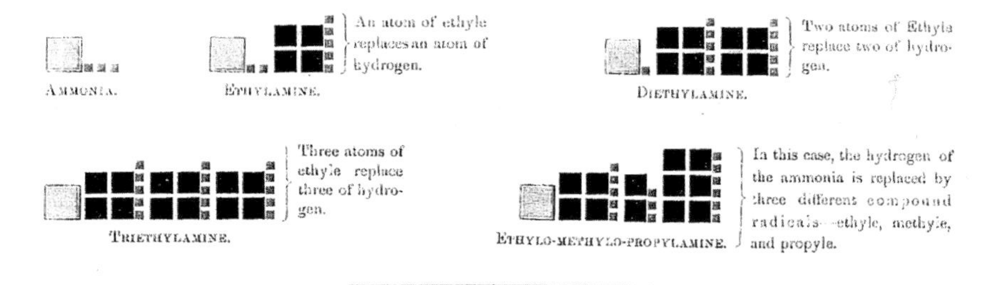

THEORY OF PAIRING—EXAMPLE OF COUPLED ACIDS.

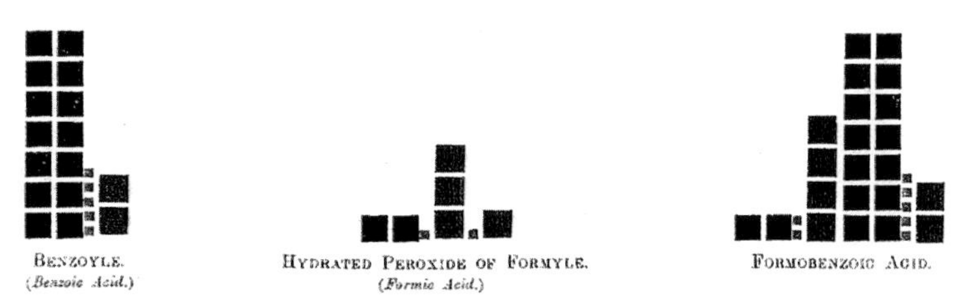

FIGURE 116 ■ For full description, see page xv.

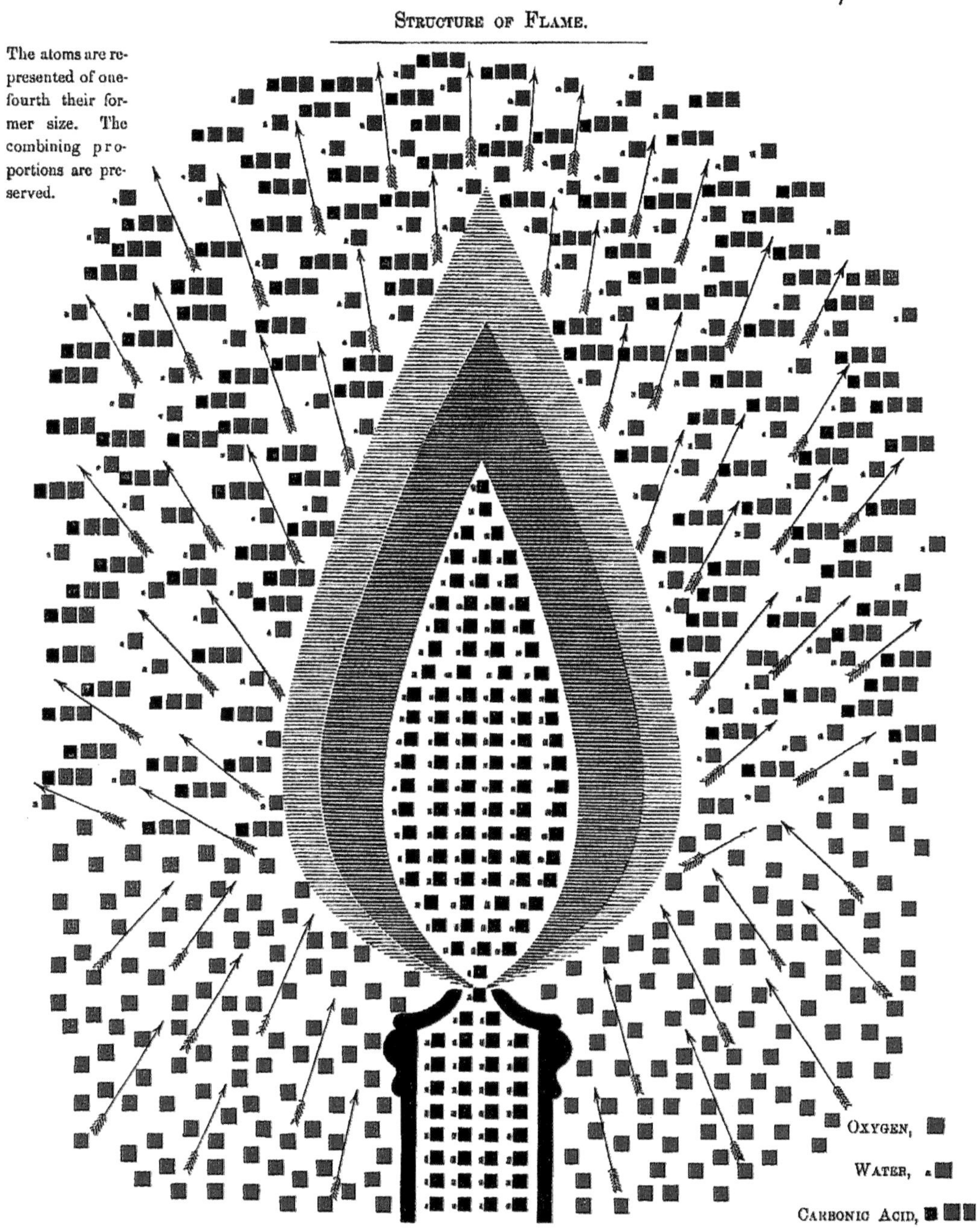

FIGURE 129 ▪ For full description, see page xv.

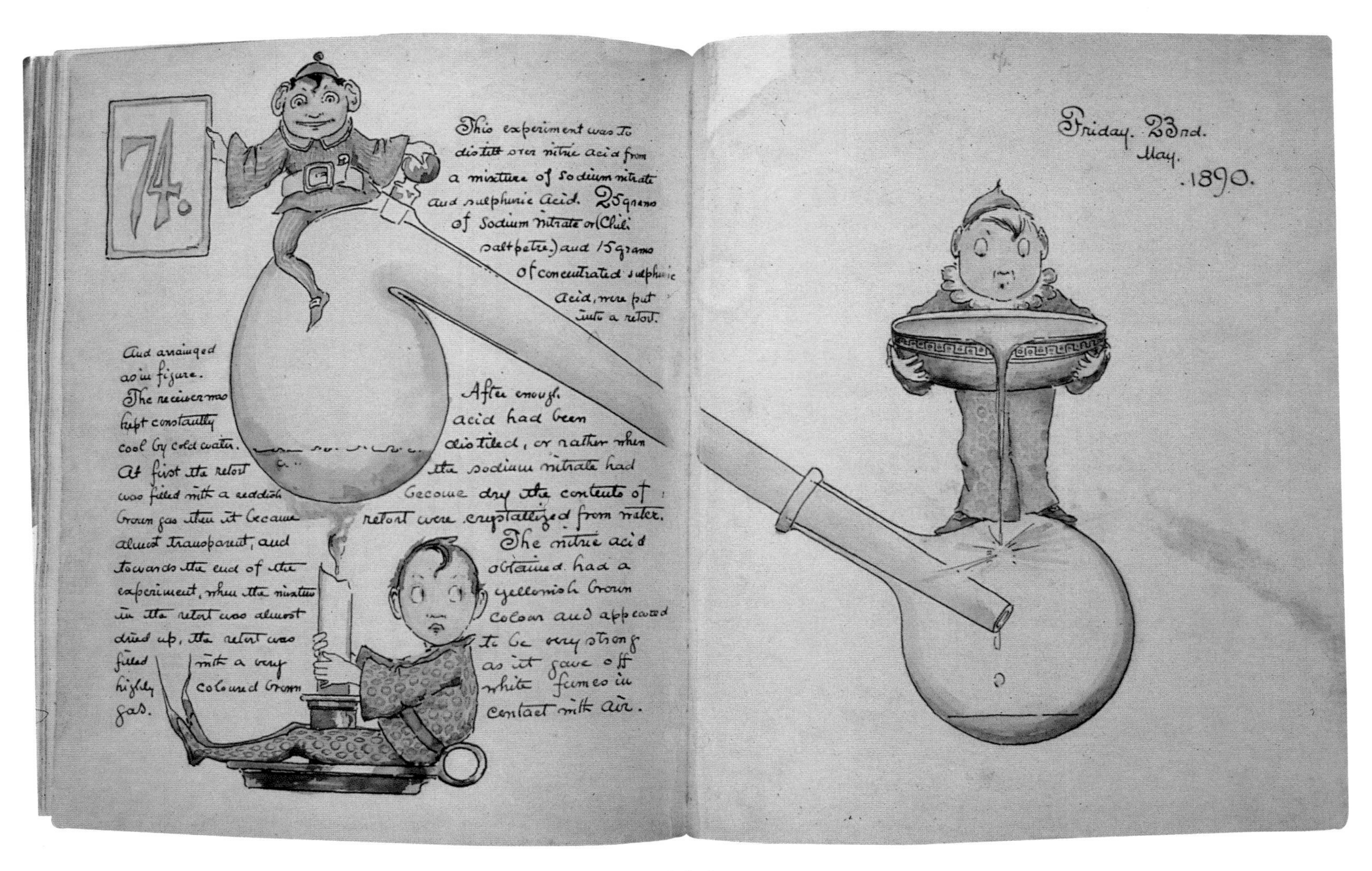

74.

Friday. 23rd.
May.
.1890.

This experiment was to distill over nitric acid from a mixture of sodium nitrate and sulphuric acid. 25 grams of Sodium nitrate or (Chili saltpetre) and 15 grams of concentrated sulphuric acid, were put into a retort.

And arranged as in figure.
The receiver was kept constantly cool by cold water.
At first the retort was filled with a reddish brown gas then it became almost transparent, and towards the end of the experiment, when the mixture in the retort was almost dried up, the retort was filled with a very highly coloured brown gas.

After enough acid had been distilled, or rather when the sodium nitrate had become dry the contents of retort were crystallized from water.
The nitric acid obtained had a yellowish brown colour and appeared to be very strong as it gave off white fumes in contact with air.

FIGURE 136 ■ For full description, see page xv.

COVER ILLUSTRATION ■ For full description, see page xv.

LA REVOLUTION CHIMIQUE COMMENCE

Antoine Laurent Lavoisier[1–3] (1743–1794) is justifiably said to be the father of modern chemistry. His greatest single contribution is the recognition that both combustion and calcination arise from combination of atmospheric oxygen with inflammable substances and metals rather than from loss of phlogiston from these substances. His greatest published work, *Traité Élémentaire de Chimie* (Paris, 1789; London, 1790; Philadelphia, 1796) is clearly a modern textbook. His contributions to chemistry, including its first systematic nomenclature, are far too numerous to mention in our brief *Chemical History Tour*. He was born to wealth, married additional wealth, lived a stylish and aristocratic life, and died by the guillotine on May 8, 1794 during the Reign of Terror. Before he died, Lavoisier experienced an angry fist-shaking crowd and perhaps stared across the River Seine to see the College Mazarin where he received his early education.[4] Twenty-eight members of the *Ferme Générale*, including Lavoisier, were executed in 35 minutes on that day.[4] The heads were placed in a wicker basket, the 28 bodies stacked on wagons and buried in large common graves dug in a wasteland named *Errancis* ("maimed person").[4] On May 9, the great mathematician Joseph Louis Lagrange commented: "It took them only an instant to cut off that head but it is unlikely that a hundred years will suffice to reproduce a similar one."[4]

The young Lavoisier showed a precocious interest in chemistry sparked by the lively demonstrations of Guillaume François Rouelle. His brilliance and his wealth gained him entry into the Académie Royale des Sciences in 1768 at the age of 25 and full membership in 1769. In 1768 he purchased a privilege to collect taxes in the Ferme Générale. Jacques Paulze, a senior member of the Ferme, had a beautiful and gifted young daughter named Marie-Anne Pierrette who was attracting unsuitably aged suitors. He introduced Antoine and Marie and they were married in 1771—just short of her fourteenth birthday. Intellectually, they were well met and Marie learned sufficient chemistry to be an effective and critical translator of texts in other languages including English, thus opening the wider chemical literature to Antoine. Her artistic talents also found some expression in the drawings used to illustrate his texts.

Lavoisier's earliest studies showed a respect for precise measurement. He demonstrated that diamonds decompose in strong heat (Boyle had proven this a century earlier) but showed that air was necessary and that the decomposition product turned lime water milky and was thus fixed air (CO_2). In 1772 his studies extended to the combustion of phosphorus and sulfur, which, like carbon, produced "acid airs" that weighed more than the solids that produced them. Similarly, he verified the observation by Jean Rey in 1630, also noted by Boyle and others, that the calxes formed by heating metals were heavier than the metals themselves. In his first great book (*Opuscules Chimiques et Physiques*, Paris, 1774; *Essays Physical and Chemical*, London, 1776), Lavoisier first offered the idea that these processes involved absorption of some "elastic fluid" present in air rather than loss of phlogiston to the air. In this book he confused this elastic fluid with fixed air.[1,2]

Figure 87 is from the 1776 *Essays Physical and Chemical*. In *Fig.* 8 (in Fig. 87) we see an apparatus for measuring the "air" absorbed during calcination of

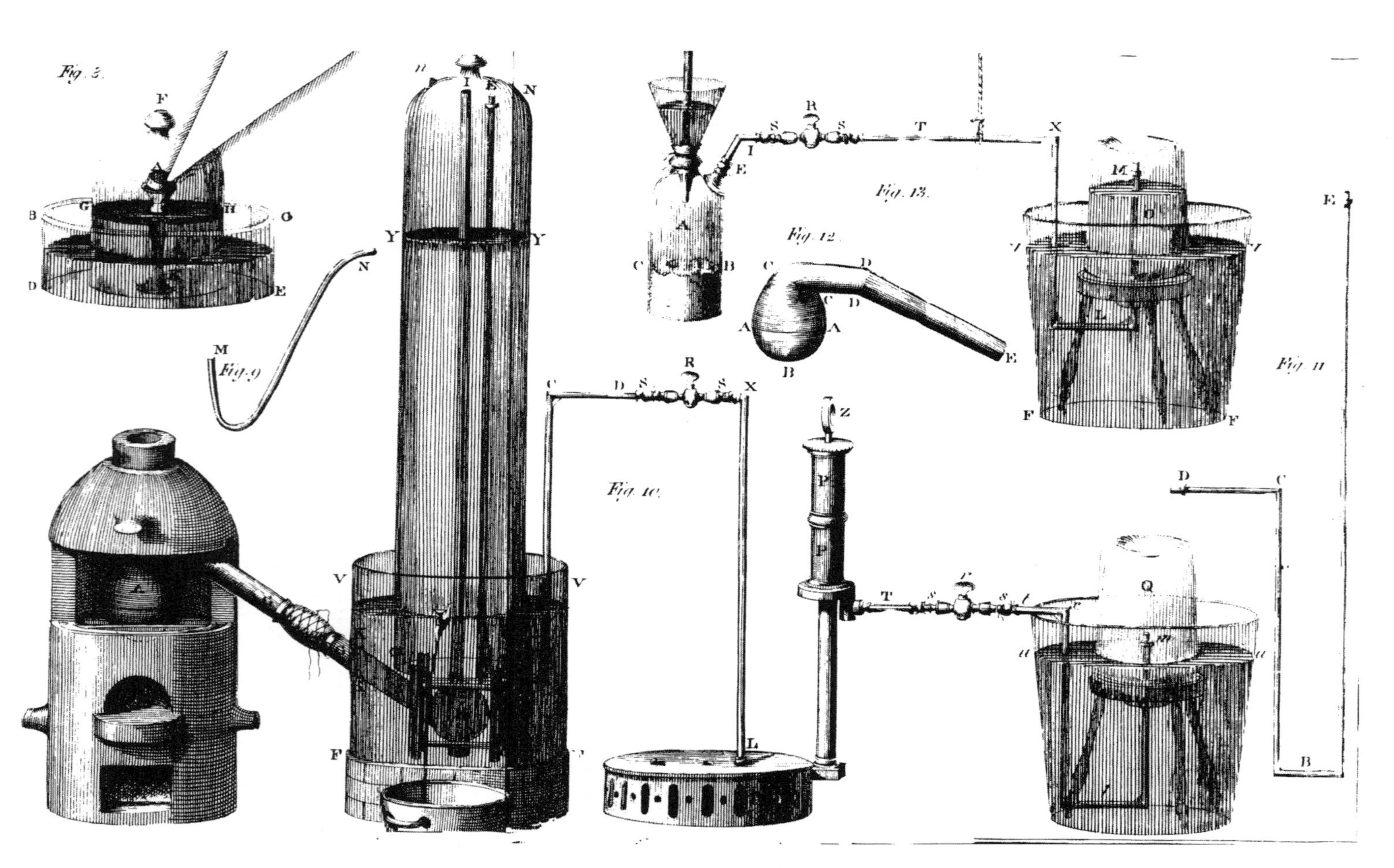

FIGURE 87 ■ Antoine Laurent Lavoisier's pneumatic experiments in *Essays Physical and Chemical*, London, 1774; the French edition was published in 1774; see text.

lead or tin under a powerful magnifying lens ("heating glass"). The inverted bell jar sits over a vessel filled with water. In the middle is a glass column with a cuplike indentation on the top. Some lead or tin is placed into a porcelain dish placed on top of the glass column. Siphon *MN* is placed under the bell jar and air withdrawn until the water level rises to the desired level. Heating of the metal should produce calx with the loss of some "aerial fluid" and a rise in the water column. Unfortunately, the heating glass was too powerful and molten metal evaporated and splashed onto the sides of the bell jar giving inconclusive results.

Figure 10 in Fig. 87 shows an apparatus for measurement of the gas (CO_2) released when minium (red-lead or litharge, Pb_3O_4) mixed with charcoal is heated in a furnace. Glass retorts were attacked by this chemical mixture, so Lavoisier fabricated an iron retort (*Fig. 12*). The tall inverted bell jar *nNoo* sits in a wooden or iron trough filled with water. A siphon inserted at *n* raises the water to *YY*. Alternatively, hand-pump *P* can connected using siphon *EBCD* (*Fig. 11*) and used to raise the column fairly high. The top of the water in jar *nNoo* is coated with a thin layer of oil. This is another way to collect a water-soluble gas such as CO_2 rather than by using mercury. To the right in *Fig. 10* we see an apparatus for transfer of the gas collected in jar *nNoo* to glass bottle *Q*. This important experiment demonstrated the release of an "aerial fluid" upon heating red-lead.

Figure 13 in Fig. 87 depicts an apparatus for generating CO_2 by adding dilute oil of vitriol onto powdered chalk. The water-soluble gas is collected over water having a layer of oil on top.

1. J.R. Partington, *A History of Chemistry*, MacMillan, London, 1962, Vol. 3, pp. 363–495.
2. A.J. Ihde, *The Development of Modern Chemistry*, Harper & Row, New York, 1964, pp. 57–88.
3. J.-P. Poirier, *Lavoisier—Chemist, Biologist, Economist*, R. Balinski, (translator), University of Pennsylvania Press, Philadelphia, 1996.
4. J.-P. Poirier, op. cit., pp. 381–382.

SIMPLIFYING THE CHEMICAL BABEL

Peter Bruegel The Elder depicted The Tower of Babel in 1563. This huge city reaching into the clouds was a human conceit and according to *Genesis* 11:9: "Therefore its name was called Babel, because there the Lord confused the language of all the earth." As the chemical edifice was erected through the eighteenth century a form of chemical babble arose in a confusing nomenclature. This was due in part to different degrees of purities of substances as well as uncontrolled neologisms. A look at the literature of the time shows that the term "mephitic air" (*mephitic* means "pestilential exhalation"), while most often used for carbon dioxide, was sometimes employed for the nitrogen that remained after "vital air" was totally consumed from common air. Eklund's useful work[1] is a helpful guide for understanding eighteenth-century nomenclature. In 1787, Lavoisier, de Morveau, Berthollet, Fourcroy, Hassenfratz, and Adet collaborated on the book *Méthode de Nomenclature Chimique* (Paris and London, 1788). Figures 88 and 89 are derived from this work but are actually taken from the second English edition (1793) of Lavoisier's *Traité*.

The work was of immense importance to the field, but let's note some interesting little flaws that prove that even Lavoisier was not infallible. First, he names "vital air" as oxygene, which means "acid maker." This was reasonable to Lavoisier since combustion of carbon, sulfur, and phosphorus in pure oxygen each produced acids. His oxygen theory of acids was well accepted. This included

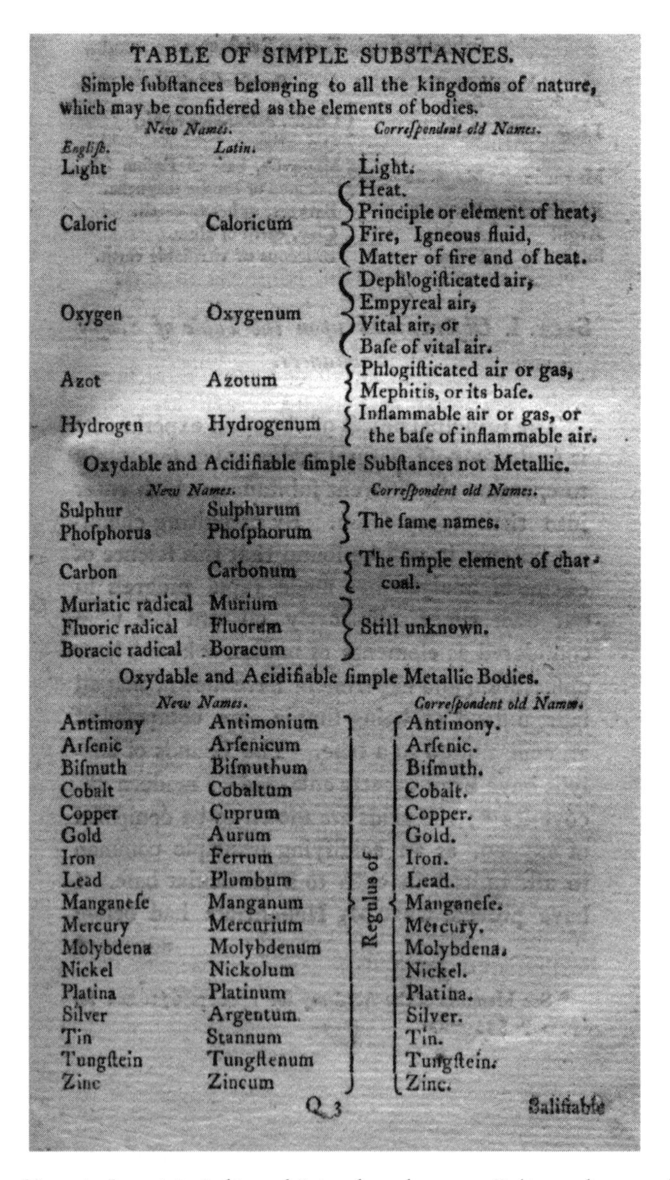

TABLE OF SIMPLE SUBSTANCES.

Simple ſubſtances belonging to all the kingdoms of nature, which may be conſidered as the elements of bodies.

New Names. English.	*Latin.*	*Correſpondent old Names.*
Light		Light.
Caloric	Caloricum	Heat. Principle or element of heat, Fire, Igneous fluid, Matter of fire and of heat.
Oxygen	Oxygenum	Dephlogiſticated air, Empyreal air, Vital air, or Baſe of vital air.
Azot	Azotum	Phlogiſticated air or gas, Mephitis, or its baſe.
Hydrogen	Hydrogenum	Inflammable air or gas, or the baſe of inflammable air.

Oxydable and Acidifiable ſimple Subſtances not Metallic.

New Names.		*Correſpondent old Names.*
Sulphur	Sulphurum	The ſame names.
Phoſphorus	Phoſphorum	The ſame names.
Carbon	Carbonum	The ſimple element of charcoal.
Muriatic radical	Murium	Still unknown.
Fluoric radical	Fluorum	Still unknown.
Boracic radical	Boracum	Still unknown.

Oxydable and Acidifiable ſimple Metallic Bodies.

New Names.			*Correſpondent old Names.*
Antimony	Antimonium		Antimony.
Arſenic	Arſenicum		Arſenic.
Biſmuth	Biſmuthum		Biſmuth.
Cobalt	Cobaltum		Cobalt.
Copper	Cuprum		Copper.
Gold	Aurum		Gold.
Iron	Ferrum		Iron.
Lead	Plumbum		Lead.
Mangaueſe	Manganum	Regulus of	Manganeſe.
Mercury	Mercurium		Mercury.
Molybdena	Molybdenum		Molybdena.
Nickel	Nickolum		Nickel.
Platina	Platinum		Platina.
Silver	Argentum		Silver.
Tin	Stannum		Tin.
Tungſtein	Tungſtenum		Tungſtein.
Zinc	Zincum		Zinc.

Q 3 Salifiable

FIGURE 88 ▪ Here is Lavoisier's list of "simple substances" (i.e., elements) in the first English edition *Elements of Chemistry* (London, 1790) of his monumental *Traité Élémentaire de Chimie* (Paris, 1789). Note caloric as an element. Count Rumford would disprove caloric about 10 years later and also marry Lavoisier's widow Marie-Anne Paulze Lavoisier.

	Names of the fimple fub-ftances.	Firft degree of oxygenation.	
		New Names.	Old Names.
Combinations of oxygen with fimple non-metallic fubftances.	Caloric -	Oxygen gas - -	Vital or dephlogifticated air - - -
	Hydrogen	Water *.	
	Azot -	Nitrous oxyd, or bafe of nitrous gas - -	Nitrous gas or air -
	Carbon -	Oxyd of carbon, or carbonic oxyd - -	Unknown - - -
	Sulphur	Oxyd of fulphur - -	Soft fulphur - -
	Phofphorus	Oxyd of phofphorus -	Refiduum from the combuftion of phofphorus
	Muriatic radical	Muriatic oxyd - -	Unknown - - -
	Fluoric radical	Fluoric oxyd - -	Unknown - - -
	Boracic radical	Boracic oxyd - -	Unknown - - -
Combinations of oxygen with the fimple metallic fubftances †.	Antimony -	Grey oxyd of antimony	Grey calx of antimony
	Silver - -	Oxyd of filver -	Calx of filver - -
	Arfenic -	Grey oxyd of arfenic -	Grey calx of arfenic -
	Bifmuth -	Grey oxyd of bifmuth	Grey calx of bifmuth -
	Cobalt -	Grey oxyd of cobalt -	Grey calx of cobalt -
	Copper -	Brown oxyd of copper	Brown calx of copper -
	Tin - -	Grey oxyd of tin -	Grey calx of tin - -
	Iron - -	Black oxyd of iron -	Martial ethiops - -
	Manganefe	Black oxyd of manganefe	Black calx of manganefe
	Mercury -	Black oxyd of mercury	Ethiops mineral‡ - -
	Molybdena	Oxyd of molybdena -	Calx of molybdena -
	Nickel -	Oxyd of nickel - -	Calx of nickel - -
	Gold - -	Yellow oxyd of gold -	Yellow calx of gold -
	Platina	Yellow oxyd of platina	Yellow calx of platina -
	Lead - -	Grey oxyd of lead -	Grey calx of lead -
	Tungftein	Oxyd of tungftein -	Calx of tungftein -
	Zinc - -	Grey oxyd of zinc -	Grey calx of zinc - -

FIGURE 89 ■ This table is also from the first English edition of Lavoisier's 1789 *Traité*. Note that oxygen gas is said to be a combination of oxygen and caloric. When a substance burns or calcines it combines with oxygen and releases caloric as heat. This has a bit of the flavor of phlogiston in it. Dephlogisticated air (oxygen) was to absorb phlogiston from burning or calcining substances.

the belief that hydrochloric acid (HCl) contained oxygen because its precursor chlorine (isolated by Scheele in 1774) must have contained oxygen. This was disproven by Humphrey Davy some 20 years after the *Nomenclature* was published.

A second problem was Lavoisier's postulation of the element "caloric"—a kind of imponderable heat fluid. In certain ways, caloric was a substitute for the phlogiston Lavoisier demolished. According to this view, gaseous oxygen contains caloric (which helps to keep it in a rarified state). When a substance burns or forms a metallic oxide, it combines with ("fixes") oxygen (thus, increasing its weight) and, in the process, frees the caloric as heat. In Fig. 88, we see caloric listed as an element ("simple substance"). In Fig. 89, we see that just as hydrogen combines with oxygen to produce water, so does caloric combine with oxygen to produce oxygen gas. Ironically (or perhaps not), it is Madame Lavoisier's second husband, Count Rumford, who eventually disproves the existence of caloric.

1. J. Eklund, *The Incompleat Chymist*, Smithsonian Institution Press, Washington, D.C., 1975.

HYDROGEN + OXYGEN → WATER. WATER → OXYGEN + HYDROGEN

Lavoisier did not invent the law of conservation of matter. It was a firm assumption in the minds of contemporary and earlier scientists. However, his careful trapping of gases in preweighed liquids and his requirement that all matter in a chemical reaction must be accounted for brought chemistry to a new level as a science—some even called it physics. If the mass at the start of a reaction and at the end could not be matched then there was not much point in analyzing the chemistry. It was as if the *Ferme Générale* were conducting an audit.

The clarity and authority of the *Traité Élémentaire de Chimie* (Paris, 1789) spelled the end for the phlogiston theory. Irish chemist Richard Kirwan[1] (1733–1812) published a book, *An Essay on Phlogiston and the Constitution of Acids* (London, 1787), that made the case forcefully for phlogiston—the English view. Madame Lavoisier translated the book into French (Paris, 1788) and it became the focus for the anti-phlogistic arguments of the French school. A second edition of Kirwan's book, including the appended 1788 essays of the French chemists, was published in London in 1789. However, by 1792 Kirwan had accepted the anti-phlogistic theory and wrote to Berthollet: "At last I lay down my arms and abandon Phlogiston."[2]

To celebrate the victorious *Traité*, Madame Lavoisier, dressed as a priestess, ceremoniously burned Stahl's works (an *auto-da-fe* of Phlogiston).[3,4] She had earlier asked one of the members of the Arsenal Laboratory, Jean Henri Hassenfratz, for suggestions for celebration of this success. In a letter dated February 20, 1788, he suggested three possibilities: a portrait of the Lavoisiers, a play involving the combat between phlogiston and oxygen, and a totally allegorical presentation about the chemical revolution.[4] The portrait was painted by the artist Jacques Louis David (who was also Mme. Lavoisier's art instructor). Hassenfratz suggested two possibilities for the play. One involved a grand battle. Oxygen's troops included carbonates, phosphates, sulfates, etc., against the allies of Phlogiston, *acidum pingue* and *acide igne*. The other was a confrontation between handsome Oxygen, with his brother-in-arms Hydrogen at his side, and the deformed Phlogiston already missing an arm. At Phlogiston's side is *acidum pingue*, already dead, and *acid igne*, ashen, defeated and dying of fear. Oxygen is poised to lop off Phlogiston's remaining arm. A play was apparently performed and reported to Crell's journal *Chemische Annalen* by a Dr. von E**. Phlogiston was placed on trial, weakly defended by Stahl, and then burned at the stake.[4] If you think about this, you will realize that if combustion releases phlogiston, then combustion of phlogiston leaves nothing.

Figure 90 is from Lavoisier's *Traité*. In *Fig. 1* we see a very sophisticated distillation apparatus designed to trap and weigh everything generated. The sample is heated in retort A, volatile and semivolatile liquids are collected in the preweighed globe C; the first preweighed three-necked bottle after C contains water and the remaining three bottles contain potash (KOH) solutions (all preweighed) to trap the acidic gases. The remaining water-insoluble, nonacidic gases (e.g., oxygen) are delivered to a bottle in a pneumatic trough or similar collection device. The tall tubes luted into the center opening of each bottle have small openings—they reach to the bottom of the liquids and will only leak if there is a pressure buildup. If a vacuum is created, the mass of air introduced is negligible compared to the mass of the glassware and their contents. Lavoisier notes that if the masses in all vessels, including residue in the retort, do not total to that of the starting material, the experiment must be redone. Figure 91 (*Fig. 1*) shows an apparatus for separation of the gaseous components arising from fermentation or putrefaction. The matrass A is connected by brass tubing and valves to glass bulb B. If frothing in A exceeds the capacity of the matrass, excess froth is collected in B and drained periodically into bottle C. Water vapor is removed in glass tube h, which contains a drying agent such as calcium chloride. Carbon dioxide from fermentation is collected in potash solutions in bottles D and E. Putrefaction sometimes produces hydrogen, collected in bell jar F in pneumatic trough GHIK.

Figure 2 in Figure 90 depicts the famous apparatus for heating metallic mercury in the presence of air. Lavoisier heated 4 ounces of mercury in the retort A.[5] After twelve days he stopped the heating and weighed the red calx (HgO) that had formed on the surface of the mercury. Its mass was 45 grains. The air volume had decreased from 50 cubic inches to 42 cubic inches (about 16%). The air remaining was "mephitic" (actually nitrogen). When the *mercurius calcinatus per se* was transferred to a small retort and heated it produced 8 to 9 cubic inches of "highly respirable air" and 41.5 grains of mercury. This was the gas that Scheele called "fire air" or "empyreal air," Priestley termed "dephlogisticated air," and Lavoisier later called "vital air" and eventually oxygen. When this oxygen was added to the "mephitic" air, normal air was reformed. The interesting apparatus in *Fig. 10* is a customized matrass [see Fig. 26(a); you may ignore the ostrich]. Its bulb has been heated in flame and flattened. The flat bottom contains mercury, which can be heated on a sand bath. The small opening at the top permits slow circulation of atmospheric air but minimizes loss of mercury vapor. Over several months, good yields of red HgO are obtained. The retort-and-bladder apparatus (*Fig. 12*) is similarly used to heat mercury in the presence of a half-bladderful of oxygen—only small amounts of red calx were formed.

In *Fig. 3* we see a small apparatus for igniting iron in a porcelain dish in a bell jar filled with oxygen over mercury. Lavoisier siphons out some air in order to raise the mercury level. He uses a red-hot iron wire (*Fig. 16*) to touch off a piece of phosphorus attached to tinder attached to the iron wire sample. *Figure 17* (upper right) depicts a fine iron wire attached to a stopper and twisted into a spiral with a small piece of tinder at point C. With the stopper and wire out, the tinder is lit and the wire lowered into the oxygen-containing bottle. As it burns, iron forms a calx that falls to the bottom, is collected, powderized, and weighed.

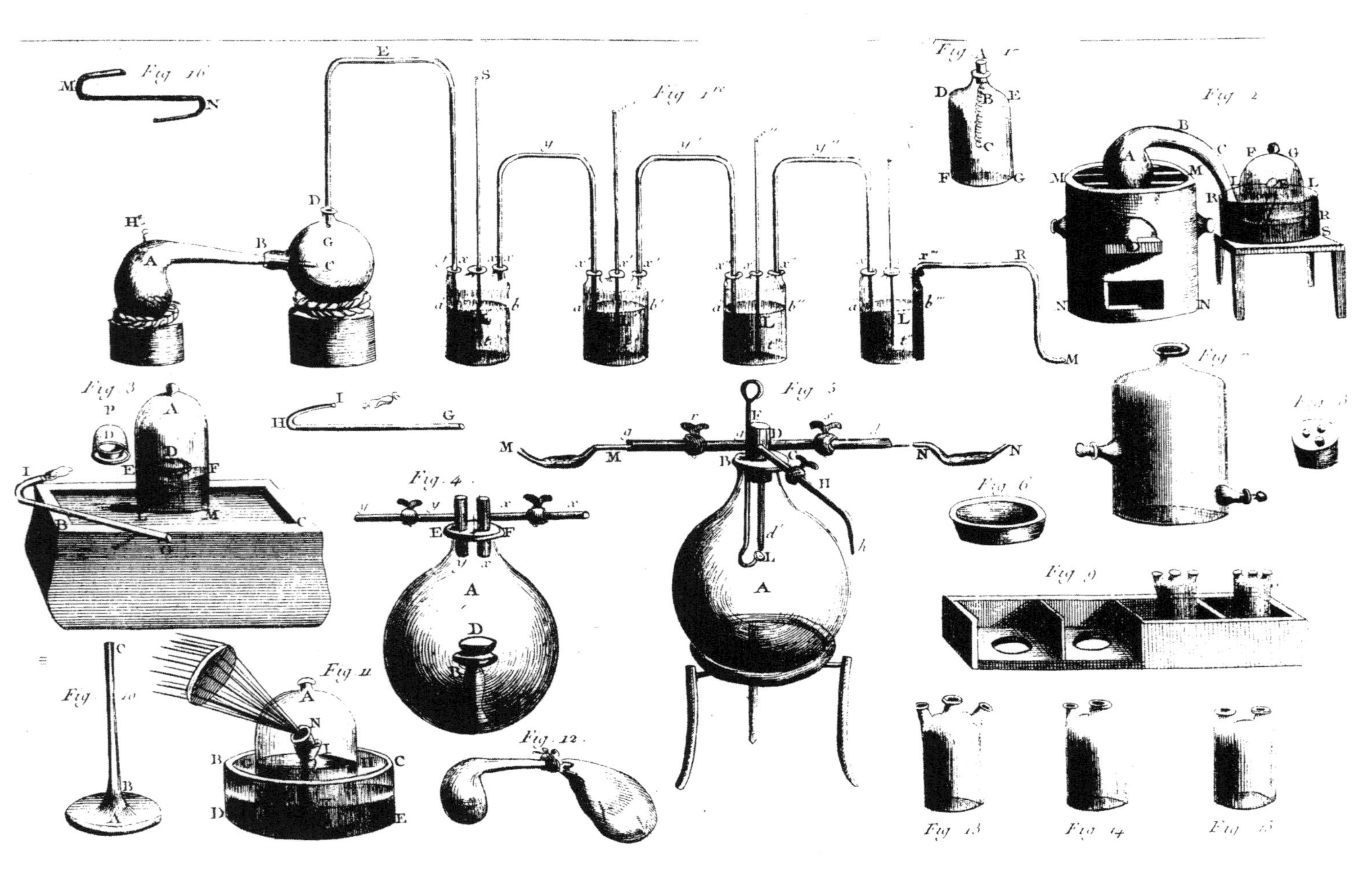

FIGURE 90 ▪ This figure is from Lavoisier's *Traité*. Madame Lavoisier, a talented artist, engraved the plates for the French editions (see text for discussion of apparatus).

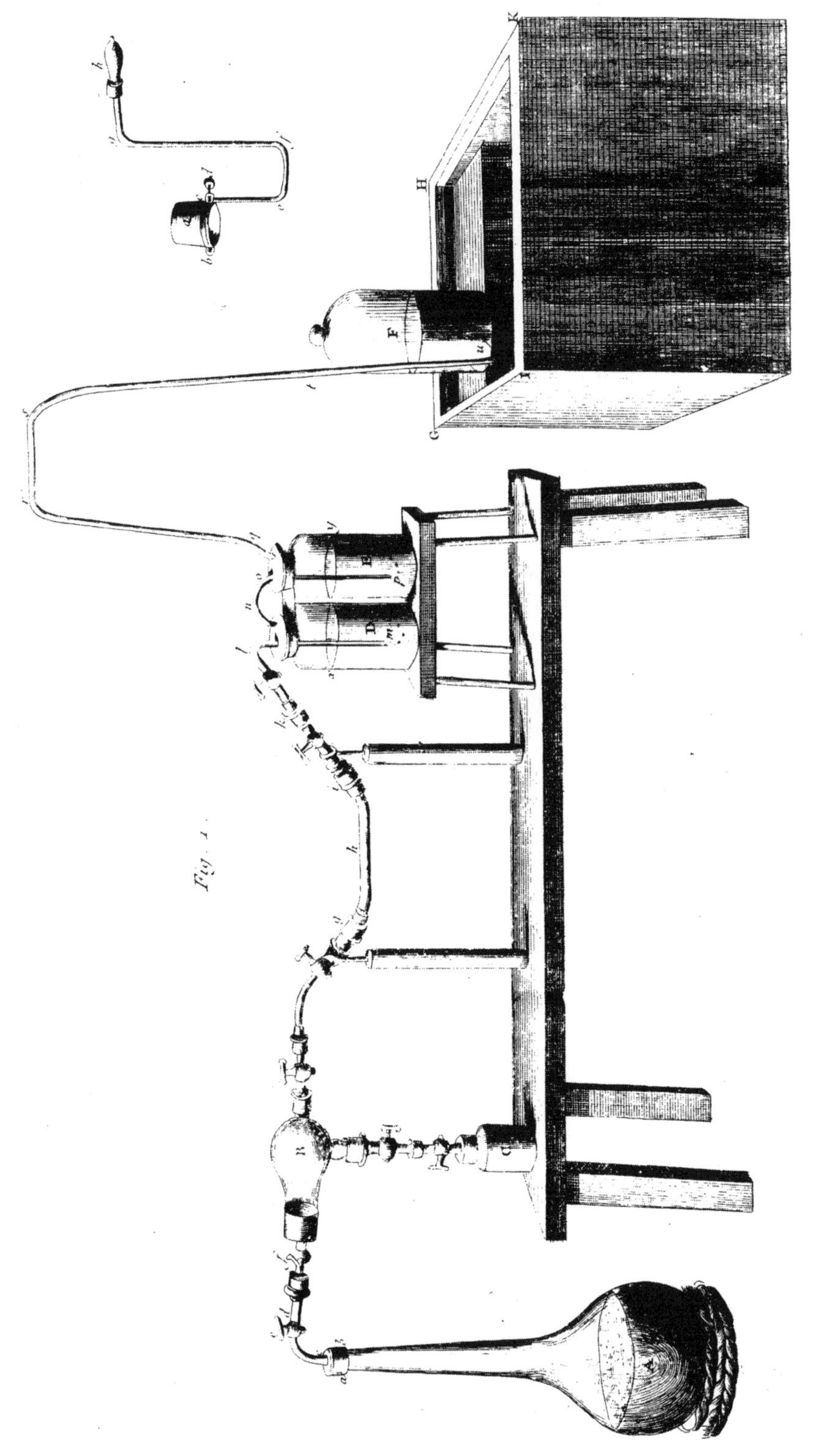

FIGURE 91 ▪ From Lavoisier's *Elements*, an apparatus for collection gases from fermentations.

Figure. 4 depicts a large vessel for combustion of phosphorus in oxygen (the opening at the top has a diameter of three inches). Phosphorus is placed in the porcelain dish *D*. Air is evacuated through one stopcock and oxygen added through the other. Combustion is started with a burning glass. In phosphorus combustion, white flakes of phosphorus pentoxide[6] (actually P_4O_{10}, which sublimes at 360°C) coat the vessel wall and interfere somewhat with the efficacy of the burning glass. This solid is extremely hydroscopic ($P_4O_{10} + 6H_2O \rightarrow 4H_3PO_4$, phosphoric acid).

Joseph Priestley may have been the first to make water by burning hydrogen in air, but he did not notice it. It was apparently Pierre Macquer who, in 1776, discovered droplets of water on a porcelain saucer and realized water was the product.[7,8] In spring of 1783, Cavendish ignited inflammable air and dephlogisticated air and weighed the water produced. Thus, he is considered to be the first person to truly synthesize water from its elements. Unfortunately, his interpretation was based upon phlogiston theory.[7,8] *Figure 5* in Figure 90 depicts the apparatus built by Lavoisier and Pierre Simon Laplace[8] (1749–1827), the famous mathematician, for making water quantitatively from oxygen introduced through tube *NN* and hydrogen introduced through tube *MM*. On June 24, 1783 the gases were added little by little and sparked with a wire ending at *L*, close to the source of hydrogen at *d'*.[8] Lavoisier and Laplace demonstrated that 85 parts of oxygen react with 15 parts of hydrogen to yield 100 parts of water.[7] In another apparatus (not shown here), Lavoisier distilled water through a glass tube containing charcoal and running through a furnace. The glass tube in the furnace was coated with clay, and an iron bar was also employed to keep it from bending. The steam exposed to charcoal formed carbon dioxide and hydrogen gas, which were duly collected and weighed. Thus, Lavoisier made water quantitatively from pure oxygen and hydrogen and also quantitatively broke down into its elements. It was, however, James Watt, who first recognized that water is a compound and not an element.

1. J.R. Partington, *A History of Chemistry*, MacMillan, London, 1962, Vol. 3, pp. 660–671.
2. A.J. Ihde, *The Development of Modern Chemistry*, Harper and Row, New York, 1964, p. 81.
3. J.R. Partington, op. cit., p. 491.
4. The author is grateful to Dr. Jean-Pierre Poirier for his correspondence, including a transcript of Hassenfratz's letter and to Professor Roald Hoffmann for making me aware of Poirier's findings concerning the Lavoisiers' play. Some discussion of the play is to be found in Poirier's book (see Ref. 8).
5. J.R. Partington, op. cit., p. 417.
6. F.A. Cotton and G. Wilkinson, *Advanced Inorganic Chemistry*, 5th ed., Wiley, New York, 1988, pp. 399–401.
7. J.R. Partington, op. cit., pp. 436–453.
8. J.-P. Poirier, *Lavoisier—Chemist, Biologist, Economist*, R. Balinski (translator), University of Pennsylvania Press, Philadelphia, 1996, pp. 140–144.

THE GUINEA PIG AS INTERNAL COMBUSTION ENGINE

Since *calorique* was a simple substance (an element), albeit imponderable, naturally Lavoisier wanted to measure it. Figure 92 shows the ice calorimeter designed by Lavoisier and Laplace. The fully assembled calorimeter is shown in

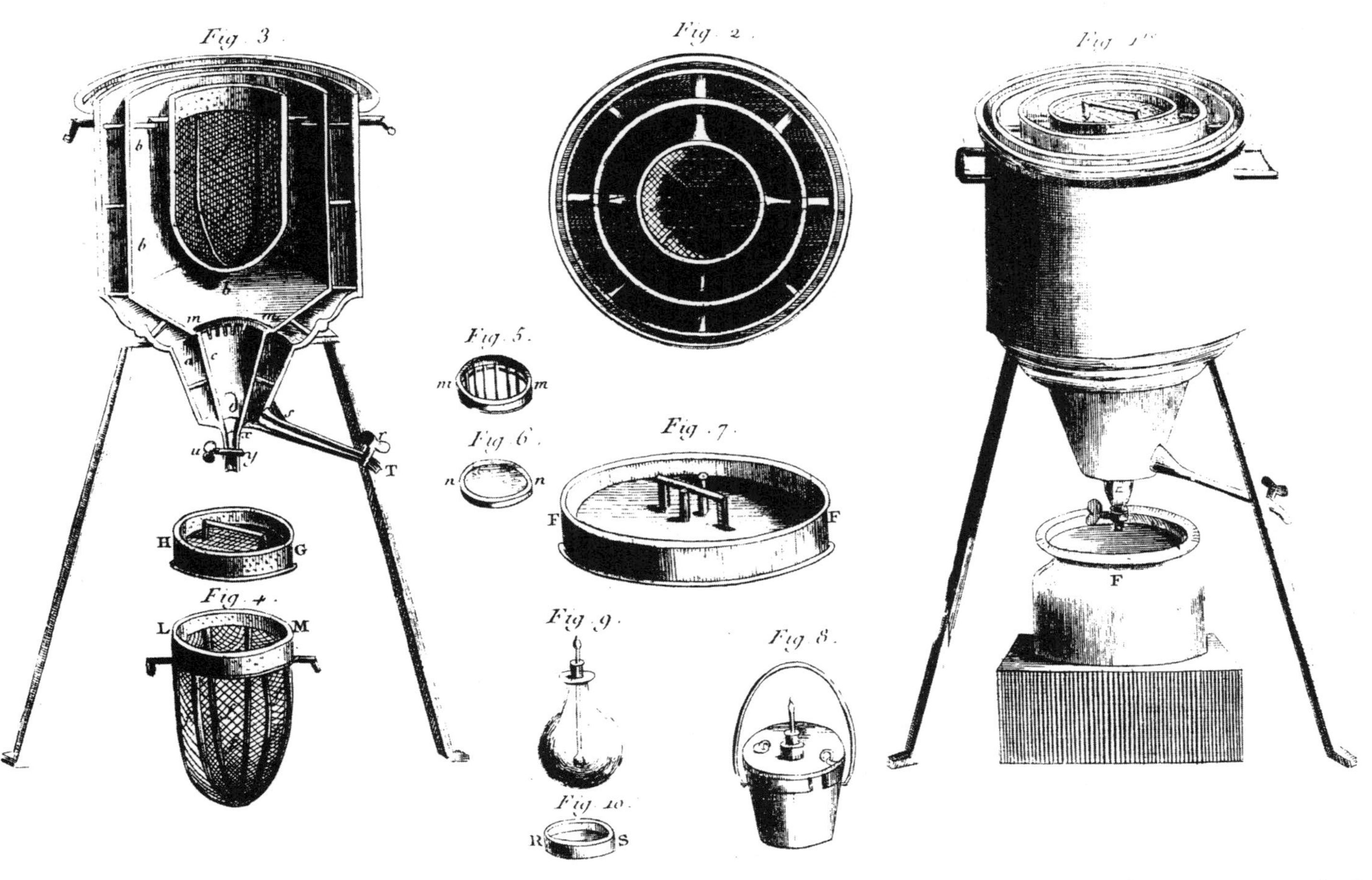

FIGURE 92 ▪ Ice calorimeter, designed by Lavoisier and the famous mathematician Laplace. Heat was defined in units of ice melted. The idea that metabolism was similar to combustion derived from the knowledge that oxygen was required, carbon dioxide and water produced and heat generated by animals. Thus, Lavoisier realized that combustion, calcination and metabolism were all related in the sense that each involved combination with oxygen.

Fig. 1 and the cutout view in *Fig. 3*. The basket *ffff*, with opening *LM*, is made of iron wire mesh and can be covered with lid *GH*. This basket holds the caloric-generating sample: hot metal, hot liquid, or chemical reaction via mixing (inside a suitable container), combustion sample, or live guinea pig. Crushed ice is placed in chamber *bbbb* as well as in jacket *aaaa*. Chamber *aaaa* insulates the apparatus—water may be tapped conveniently through *sT*. The ice in chamber *bbbb*, supported by screen *mm* and sieve *nn*, absorbs the heat from basket *ffff*. The resulting water is tapped through *xy* and weighed. Prior to the experiment, crushed ice is tightly packed into chambers *aaaa* and *bbbb*, into lid *GH*, and the apparatus cover (*Fig. 7*) and allowed to attain equilibrium. These experiments are best done in rooms not much warmer than 50°F and definitely not colder than 32°F (since ice must be at this temperature and not colder). A large sample is placed in a metallic bucket equipped with a thermometer (*Fig.* 8; a corrosive liquid would be placed in a glass vessel equipped with a thermometer, *Fig.* 9). The bucket or glass vessel is placed in a bath of boiling water. Just prior to transfer, the last drops of water are tapped through *xy* and discarded. Quick transfer of the hot sample is performed. It takes typically 10 to 12 hours for the entire internal calorimeter to return to 32°F. The water from chamber *bbbb* is then tapped and carefully weighed. Lavoisier and Laplace realized that there had to be heat losses that limited the accuracy of their determinations.

Lavoisier and Laplace defined their heat unit as the quantity required to melt one pound of ice (at 32°F). They demonstrated that it requires one pound of water starting at 167°F and cooling by 135°F (to 32°F) to melt this ice. Thus, they took 7.707 pounds of iron strips heated in a boiling water bath to 207.5°F and added the metal quickly into basket *ffff* and closed the calorimeter. After eleven hours, 1.109795 pounds of ice had melted. The iron had thus cooled by 175.5°F. Using the ratio $175.5/1.109975 = 135/x$, they found that $x = 0.85384$. Dividing this by 7.707, the quotient 0.1109 is the quantity of ice melted by one pound of iron cooling through 135°F. Other caloric-generating processes could be placed on this arbitrary scale.

Guinea pigs "thrive" in ice calorimeters better than mice. The air entering and leaving basket *ffff* had to, of course, pass through tubing immersed in the crushed ice. The realization gained over the previous decade that both respiration and combustion required oxygen "jelled" with the earlier observations that both processes produced carbon dioxide. It was thus a relatively small creative leap to equate the two and try to measure the slow internal combustion recognized as animal heat.

Imagining a mouse shivering inside the ice calorimeter (Figure 92), enduring "mephitic air" in the apparatus of Priestley (Figure 86), Scheele, Lavoisier or Mayow (Figure 67) inspires respect for the role this hardy and courageous mammal has played and continues to play in science. And while we previously noted that Priestley was "nice to his mice" (p. 137), Franklin wrote to him suggesting, in effect, that he ". . . repent of having murdered in mephitic air so many honest, harmless mice . . .".[1] Perhaps a statue should be erected honoring the mouse at the Royal Institute in Stockholm.

1. W. C. Bruce, *Benjamin Franklin Self-Revealed*, Second Revised Edition, Vol. I, Putnam, New York, 1923, pp 106–107. I thank Professor Roald Hoffmann for bringing this material to my attention and Professor Susan Gardner for suggesting homage to mice.

A SINGLE ELECTIVE ATTRACTION (SINGLE DISPLACEMENT)

The Swedish chemist Torbern Bergman (1735–1784)[1] systemized chemical affinities and displacements (single or double)[2] in the "wet way" or "dry way" in his book *A Dissertation on Elective Attractions*.[2] See Figure 93 and the enlargement of item 20 in Figure 94(a). Calcium sulfide [CaS or (1)] will be decomposed by sulfuric acid [H_2SO_4 or (2)] in water (3) to produce elemental sulfur (4), which precipitates (downward half-bracket) and calcium sulfate [gypsum, $CaSO_4$, or (5)], which also precipitates (downward bracket). Thus, sulfuric acid (2) has a higher affinity for pure calcareous lime (6)—really the source of calcium in (1)—than does sulfur (4).

1. J.R. Partington, *A History of Chemistry*, MacMillan, London, 1962, Vol. 3, pp. 179–199.
2. T. Bergman, *A Dissertation on Elective Attractions*, Edinburgh, 1785.

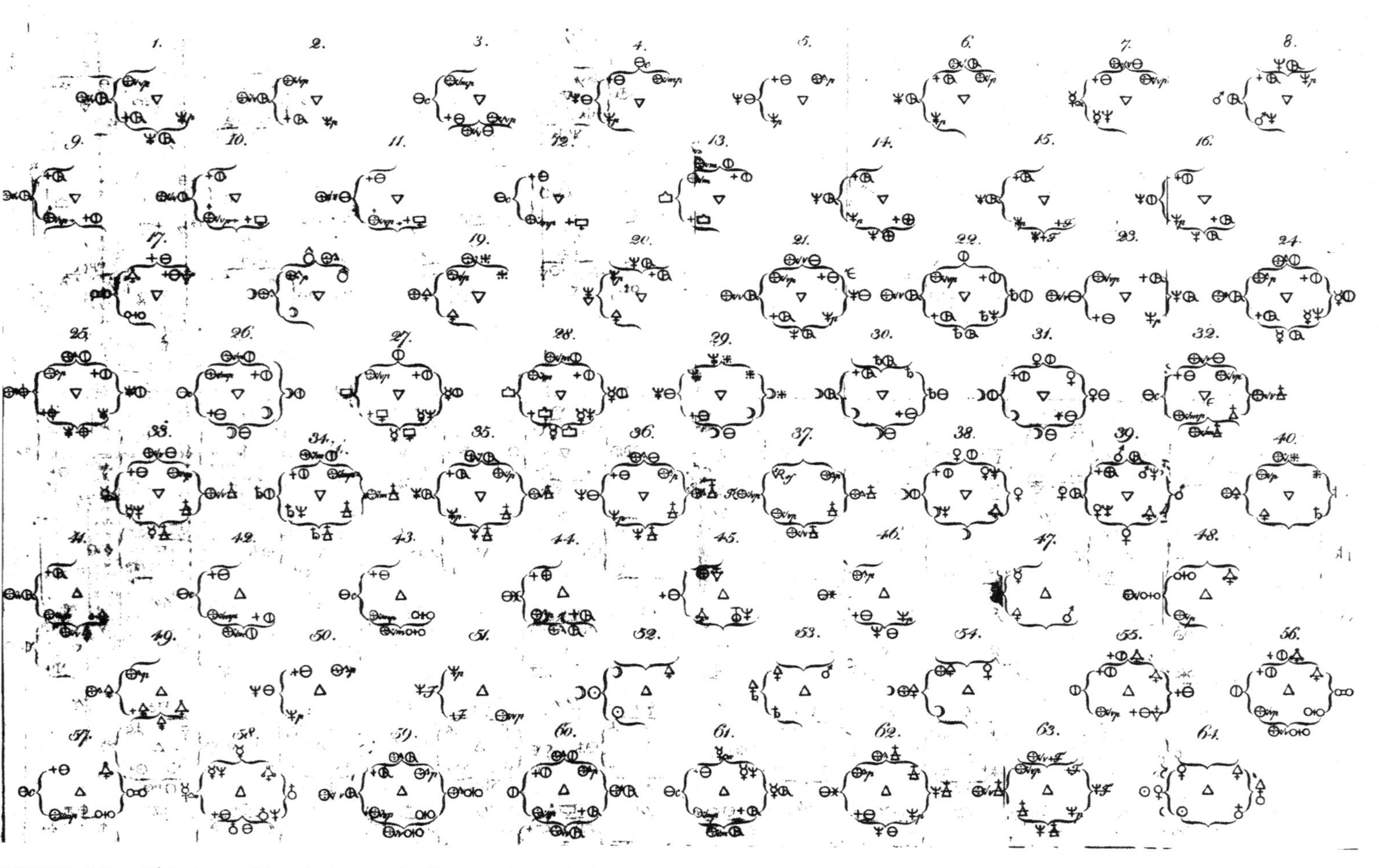

FIGURE 93 ■ This is a table of chemical affinities from Torbern Bergman's *A Dissertation on Elective Attractions* (London, 1785) [see Figs. 94(a) and 94(b) and text for discussion].

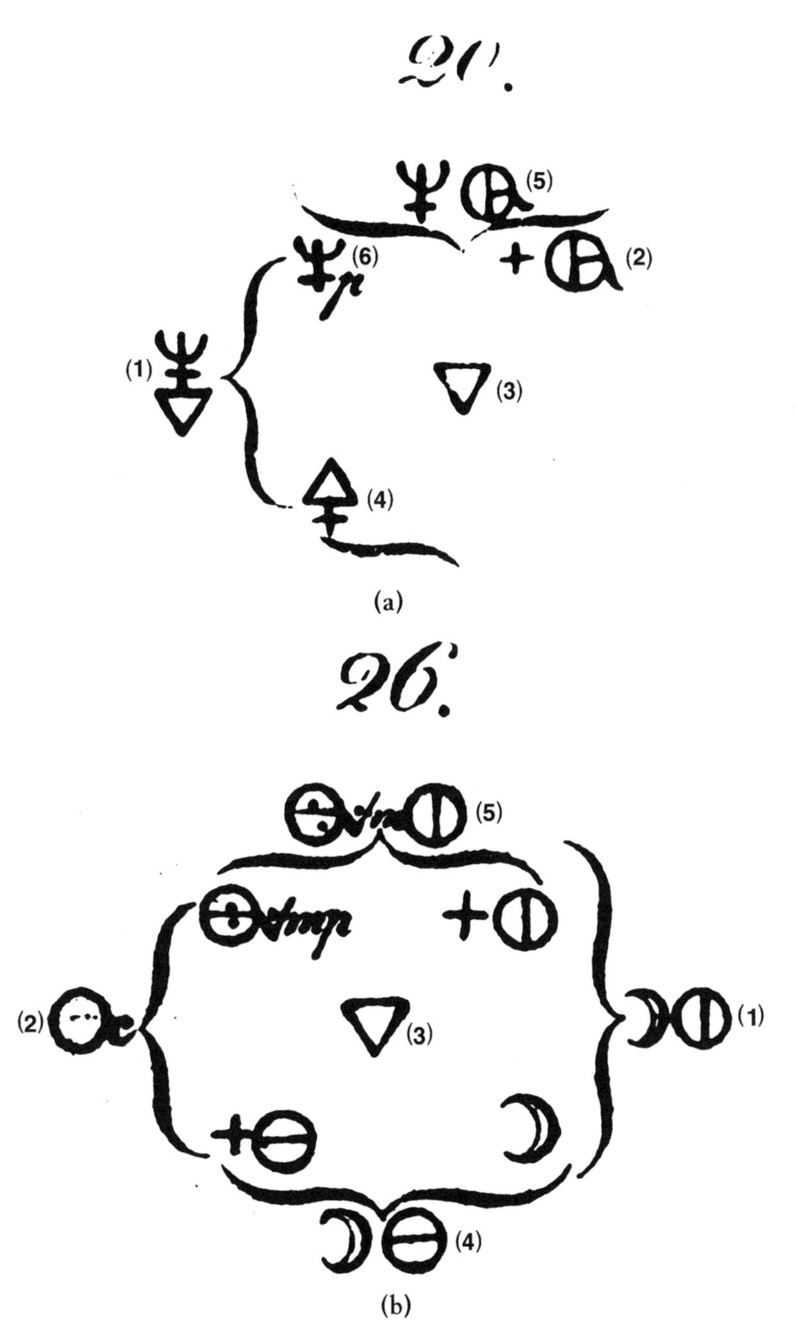

FIGURE 94 ▪ (a) Single-elective attraction (decomposition) of calcium sulfide from Bergman's tables (see Fig. 93). (b) Double elective attraction between silver nitrate and sodium chloride (see text).

A DOUBLE ELECTIVE ATTRACTION (DOUBLE DISPLACEMENT)

See Figure 93 and the enlargement of item 26 in Figure 94(b). Silver nitrate (1) and sodium chloride [(2), table salt] decompose each other in water (3) to produce silver chloride (4), which precipitates (downward bracket), and sodium nitrate (5), which remains in solution (upward bracket).

THE PHOENIX IS A "HER"?

Mrs. Elizabeth Fulhame, whom Laidler calls a "forgotten genius,"[1,2] authored a remarkable book (Fig. 95) published in 1794 (German translation, 1798; American edition, 1810; see Fig. 96). Women were not only not encouraged, they were actively discouraged from pursuing scientific interests. The 1794 edition was published privately for the author, presumably with the support of her husband Dr. Thomas Fulhame. From her preface to this book:[3]

> It may appear presuming to *some*, that I should engage in pursuits of this nature; but averse from indolence, and having much leisure, my mind led me to this mode of amusement, which I found entertaining and will I hope be thought inoffensive by the liberal and the learned. But censure is perhaps inevitable; for some are so ignorant, that they grow sullen and silent, and are chilled with horror at the sight of anything that bears the semblance of learning, in whatever shape it may appear; and should the *spectre* appear in the shape of a *woman*, the pangs which they suffer are truly dismal.

Mrs. Fulhame made two, probably three, great discoveries. She was the first to demonstrate photoimaging and used salts of gold and other metals. The famous Count Rumford (Benjamin Thompson—see pp 161–163) differed with her chemical interpretation as opposed to a purely physical one.[4] He was wrong—the photochemical reduction of gold or silver ions to the respective metals is considered to be the first demonstration that ambient aqueous chemistry can accomplish the work of high-temperature smelting.[3]

Her work on the participation of water as a catalyst in the oxidation of charcoal to carbon dioxide, later proven,[3] was of great importance and anticipated the concept of *catalysis* (term introduced by Berzelius in 1836—"wholly loosening" from the Greek[2]). Implicit in this is also the modern concept of the chemical mechanism: a stepwise, "blow-by-blow" account of a chemical reaction. We will illustrate this briefly with the rusting of iron—it was the Irish chemist William Higgins who first discovered the role of water in this process and he accused Mrs. Fulhame of plagiarism (but he also accused John Dalton of plagiarizing the Atomic Theory from him).[4] Mrs. Fulhame's concept was more general; she clearly overextended it.[2,4]

Although the rusting of iron involves reaction of the metal with oxygen to form red-brown iron(III) oxide (Fe_2O_3), we know iron doesn't just rust in the

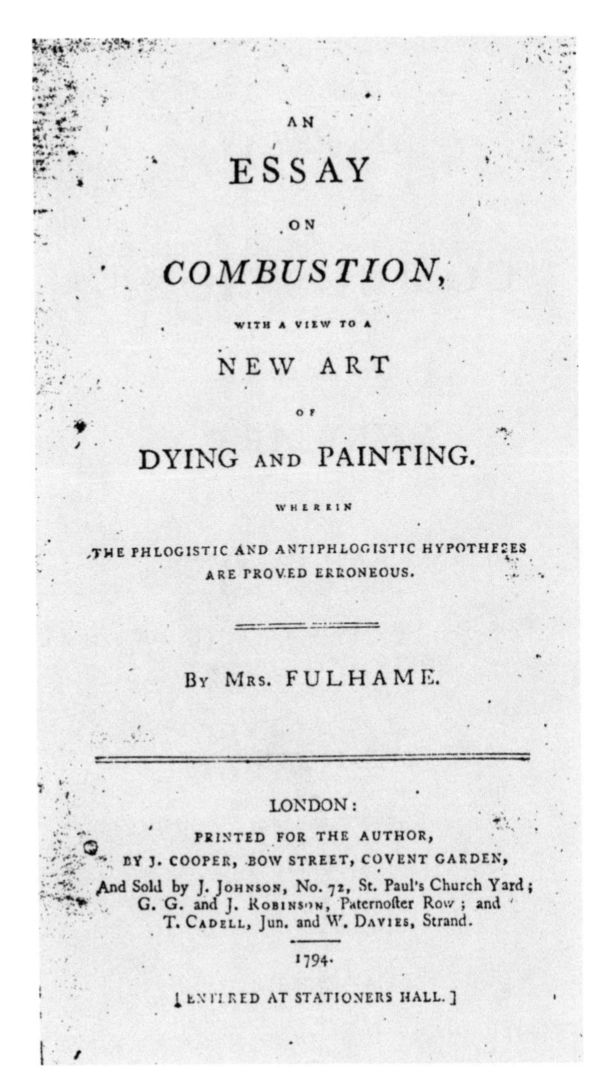
AN

ESSAY

ON

COMBUSTION,

WITH A VIEW TO A

NEW ART

OF

DYING AND PAINTING.

WHEREIN

THE PHLOGISTIC AND ANTIPHLOGISTIC HYPOTHESES
ARE PROVED ERRONEOUS.

BY MRS. FULHAME.

LONDON:
PRINTED FOR THE AUTHOR,
BY J. COOPER, BOW STREET, COVENT GARDEN,
And Sold by J. JOHNSON, No. 72, St. Paul's Church Yard;
G. G. and J. ROBINSON, Paternofter Row; and
T. CADELL, Jun. and W. DAVIES, Strand.

1794.

[ENTERED AT STATIONERS HALL.]

FIGURE 95 ■ Title page of Mrs. Elizabeth Fulhame's book on the theory of combustion. Laidler calls her a "forgotten genius" who first demonstrated photoimaging and may rightly be called the "mother of mechanistic chemistry (courtesy of distinguished chemist/book collector Dr. Roy G. Neville).

open air if it is kept dry. Water plays the roles of electrolytic solvent and catalyst. If iron is wet and exposed to an ample supply of oxygen, the following reactions occur.[5] Reaction 1:

$$4Fe(s) + 4H_2O(l) + 2O_2(g) \rightarrow 4Fe(OH)_2(s)$$

Reaction 2:

$$4Fe(OH)_2(s) + O_2(g) \rightarrow 2Fe_2O_3 \cdot H_2O(s) + 2H_2O(l)$$

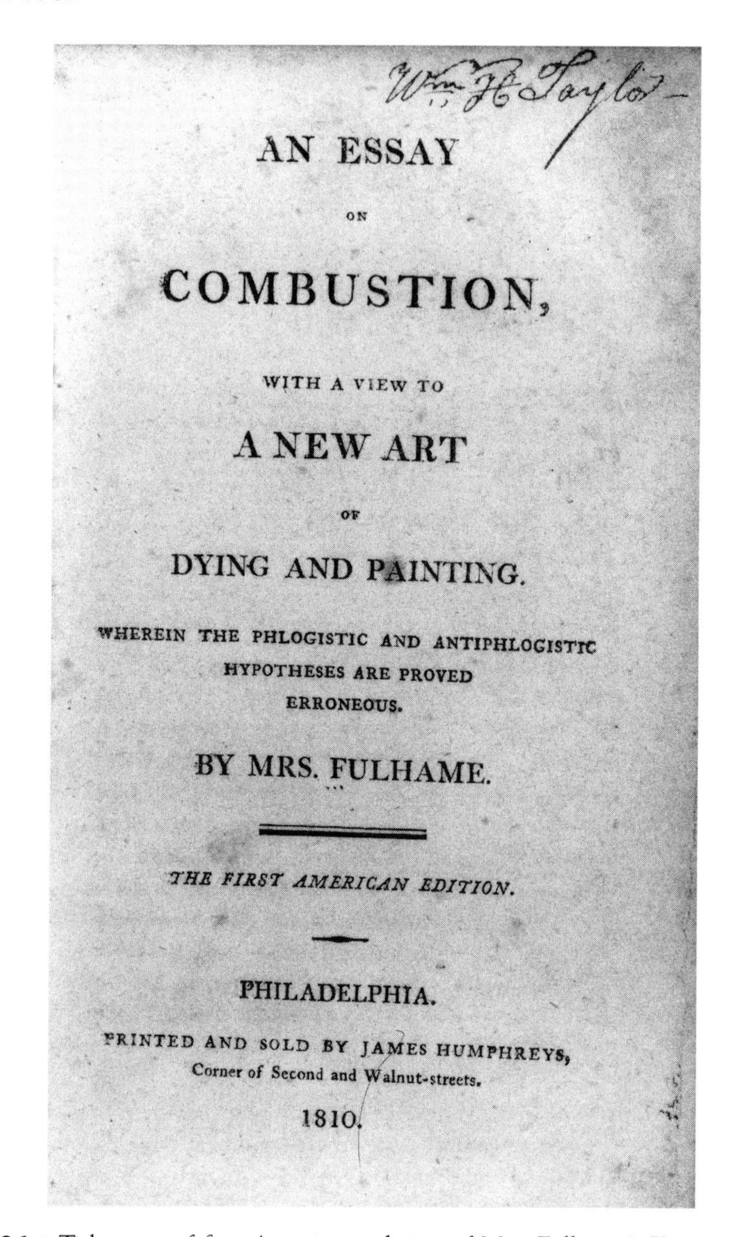
AN ESSAY

ON

COMBUSTION,

WITH A VIEW TO

A NEW ART

OF

DYING AND PAINTING.

WHEREIN THE PHLOGISTIC AND ANTIPHLOGISTIC
HYPOTHESES ARE PROVED
ERRONEOUS.

BY MRS. FULHAME.

THE FIRST AMERICAN EDITION.

PHILADELPHIA.

PRINTED AND SOLD BY JAMES HUMPHREYS,
Corner of Second and Walnut-streets.

1810.

FIGURE 96 ▪ Title page of first American edition of Mrs. Fulhame's Treatise. She was an early member of the Philadelphia Chemical Society, possibly nominated by Joseph Priestley with whom she differed about phlogiston (courtesy of Edgar Fahs Smith Collection, Rare Book & Manuscript Library, University of Pennsylvania).

Net Reaction:

$$4Fe(s) + 3O_2(g) + 2H_2O(l) \rightarrow 2Fe_2O_3 \cdot H_2O(s)$$

It is clear that two of the water molecules in Reaction 1 were regenerated in Reaction 2 and therefore do not appear in the net reaction. They were "temporarily tied-up" in the $Fe(OH)_2$ *intermediate* but then regenerated when the intermediate reacted.

Although Mrs. Fulhame was much more anti-phlogistonist than phlogistonist and thus closer to Lavoisier, Laidler speculates that it may have been Priestley who nominated her for the Philadelphia Chemical Society.[2] Mrs. Fulhame ends her book with an ebullient reference to the Phoenix,[4] a majestic bird symbolizing renewal or rebirth from the ashes—it adorns the symbol for the American Chemical Society:

> This view of combustion may serve to show how nature is always the same, and maintains her equilibrium by preserving the same quantities of air and water on the surface of our Globe: for as fast as these are consumed in the various Processes of combustion, equal quantities are formed, and Regenerated like the Phoenix from her ashes.

Partington,[4] who was a great authority if ever there was one, cautiously avers: "The phoenix, it may be noted, was a fabulous bird regarded as sexless."

1. K.J. Laidler, *Accounts of Chemical Research*, 1995, Vol. 28, pp. 187–192.
2. K.J. Laidler, *The World of Physical Chemistry*, Oxford University Press, Oxford, 1993, pp. 250–252; 277–278.
3. M. Rayner-Canham and G. Rayner-Canham, *Women in Chemistry: Their Changing Roles From Alchemical Times to the Mid-Twentieth Century*, American Chemical Society and Chemical Heritage Foundation, Washington, D.C. and Philadelphia, 1998, pp. 28–31.
4. J.R. Partington, *A History of Chemistry*, MacMillan, London, 1962, Vol. 3, pp. 708–709.
5. J.C. Kotz and P. Treichel, Jr., *Chemistry and Chemical Reactivity*, 3rd ed., Saunders, Fort Worth, 1996, pp. 980–983.

CHEMISTRY IN THE BARREL OF A GUN

In the wrong circumstances, charcoal can be dangerous. Just ask Johann Baptist Von Helmont, who coined the term *gas* and then almost "gassed" himself by burning charcoal indoors.[1]

Carbon monoxide (CO) is formed as a by-product of combustion in oxygen-poor environments. In oxygen-rich environments, typical flammable materials burn (oxidize) completely to form carbon dioxide (CO_2) and water (H_2O). Under these conditions, CO is a short-lived *intermediate* that reacts as quickly as it is formed. [$Fe(OH)_2$, described in the previous essay, is a longer-lived intermediate en route to rust.] Mrs. Fulhame correctly concluded that water accelerates charcoal[2] combustion. The reason is, once again, clarified by *parts* of the very complex reaction mechanism for combustion.[3] In the absence of hydrogen-containing substances, the key chain-initiating reaction is

$$CO + O_2 \rightarrow CO_2 + O \qquad (1)$$

This is followed by many other reactions that keep the chain going. One is thought to be (where M is a molecule or atom for collision):

$$M + CO + O \rightarrow CO_2 + M \quad (2)$$

However, where sources of hydrogen are present [water, methane (CH_4), etc.], a very different and even faster chemistry occurs:[3]

$$CO + H_2O \rightarrow CO_2 + H_2 \quad (3)$$

$$H_2 + O_2 \rightarrow H_2O_2 \quad (4)$$

$$H_2O_2 \rightarrow 2OH \quad (5)$$

$$OH + CO \rightarrow CO_2 + H \quad (6)$$

$$H + O_2 \rightarrow OH + O \quad (7)$$

Charcoal does not burn rapidly because of its solid structure and the absence of sources of hydrogen; the latter also explains the relative abundance of carbon monoxide in its emissions. To trap and observe CO, it needs to be made outside of normal combustion conditions. Imagine newborn guppies amidst a tank of voracious fish. Now imagine the same newborn guppies deposited by their mother directly into an incubator tank so they may be studied.

Carbon monoxide was a "puzzlement" during the late eighteenth century when it was discovered independently by Torbern Bergman, Joseph Marie François de Lassone, and Joseph Priestley.[4] Steam passed over red-hot charcoal produces "water gas," which is useful for combustion energy but highly toxic. (We understand today that it is a mixture of CO, H_2, and CO_2). Priestley observed that when chalk ($CaCO_3$) was heated in a red-hot gun barrel, the result was "inflammable air" of a relatively "heavy" nature that burned with a blue flame to form "fixed air" (CO_2). When slaked lime [$Ca(OH)_2$] was heated in a red-hot gun barrel, the result was "light inflammable air" that burned explosively. Reactions (8) and (9) correspond to the first case, wherein CO is formed. Reactions (10) and (11) correspond to the second case wherein H_2 is formed. Of course, to add to the confusion, water gas contained both "light" and "heavy inflammable airs."

$$CaCO_3 \rightarrow CaO + CO_2 \quad (8)$$

$$3CO_2 + 2Fe \rightarrow Fe_2O_3 + 3CO \quad (9)$$

Here the gaseous product is "heavy inflammable air."

$$Ca(OH)_2 \rightarrow CaO + H_2O \quad (10)$$

$$3H_2O + 2Fe \rightarrow Fe_2O_3 + 3H_2 \quad (11)$$

Here the gaseous product is "light inflammable air."

Now, if you are Joseph Priestley and firmly wedded to the phlogiston theory, the conclusion is obvious—both "fixed air" (CO_2) and steam are releasing the phlogiston from iron in the gun barrel although to different extents. In 1801, William Cruickshank, Ordnance Chemist, Lecturer in Chemistry in the Royal Artillery Academy, Surgeon of Artillery and Surgeon to the Ordnance Metal Department finally succeeded in differentiating hydrogen from carbon monoxide.[4] As we will soon see, Count Rumford also used artillery to do science. Perhaps England's eighteenth-century wartime economy produced a surplus of weapons to be later exploited as scientific apparatus.

1. J.R. Partington, *A History of Chemistry*, MacMillan, London, 1962, Vol. 2, p. 229.
2. Charcoal is formed from slowly heating wood to rather high temperatures. The result is a mass of about 75% carbon, 20% volatiles (boiled away in red-hot charcoal), and 5% ash.
3. K.K. Kuo, *Principles of Combustion*, Wiley, New York, 1986, pp. 148–149.
4. J.R. Partington, op. cit., Vol. 3, pp. 271–276.

A BORING EXPERIMENT

Even as Lavoisier demolished phlogiston, he postulated a new gaseous "simple substance" or "element" called *caloric*—the element of heat (see Lavoisier's Table of Elements, Fig. 88). Caloric could be transferred from a warmer body to a cooler body without chemical change. However, Lavoisier also posited that oxygen gas contained caloric, released as heat and light when a substance burned (see Figure 89). The similarity between the caloric concept and the phlogiston concept is almost obvious.

Figure 97 is from Volume 2 of *Essays, Political, Economical and Philosophical* (3rd ed., 1798) by Benjamin Thompson (Count Rumford) (1753–1814). He demonstrated that the mechanical work involved in boring a brass cannon was sufficient to boil water and that the heat capacity of the chips produced by boring was the same as when these chips were part of the cannon.[1] One would have expected a loss in caloric to be manifested in a loss of mass and/or heating capacity. In effect, Rumford showed that there was no limit to the amount of caloric that could be released as the result of mechanical friction. This was, of course, impossible. He also carefully established that there is no change in mass upon freezing water. At the time, Rumford's work had little impact: Explanations offered were that the quantity of caloric present in the cannon was incredibly large and hardly any had been released in Rumford's experiments and that caloric was exceedingly light.

This study was a first quantitative step toward establishing the *First Law of Thermodynamics* in terms of the mechanical equivalent of heat:

$$\text{Energy}_{\text{System}} = (\text{Heat added})_{\text{System}} - (\text{Work on surroundings})_{\text{System}}$$

In the boring experiment, work is done by the surroundings on the system (the brass cannon), the energy of the system rises and heat is also released to the surroundings (water bath).

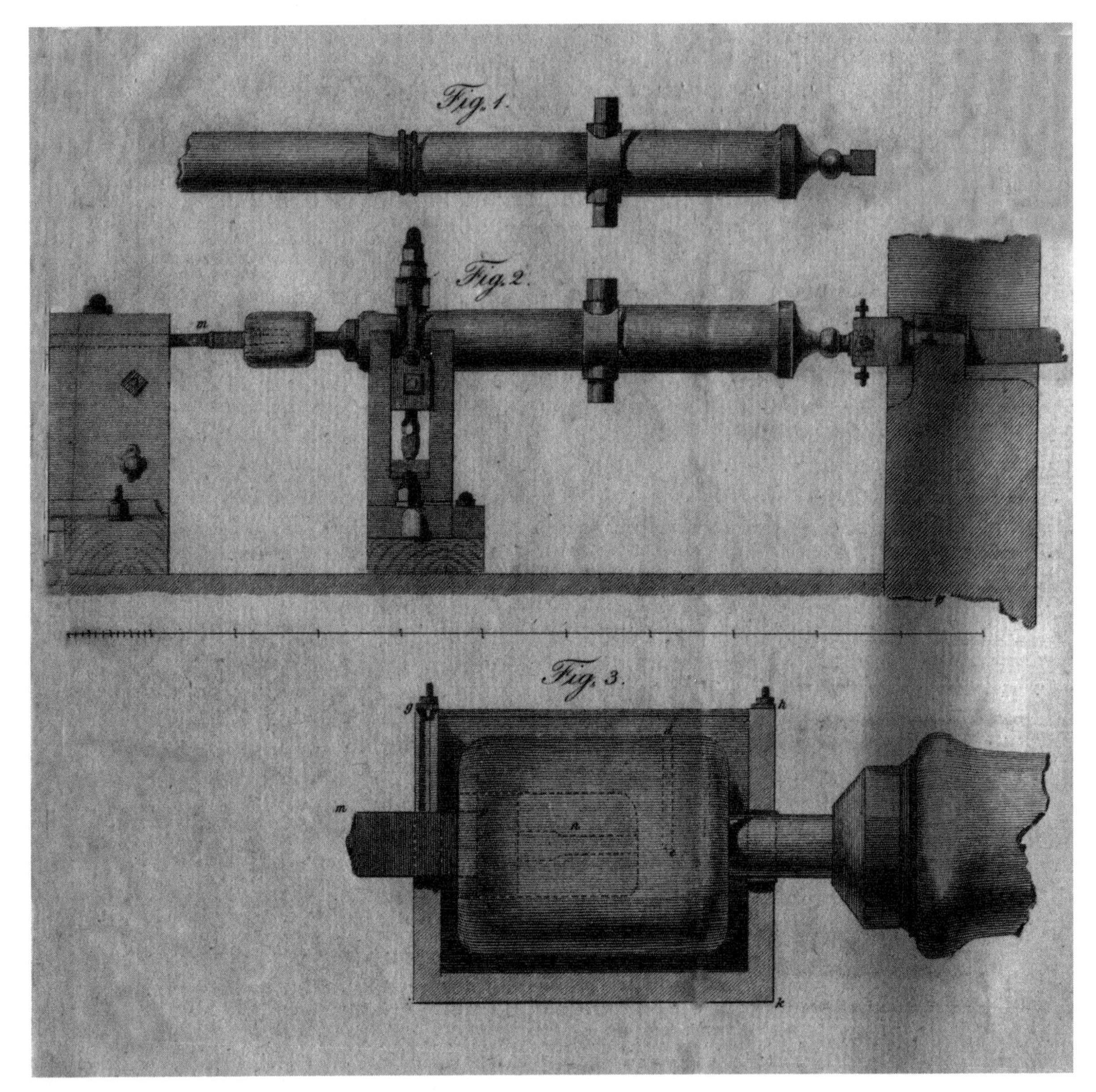

FIGURE 97 ■ The cannon-boring experiment of Benjamin Thompson (Count Rumford), which disproved the caloric theory (from Vol. 2, *Essays, Political, Economical and Philosophical*, 3rd ed., London, 1798).

Benjamin Thompson was born to a modest farming family in the Colony of Massachusetts in 1753.[2] He received little formal education, was largely self-taught, moved to Concord, New Hampshire to teach school and married a wealthy widow 14 years his senior when he was 19. They separated permanently in 1775 as the American Revolution began and Thompson worked as a spy for the English, eventually fleeing to England. He retired from the British Army and was Knighted by George III in 1784 and moved to Germany, became head of the Bavarian Army, and was appointed Count Rumford of the Holy Roman Empire in 1793. The early thermodynamics studies grew out of this military experience in Germany. Count Rumford returned to England in 1798, helped found the Royal Institution in 1799, and he appointed Humphrey Davy Lecturer in Chemistry in 1801 following publication of his work on laughing gas.

Count Rumford successfully courted Madame Marie-Paulze Lavoisier over a four-year period and they were married in 1805. However, according to the Rayner-Canham's, "he was a rather conceited, boring individual, who was expecting to live well on Paulze-Lavoisier's finances, while pursuing his researches alone" and their marriage apparently deteriorated in two months with separation occurring in 1809.[3]

1. W. Kauzmann, *Thermodynamics and Statistics: With Applications To Gases*, Vol. II of *Thermal Properties of Matter*, Benjamin, New York, 1967, pp. 34–35.
2. *Dictionary of Scientific Biography*, Scribner, New York, 1970, Vol. 13.
3. M. Rayner-Canham and G. Rayner-Canham, *Women in Chemistry: Their Changing Roles from Alchemical Times to the Mid-Twentieth Century*, American Chemical Society and the Chemical Heritage Foundation, Washington, D.C. and Philadelphia, 1998, pp. 17–22.

LAUGHING GAS FOR EVERYBODY!

Humphrey Davy[1] (1778–1829) was apprenticed to a surgeon in Penzance in 1795 but started reading Lavoisier's *Elements of Chemistry* and Nicholson's *Dictionary of Chemistry*, which still retained some phlogistic influences. His early investigations caught the attention of Thomas Beddoes and he was appointed to Beddoes's Pneumatic Institution in 1798. The Institution's purpose was to use inhalable gases to cure diseases.

Priestley's work on different kinds of air in 1772 produced impure nitrous oxide (N_2O). In 1799 Davy heated ammonium nitrate in the retort depicted in Figure 98 (*Fig. 2*) and obtained the pure gas, collected over water. His experimental and physiological studies were published in *Researches, Chemical and Philosophical, Chiefly Concerning Nitrous Oxide, Or Dephlogisticated Nitrous Air And Its Respiration* (1800). The plate shown is the frontispiece from the 1839 reprint of this exceedingly rare book and depicts a gas holder and breathing apparatus. Davy's reckless breathing of the newly discovered gases of the period were, for once, rewarded with nitrous oxide (laughing gas):

> On April 16th, Dr. Kinglake being accidentally present, I breathed three quarts of nitrous oxide from and into a silk bag for more than half a minute, without previously closing my nose or exhausting my lungs. The first inspirations occasioned a slight degree of giddiness. This was succeeded by an uncommon sense of fulness of the head, accompanied by loss of distinct sensation and voluntary power, a feeling analogous to that produced in the first stage of intoxication; but unattended by pleasurable sensation. Dr. Kinglake, who felt my pulse, informed that it was rendered quicker and fuller.

Davy, who wrote good poetry and was an avid fisherman,[1] had a wide variety of friends and correspondents who sampled nitrous oxide: these included Dr. Peter Mark Roget, future physician and author of the *Thesaurus*, but only 20 years old

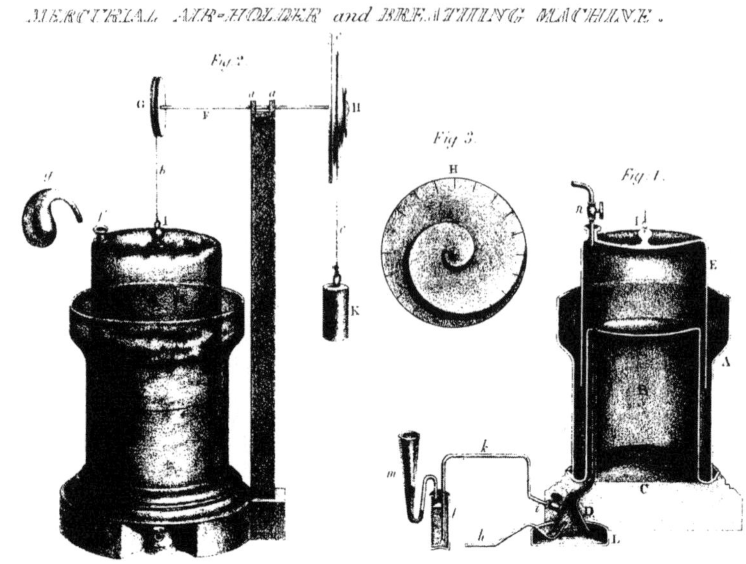

FIGURE 98 ■ Diagram of Humphrey Davy's apparatus for storing and breathing nitrous oxide (from *Researches, Chemical and Philosophical, chiefly concerning Nitrous Oxide, or Dephlogisticated Nitrous Air, and its Respiration*; the original edition, published in 1800, is of very great rarity). Davy's artistic circle of friends included poet Samuel Taylor Coleridge and Dr. Mark Roget (*Thesaurus* fame) who sampled laughing gas with Davy and recorded their scientific observations.

at the time, and Samuel Taylor Coleridge, one year after composing *The Rime of the Ancient Mariner*. Coleridge's description is the more poetic:

> The first time I inspired the nitrous oxide, I felt a highly pleasurable sensation of warmth over my whole frame, resembling that which I remember once to have experienced after returning from a walk in the snow into a warm room. The only motion which I felt inclined to make, was that of laughing at those who were looking at me. My eyes felt distended, and towards the last, my heart beat as if it were leaping up and down. On removing the mouth-piece, the whole sensation went off almost instantly.

Nitrous oxide was first used as an anesthetic in 1846 but not before it had caused a stir in college dorms of the period. And James Gillray's 1802 carricature (Figure 99) shows Davy holding bellows and assisting a lecture-hall laughing gas demonstration. And at the right standing, that *is* Count Rumford smiling with approval.[2]

1. J.R. Partington, *A History of Chemistry*, MacMillan, London, 1964, Vol. 4, pp. 29–73.
2. J. Read, *Humour and Humanism in Chemistry*, G. Bell, London, 1947, p. 207.

FIGURE 99 ▪ Artist James Gillray depicted a Royal Institution Lecture in 1802 in which Humphrey Davy, holding the bellows, delights as a subject receives a dose of laughing gas (see *Chemical Heritage*, Vol. 17, No. 2. p. 45, 1999). The tall standing figure on the right is Count Rumford. (Courtesy of Edgar Fahs Smith Collection, Rare Book & Manuscript Library, University of Pennsylvania.)

"LAVOISIER IN LOVE"

Draft for a screenplay: "Lavoisier In Love." Noting the great critical (*and* commercial) success of the 1998 film "Shakespeare In Love," we[1] feel that the roughly 30-year period between 1772 and 1805 that witnessed the chemical revolution could furnish a blockbuster.[2] Although we offer the idea later in a humorous vein, we honestly feel that an epic of more appropriate title could really be quite good. We see Kenneth Branagh directing the screenplay and playing Antoine; Gwyneth Paltrow as the young Marie; Judi Dench as Marie in her later years. Are there any financial backers out there?

We are not even wedded to the above title—"Antoine and Marie, The Tax Collector's Daughter" is another possibility.[3] Where "Shakespeare in Love" offers mere swordplay, we offer certified gunpowder and real pyrotechnics in the laboratory, in the streets and on the high seas. The film will be a period piece centered around the lives of Antoine Laurent Lavoisier, the father of modern chemistry, and his wife Marie-Anne Pierette Paulze-Lavoisier, one of the most sophisticated and alluring women of the age. The voice-over narration[4] is that of Madame Paulze-Lavoisier, the nexus of our drama. The background is the American Revolution, the French Revolution, and the fearful and violent reaction in England during and after the loss of the jewel in the crown of its North American empire. ("The Madness of King George" did pretty well in 1995 but it was aimed at the brie-and-merlot crowd, not the masses.) There's sex, violence, adultery, abandonment, lechery, espionage, treason—it will make "Les Liaisons Dangereuses" seem like "Sesame Street."

It is 1766 and phlogiston, a last remnant of alchemical thought, has held sway for nearly 100 years. In England, Henry Cavendish, an eccentric genius millionaire, thinks he has isolated the elusive phlogiston but has really made explosive hydrogen. (FLASH-FORWARD 20 years—French exploring air travel in hydrogen-filled balloons—a fiery disaster occurs). CUT TO Birmingham, England in the early 1770s: Joseph Priestley, a rather stiff-necked Unitarian clergyman discovers oxygen, finding that it sustains animals five times as long as regular air. He is a friend and correspondent of Benjamin Franklin as the American Revolution commences. JUMP CUT TO Fall, 1775; Ben advises:[5]

> "Joseph, Britain, at the expense of three millions, has killed one hundred and fifty Yankees this campaign, which is twenty thousand pounds a head . . . During the same time sixty thousand children have been born in America."

Enter the wealthy and brilliant 28-year-old Antoine to rescue 14-year-old Marie-Anne from the doddering lechers who work with him and Monsieur Paulze at La Ferme Générale. (La Ferme was a private finance company employed by the government to collect taxes—more on that later.) However, there is a foreshadowing as the shadow of a blade falls on a sausage for the Company Christmas party *choucroutes garnie*.[6] Marie and Antoine marry in 1771 and form a partnership that any couple would envy. Antoine begins his scientific studies in their residence. Marie's facility with languages brings Antoine access to the foreign chemical literature. He doesn't like what he reads and decides to change *everything.* Marie learns enough chemistry to be able to translate and comment critically on foreign texts. A gifted artist, also she engraves the plates for his monumental *Traité Élémentaire de Chimie* and paints a portrait of Franklin that Ben treasures highly. Saturdays are spent in the Salon with Antoine and Marie discussing the week's experiments with the *cogniscenti.*

Enter Pierre du Pont and his son Eleuthère Irénée who will later find a small measure of success with a start-up chemical company in wild and remote Delaware. Pierre is dashing and ebullient. Antoine is analytical and totally devoted to his beakers, flasks, and balances. Pierre and Marie begin an affair starting in 1781 that will last over 10 years without damaging the friendship between Pierre and Antoine (or, for that matter, Marie and Antoine)—ah, the French. From a scene:

Antoine: There are 26 of Pierre's robes in my armoire. I can't seem to find my lab coat!

Marie: It's in the laboratory with your clean underwear. I'll see you next Saturday in the Salon, Cheri.

FLASHBACK TO Benjamin Thompson, born to a family of modest means in the Colony of Massachusetts, who marries at the age of 19 a wealthy widow some 14 years his senior. During the American Revolution he spies for the British, is almost caught, abandons his wife, takes a fortune, and flees to England where he is knighted by George III in 1784. He will return.

Meanwhile, the English and French have been fighting for global dominance directly and by proxy for over 100 years. The phlogiston controversy gives them a fresh field for rivalry. Volleys of rhetoric fly back and forth across the Channel. Richard Kirwan, a viriolic Irishman, attacks Antoine (in print). Marie translates Kirwan's work—it provides Antoine with just the ammunition he needs and he appends his own notes to Marie's published translation. Hoist on his own petard and mortally wounded, Kirwan abandons phlogiston. Priestley holds fast—he never abandons phlogiston! The American Revolution triumphs. England is full of fear and anger. The excesses of the French Revolution add to this fear. Lavoisier, the tax collector, is guillotined (remember the foreshadowing?). Priestley escapes England as an angry rabble vows to "shake the powder from his wig" and burns his church to the ground. (Future portraits will show Priestley *sans* wig.)[7]

After Lavoisier is executed, the most eligible, wealthy, and brilliant suitors in Europe court Marie. REENTER Benjamin Thompson, now Count Rumford of the Holy Roman Empire, retired Head of the Bavarian Army, and vanquisher of Antoine's caloric theory, who emerges from the pack. The happily unmarried couple tour Europe together for four years. He is once again willing to endure the consequences of marrying a wealthy woman. They marry in 1805, but the marriage is on the rocks in two months—it seems that Rumford locked the front gates on Marie's guests one day and she responded by pouring boiling water on his prize flowers (lots of doctoral theses by students of cinema on the symbolism of these two actions). To this interesting cast we can add the laughing-gas-sniffing parties of Humphrey Davy and his artistic friends.

In our advertising trailer—COMING TO A THEATRE NEAR YOU:

SEE! Franklin Cruise The Salons of Paris!
SEE! Cavendish Make Water!
SEE! Marie Paulze-Lavoisier Scald Rumford's Flowers!

1. Ideas were contributed to this essay by Professor Susan Gardner, University of North Carolina at Charlotte.
2. Professor Roald Hoffmann was kind enough to share with me his earlier ideas about dramatizing the Lavoisiers. As this book is being completed a play titled "Oxygen" is being written by Professors Carl Djerassi and Roald Hoffmann organized around the idea of a Retro-Nobel Prize in 2001 (see *Chemical & Engineering News*, October 11, 1999, p. 7).
3. In the movie, Shakespeare's original title appears to be: "Romeo And Ethel, The Pirate's Daughter."
4. Professor Susan Gardner proposed Mme. Lavoisier as the narrative voice.
5. E. Wright, *Franklin of Philadelphia*, Belknap Press of Harvard University Press, Cambridge, 1986, p. 239.
6. Homage to Jane Campion's movie "The Piano."
7. Well, at least Rembrandt Peale's 1801 oil-on-canvas portrait is wigless. See B.B. Fortune and D.J. Warner, *Franklin and His Friends: Portraying The Man of Science in Eighteenth-Century America*, Smithsonian Portrait Gallery and University of Pennsylvania, Washington, D.C. and Philadelphia, 1999, p. 151.

SOME LAST-MINUTE GLITSCHES BEFORE THE DAWN OF THE ATOMIC THEORY

Introductory chemistry books paint a fairly neat picture of the orderly march toward Dalton's atomic theory: Discovery of the Laws of Conservation of Matter, Definite Composition and Multiple Proportions, and thence Atomic Theory. It was never quite so neat.

Chemists who preceded Lavoisier for decades if not centuries implicitly assumed that matter could not be created nor destroyed.[1] Why else would they postulate the addition of effluviums of fire (see Becher, Boyle, or Freind) to explain the increase in mass when metals form calxes, or the need to postulate buoyancy (or negative mass) for phlogiston, to explain the same phenomena? However, Lavoisier's careful work with chemical balances and pneumatic chemistry established the Law of Conservation of Matter on firm scientific ground.[1] Similarly, the Law of Definite Composition had long been assumed—that the back oxide of copper, for example, would always be 80% by weight copper and 20% by weight oxygen no matter the country, chemist, or method of origin. The studies of Joseph Louis Proust (1754–1826) established this and helped to solidify the principles of chemical composition (*stoichiometry*).

However, Claude Louis Berthollet (1748–1822), one of the great collaborators with Lavoisier on the *Nomenclature Chimiques*, raised some difficult questions in his book *Essai de Statique–Chimique* published in 1803 (Fig. 100).[2] Although there was some confusion about mixtures and compounds, he noted that there were some crystalline compounds having indefinite and varying compositions. He was correct. For example, the iron ore wustite is typically given the formula FeO although it really ranges from $Fe_{0.95}O$ (76.8% iron) to $Fe_{0.85}O$ (74.8% iron) depending, as we know today, on the balance between Fe^{2+} and Fe^{3+} ions to balance the O^{2-} ions in the ionic salt.[3] Since two Fe^{3+} ions will be

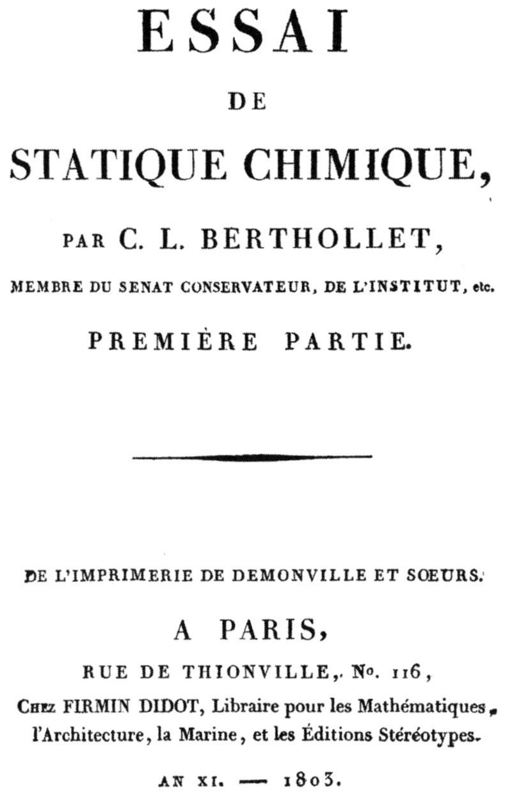

ESSAI

DE

STATIQUE CHIMIQUE,

PAR C. L. BERTHOLLET,

MEMBRE DU SENAT CONSERVATEUR, DE L'INSTITUT, etc.

PREMIÈRE PARTIE.

DE L'IMPRIMERIE DE DEMONVILLE ET SOEURS.

A PARIS,

RUE DE THIONVILLE, No. 116,

CHEZ FIRMIN DIDOT, Libraire pour les Mathématiques, l'Architecture, la Marine, et les Éditions Stéréotypes.

AN XI. — 1803.

FIGURE 100 ■ Title page of Claude Louis Berthollet's book, published just before Dalton's Atomic Theory. Berthollet discovered that chemical compositions were not always "definite" but often depended upon reaction conditions. He had really discovered the law of mass action.

equivalent to three Fe^{2+} ions in neutralizing three O^{2-} ions, replacement of Fe^{2+} by Fe^{3+} ions will produce gaps in the crystalline lattice and cause the Fe/O ratio to be less than 1:1 and slightly variable. Wustite is an example of a *nonstoichiometric compound* and such compounds are sometimes called *berthollides.*

Even more serious was Berthollet's finding that in some cases the products obtained in a chemical reaction depended upon reaction conditions. For example, a well-known laboratory chemical reaction is:

$$CaCl_2 + Na_2CO_3 \rightarrow CaCO_3 + 2NaCl$$

where $CaCl_2$ is a muriate of lime, Na_2CO_3 is soda, $CaCO_3$ is limestone, and NaCl is salt. The precipitation of solid limestone drives this "double elective attraction." However, accompanying Napolean on a trip to Egypt in 1798, Berthollet was surprised to discover deposits of soda on the shores of the salt lakes.[4] He reasoned that high concentrations of salt in the lakes could reverse the normal affinities, and thus the products of the reaction depended upon conditions. In fact, he had discovered the reversibility of chemical reactions and the Law of Mass Action, but this was only understood later.

$$CaCl_2 + Na_2CO_3 \rightleftarrows CaCO_3 + 2NaCl$$

There is something here for us to learn about the Scientific Method. To borrow the oft-cited example given by the philosopher Karl Popper: *If* one observes only white swans for decades, then the hypothesis "All swans are white" appears reasonable and as it continues to be verified over decades it assumes the status of a confirmed theory and possibly even a law. It can never be proven *true* since all possible future cases cannot be tested. However, the confirmed scientific observation of a black swan will overturn the theory. Now Berthollet's scientific observations might have been taken as invalidating the Law of Definite Composition and seriously undermining the Atomic Theory. However, rather than tossing them out due to the observation of a few "black swans," chemists retained these explanations, correctly anticipating that the inconsistencies would be explained in the future.

1. F.L. Holmes, *Chemical and Engineering News*, **72** (37): 38–45, 1994.
2. H.M. Leicester and H.S. Klickstein, *A Source Book in Chemistry 1400–1900*, McGraw-Hill, New York, 1952, pp. 192–201.
3. D.W. Oxtoby and N.H. Nachtreib, *Principles of Modern Chemistry*, 3rd ed., Saunders College Publishing, Fort Worth, 1996, p. 9.
4. W.H. Brock, *The Norton History of Chemistry*, Norton, New York, 1993, p. 144.

THE ATOMIC PARADIGM

Paradigm is a much overused word. However, the existence of atoms is so fundamental to the very fabric of chemical understanding that we can say virtually nothing scientifically sensible without it. *This* is a paradigm! Figures 101 and 102 are derived from John Dalton's 1808–1810 *A New System of Chemical Philosophy* (Vol. I, Parts I and II; the third volume, Vol. II, Part I appeared in 1827; although less important than the earlier volumes it is of utmost rarity and pricy indeed). Dalton (1766–1844), born to Quaker parents of modest means,[1] was largely self-educated. He taught school at the age of 12 and in 1793 moved to Manchester where he was for a period Professor of Mathematics and Philosophy at New College. This college moved from Manchester in 1803 and, after a variety of incarnations, became Manchester College in Oxford in 1889.[1] Dalton, however, remained in Manchester where he earned a modest living tutoring, lecturing, and consulting while performing his research.[1] Partington conjectures that the "robust and muscular" Dalton inherited his nature largely from his "energetic and lively" mother.[1] He never married but was attracted briefly to a widow of "great intellectual ability and personal charm": "During my captivity, which lasted about one week, I lost my appetite and had other symptoms of bondage about me as incoherent discourse, etc., but have now happily regained my freedom."[1]

Dalton had a lifelong interest in meteorology.[1,2] He published a book on this topic in 1793. However, his studies of the composition of the atmosphere

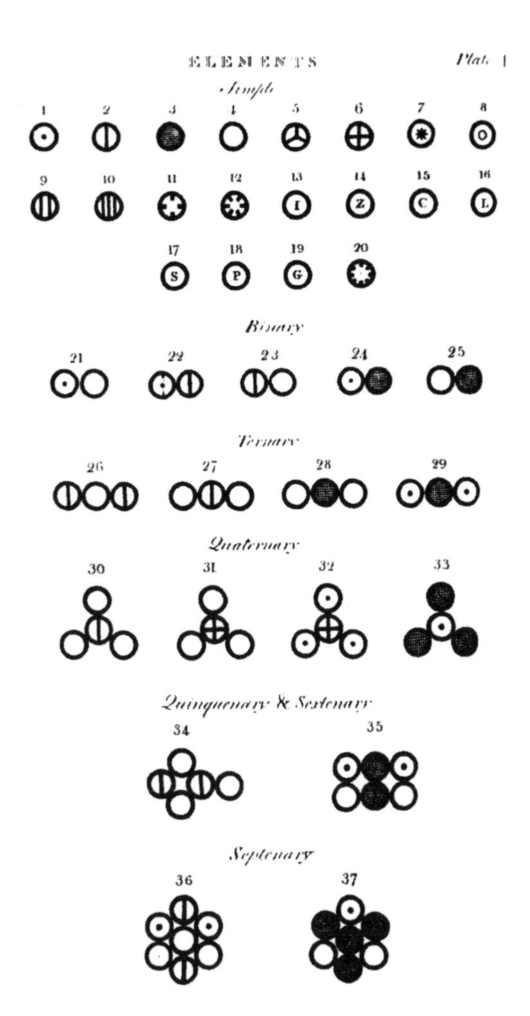

FIGURE 101 ■ Plate depicting atoms in Dalton's *A New System of Chemical Philosophy* (Manchester, 1808–1810). Dalton's "rule of greatest simplicity" (perhaps a Quaker style of science as well as lifestyle) has the formula of water as HO rather than H_2O (hydrogen peroxide, H_2O_2, would be discovered by Gay-Lussac and Thenard in 1815).

gave him the first clues leading to his atomic theory. Dalton realized that the composition of air was independent of altitude. Although oxygen and nitrogen differed in density, they did not form layers. His thoughts at this time included the idea that individual atoms were surrounded by envelopes (atmospheres) of caloric that repelled like atoms and attracted different atoms, thus explaining atmospheric mixing. During the period 1799–1801 he defined the vapor pressure of water and realized that when water was added to dry air, the total pressure was the sum of the dry air pressure and water's vapor pressure—the gases mixed yet acted totally independently (Dalton's Law of Partial Pressures). He also showed, as had Charles earlier, that air expands its volume linearly upon heating.

Although evidence suggests that chemists of the late eighteenth century assumed that specific substances had definite compositions,[2] Berthollet's studies (see Fig. 100) showed that compositions of "substances" often depended upon starting conditions. We now understand that Berthollet was observing mixtures whose proportions changed with conditions prevailing at equilibrium. Joseph

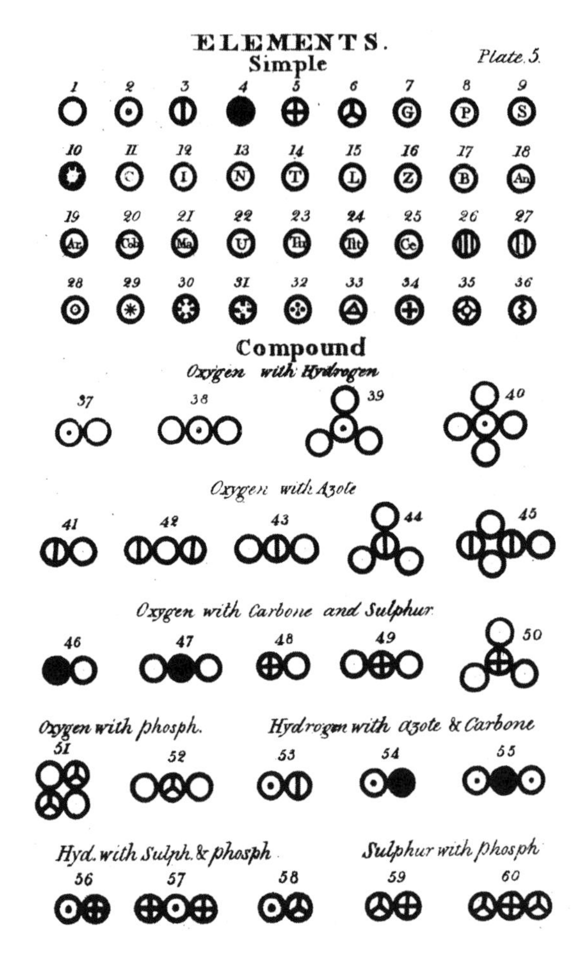

FIGURE 102 ■ Plate depicting atoms in Dalton's *A New System of Chemical Philosophy* (Manchester, 1808–1810).

Louis Proust (1754–1826) was educated in Paris but moved to Madrid where he held academic positions.[2] He engaged in a respectful debate with Berthollet over the course of a number of years and eventually prevailed. Proust demonstrated that there were two distinct oxides of tin and two distinct sulfides of iron—each with their own definite composition. Previous uncertainties were the result of mixtures of each pair of binary compounds.[2]

Dalton applied the Laws of Conservation of Mass and Definite Composition to explain his Atomic Theory.[3] He developed the chemical theory in 1803 and told Thomas Thomson, University of Edinburgh, about it in 1804. The third edition of Thomson's multivolume *A System of Chemistry* (Edinburgh, 1807) was the first book to include Dalton's Atomic Theory. From it, Dalton developed a third law, the Law of Multiple Proportions, to explain different formulas for binary compounds. For example, look at the binary compounds (*41* to *45*) formed between oxygen and nitrogen (Fig. 102). In comparing nitric oxide (*41*, NO) with nitrous oxide (*42*, N_2O), we see that the mass of nitrogen combining with a mass of oxygen in compound *42* is twice that in *41*. Dalton's atoms were real, indestructible, and unique for each of Lavoisier's ponderable elements. They were a total denial of alchemical transmutation. Dalton even built molecular models.

Dalton, a Quaker in science style as well as lifestyle, also assumed a "rule of greatest simplicity."[2] He originated the concept of atomic weights but could neither measure them nor even understand their basis. He chose to assign a relative weight of 1 to hydrogen, the lightest element, and to assume that combinations were the simplest possible. For example, we know water to be H_2O, ammonia to be NH_3, and methane to be CH_4. Dalton assumed they were HO, NH, and CH, respectively. Based upon the chemical analyses of 1803, which were good but far from perfect, he derived atomic weights as follows: hydrogen, 1.0 (assumed); oxygen, 5.5; nitrogen, 4.2; carbon, 4.3. By 1808, he had modified the values to include the latest data and rounded them off to whole numbers: hydrogen, 1 (assumed); nitrogen, 5; carbon, 5; oxygen, 7.[3,4] These assumptions would continue to cause confusion for decades. The assumption of 1.0 for hydrogen appears to be prescient although there was no basis at all for its assumed "oneness." We now know that hydrogen's "oneness" derives from its nucleus that has one proton only. Although hydrogen gas is actually H_2, oxygen (O_2), nitrogen (N_2), and some others are also diatomic and their relative densities are direct reflections of atomic masses. It remained for physician William Prout (1785–1850) to hypothesize in 1815–1816 that atomic weights are whole-number multiples of the atomic weight of hydrogen.[3,4]

Partington[1] notes that "Dalton never pretended that his teaching work interfered with his research, saying that 'teaching was a kind of recreation, and if richer he would not probably spend more time in investigation than he was accustomed to do.'" This is food for thought for those occasional self-important professors whose accomplishments do not include discovering their discipline's paradigm. Of course, one of Dalton's students was the renowned physicist James Prescott Joule—now there is a "career student" for any dedicated teacher!

1. J.R. Partington, *A History of Chemistry*, MacMillan, London, 1962, Vol. 3, pp. 755–822.
2. A.J. Ihde, *The Development of Modern Chemistry*, Harper and Row, New York, 1964, pp. 98–111.
3. J.R. Partington, op. cit., pp. 713–714.
4. W.H. Brock, *The Norton History of Chemistry*, Norton, New York, 1993, pp. 133–147; 160–162.

"WE ARE HERE! WE ARE HERE! WE ARE *HERE*!"

Dr. Seuss's wonderful book *Horton Hears A Who*[1] tells of the Whos of Whoville, a town on a speck of dust. They are too small to be seen and only Horton the elephant can hear them. Before the dust speck is boiled in oil, Horton exhorts the entire town to make a loud unified noise to announce the Whos' existence and save their lives: "We are here! We are here! We are here!"

In many ways the invisible (and voiceless) atoms were calling attention to themselves early in the nineteenth century. Figure 103 is from Dalton's 1808 *A New System of Chemical Philosophy*. In illustration 1 we see Dalton's depiction of the structure of liquid water. Dalton postulated that when water freezes, the atoms in a layer move from the square to rhomboid arrangement of illustration

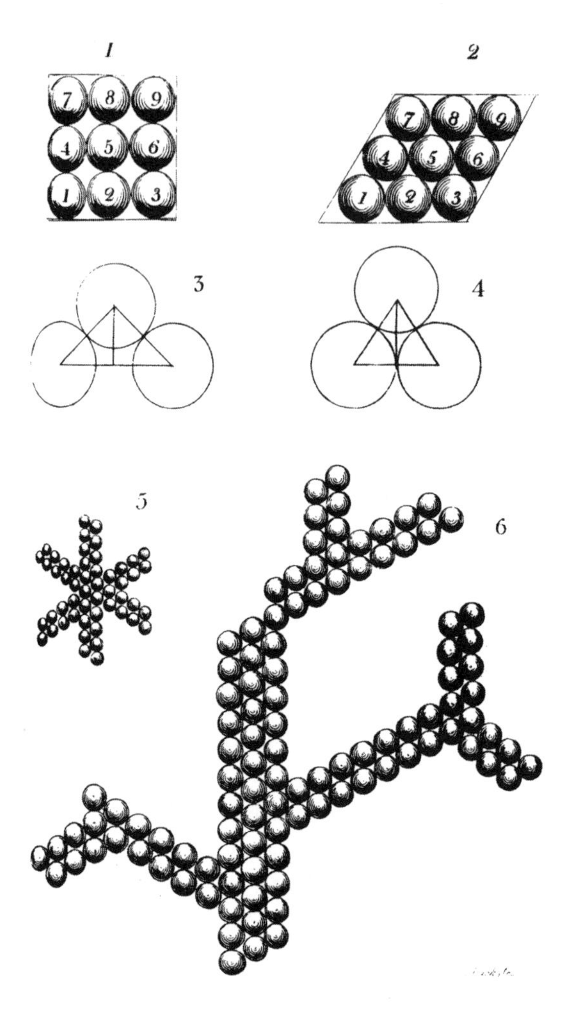

FIGURE 103 ■ A plate from Dalton's *A New System of Chemical Philosophy* explaining why ice is less dense than water using Atomic Theory. Although the details are not correct, Dalton's explanation of the sixfold symmetry of snowflakes and ice crystals originating at the molecular level was incredibly insightful. Berzelius and others noted the trigonometric error in his explanation of the decrease in density when water (*1* in this figure) becomes ice (*2*).

2. It is this hexagonal arrangement that Dalton perceived to be responsible for the well-known hexagonal symmetry of snow flakes and ice crystals (see illustration 5). He also tried to use these structures to explain the known fact that ice is less dense than water (ice floats). His arguments (using illustrations 3 and 4) were incorrect (this was noted with disapproval by Berzelius in 1812).[2] Also, liquid water is not an orderly array as depicted in diagram 1. Nevertheless, the core idea about ice structure was correct. We understand that water molecules are not perfect spheres and that ice is less dense than water to allow full hydrogen bonding between water molecules. The overall molecular lattice of ice has sixfold symmetry and this is indeed reflected in the snowflake.

In 1808 Jean Louis Gay-Lussac[3] (1778–1850) summarized the results of experiments of others and a few of his own and discovered the law of combining volumes of gases. He realized that volumes of gases could only be compared if their pressures and temperatures were equal. (The first statement is Boyle's Law; the second is sometimes called Charles Law after the first discoverer or Gay-Lussac's Law after the person who first published it.) Thus, equal volumes of ammonia (NH_3) and muriatic acid (HCl) combine perfectly to form a solid salt; one volume of nitrogen and one volume of oxygen form two volumes of "nitrous gas" (NO). One volume of nitrogen and three volumes of hydrogen form two volumes of ammonia.

Dalton resisted Gay-Lussac's findings. They caused problems for his rule of greatest simplicity. Since an equal volume of hydrogen reacts with an equal volume of chlorine, it is reasonable that those volumes contain equal numbers of elementary particles. However, if two volumes of hydrogen react with one volume of oxygen, as observed by Gay-Lussac, this would not be consistent with Dalton's formulation of water as HO. (Note that hydrogen peroxide, H_2O_2, was not discovered until 1815.)

In August, 1804 Biot and Gay-Lussac, like Charles in 1783, ascended in a hydrogen-filled balloon to make measurements of the earth's magnetic field. In September, 1804, Gay-Lussac ascended to 23,000 ft above Paris, collected air samples and found them to have the same composition as air at sea level.[3] You have to admire Charles's and Gay-Lussac's confidence in the gas laws.

In 1811, Amedeo Avogadro[4] (in full, Lorenzo Romano Amedeo Carlo Avogadro di Quaregua e di Cerreto, 1776–1856) used Gay-Lussac's, Dalton's, and others' works to make his hypothesis: equal volumes of gases (at same temperature and pressure) have equal numbers of molecules. Interestingly, this contribution was largely forgotten until resurrected by Cannizaro in 1858.[5]

Another piece of macroscopic evidence favoring atoms was the concept of Isomorphism enunciated by Eilhardt Mitscherlich (1794–1863) around 1818–1819.[6,7] He related atomic composition to observable crystal structures. Thus, phosphates and arsenates (e.g., $Na_2HPO_4 \cdot 12H_20$ and $Na_2HAsO_4 \cdot 12H_20$) as well as sulfates and selenates (e.g., Na_2SO_4 and Na_2SeO_4) had identical or very similar crystal structures because their atomic compositions were so similar. (Can you hear the pendulum of the future Periodic Table swinging here?) Berzelius used these relationships to help in his assignments of atomic weights.

Calorimetric studies of the type done by Lavoisier and Laplace (Fig. 92) were continued by others including Pierre Louis Dulong (1785–1838) and Alexis Therese Petit (1791–1820). They discovered the law that bears their names: the product of the specific heat and the atomic weight of solid elements (e.g., lead, gold, tin, silver, and sulfur) is constant. This really implies that all atoms (independent of their identities) have the same capacity for heat. This result was later extended to solid compounds and ultimately cleared up confusions such as whether the binary oxides of copper were really CuO and CuO_2 or Cu_2O and CuO.

1. T.S. Geisel, *Horton Hears A Who* (By Dr. Suess), Random House, New York, 1954.
2. W.H. Brock, *The Norton History of Chemistry*, Norton, New York, 1993, pp. 158–162.
3. J.R. Partington, *A History of Chemistry*, MacMillan, London, 1964, Vol. 4, pp. 77–90.
4. J.R. Partington, op. cit., pp. 213–217.
5. J.R. Partington, op. cit., pp. 489–494.
6. J.R. Partington, op. cit., pp. 207–212.
7. A.J. Ihde, *The Development of Modern Chemistry*, Harper and Row, New York, 1964, pp. 147–149.

WAS AVOGADRO'S HYPOTHESIS A PREMATURE DISCOVERY?

Avogadro got it right in 1811 when, combining Dalton's Atomic Theory and Gay-Lussac's Law of Combining Volumes, he concluded that equal volumes of gas (same temperature and pressure) have equal numbers of ultimate particles. One critical aspect was Avogadro's term *half-molecules*, which were really atoms for diatomic molecules such as H_2, O_2, N_2, and Cl_2. Dalton had never accepted the combining volumes law. He was bothered, for example, by the "sesquioxide" of nitrogen (1 volume of nitrogen atoms; 1.5 volumes of oxygen atoms). In his Quaker style of speech he said:[1] "Thou knows . . . no man can split an atom." Avogadro's nomenclature was somewhat confusing. For example, the "integrant molecule" of water contains half a molecule of oxygen and one molecule (or two half molecules) of hydrogen:[2] $1H_2 + {}^1/_2O_2 \rightarrow 2H + 1O \rightarrow 1H_2O$. Avogadro's Hypothesis also vexed Dalton: how could nitrogen gas (N) be more dense than ammonia gas (NH) if the two gases had equal numbers of molecules in a unit volume? Of course the answer is that nitrogen gas is N_2 while gaseous ammonia is NH_3.

Andre Marie Ampere, Jean Baptiste Andre Dumas, and their student Marc Augustin Gaudin adopted Avogadro's Hypothesis in their work during the next three decades.[3] Until Dumas, gas densities could be measured only for permanent gases. Dumas developed a technique to measure densities for volatile liquids and solids, thus extending the range of molecular (and atomic) weights determinable by the Ideal Gas Law ($PV = nRT$). In this way, Gaudin discovered that elemental (white) phosphorus is actually P_4.[2] Still, it remained for Stanislao Cannizzaro to reintroduce Avogadro's hypothesis in 1858.

Why did it take almost 50 years to achieve widespread acceptance of Avogadro's hypothesis? Here it may be of some value to refer to Gunther S. Stent's concept of a *premature discovery*, defined as follows: "A discovery is premature if its implications cannot be connected by a series of simple logical steps to canonical or generally accepted knowledge."[3] Stent exemplifies a premature discovery using Oswald T. Avery's unambiguous experimental identification of DNA as the genetic material in 1944.[3] What was the conceptual problem? It was known that DNA was composed of only four different nucleotides. How could such a simple "alphabet" code for an unimaginably vast store of genetic information? Proteins, by contrast, had a 20-amino-acid "alphabet" and were obviously the better choice for information storage. Thus, although Avery's experimental conclusions were solid and unambiguous, scientists did not immediately accept the conceptual framework to understand them and they remained relatively unnoticed for about five or six years. Similarly, atoms were not universally accepted as real and Avogadro's nomenclature was somewhat confusing. Moreover, Avogadro, who practiced law and was Professor of Mathematics at Turin, although "a man of great learning and modesty," was said to be "little known in Italy."[2] Similarly, Avery was "a quiet, self-effacing, non-disputatious gentleman."[3] Had Avogadro access to a good Madison Avenue public relations firm, perhaps the Periodic Table would have been discovered a decade or two earlier.

1. J.R. Partington, *A History of Chemistry*, MacMillan, London, 1962, Vol. 3, p. 806.
2. J.R. Partington, *A History of Chemistry*, MacMillan, London, 1964, Vol. 4, p. 213–222.
3. G.S. Stent, *Scientific American*, Vol. 227, No. 6 84–93, 1972.

CHEMISTRY IS *NOT* APPLIED PHYSICS

Dalton's atoms were derived from chemical experiments and explained chemical laws. Atoms were "adopted" by physicists only after many decades passed.

Indeed, attempts 100 years earlier to apply the physics of the age—Newton's great work—to chemistry failed. Among the first to attempt these applications were mathematician John Keill (1671–1721) and physician John Freind (1675–1728).[1] Newton had expressed the force arising from gravitational attraction between two bodies with the formula:

$$F = \frac{Gm_1m_2}{d^2}$$

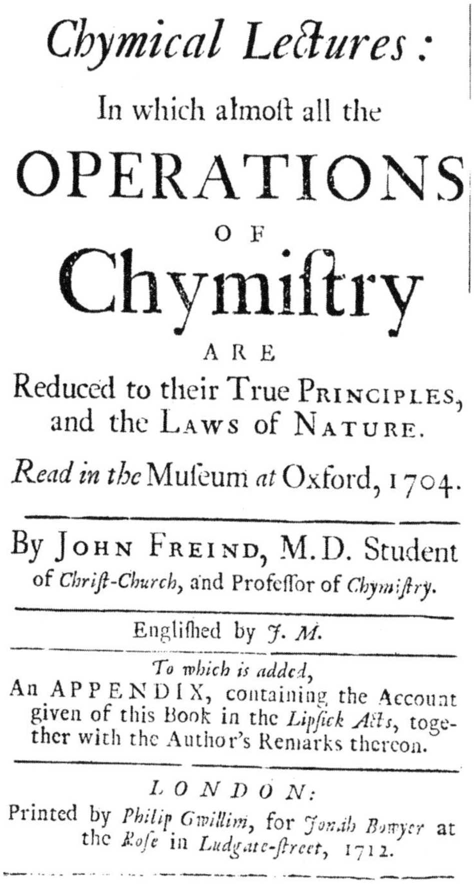

Chymical Lectures:

In which almoſt all the

OPERATIONS

OF

Chymiſtry

ARE

Reduced to their True PRINCIPLES, and the LAWS of NATURE.

Read in the Muſeum *at* Oxford, 1704.

By JOHN FREIND, M.D. Student of *Chriſt-Church*, and Profeſſor of *Chymiſtry*.

Engliſhed by *J. M.*

To which is added,
An APPENDIX, containing the Account given of this Book in the *Lipſick Acts*, together with the Author's Remarks thereon.

LONDON:
Printed by *Philip Gwillim*, for *Jonah Bowyer* at the *Roſe* in *Ludgate-ſtreet*, 1712.

FIGURE 104 ▪ Title page from Dr. John Freind's 1712 book in which he attempted to use Newtonian physics to explain physical and chemical properties of matter. Newton suspected that the forces holding matter together were electrical and magnetical.

The distance (d) was calculated from the centers of mass (center of the earth, mass m_1; center of the apple, mass m_2), and the weakening of the force with the square of the distance ($1/d^2$) meant that if the distance doubled, the force was only one-quarter of the original.

Keill and Freind both recognized gravity as a weak force unless a planet was involved. In the very rare *Chymical Lectures* by Friend (Figs. 104 and 105), he describes another similar attractive force, extremely strong at exceedingly minute distances and present on the surfaces of particles (points c and d) and having a higher-order relationship with distance ($1/d^{10}$, $1/d^{100}$, . . . , ?) and thus vanishing when the distance between c and d remains tiny.

Freind too recognized that a metal was lighter than its calx. He explained this observation by postulating the incorporation of igneous particles (particles of fire, see Boyle's "effluviums"), further separating the particles of metal; therefore weakening the forces between them. Thus, it is understandable that the melting point of silver metal is 962°C while its calx (Ag_2O) decomposes at only 230°C. Metal calxes, though not terribly water soluble, were more soluble than the metals themselves. However, lead melts at 327°C, while the white pigment litharge (PbO) melts at 886°C; mercury is a liquid, while HgO is a solid, albeit

Velocity they approach each other For the Attractive Force exerts it ſelf only in thoſe Particles which are very near one another; as for inſtance, in d *and* c*; The Force of ſuch as are remote is next to nothing.*

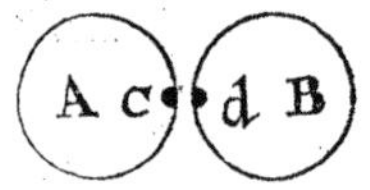

A c d B

Therefore no greater Force is requir'd to move the Bodies A *and* B*, than what would put into motion the Particles* d *and* c*, when diſengag'd from the reſt. But the Velocities of Bodies moving with the ſame Force are reciprocally, as the Bodies themſelves. Therefore the more the Body* A *exceeds the Particle* d *in Magnitude, the leſs is its Velocity; and this Motion is ſo languid, that oftentimes 'tis overcome by the Circumambient Medium, and other Bodies. Hence it is that this Attractive Force does ſcarce exert it ſelf, unleſs in the ſmalleſt Particles, ſeparated from the reſt.*

FIGURE 105 ▪ A page from Freind's book depicting gravitational attraction between atoms.

very slightly more water soluble than the metal. Adding to the confusion were calxes that were actually mixtures having components that readily decomposed.

During the twentieth century we have come to recognize that Newtonian physics explains the behavior of large, slow-moving objects like Nolan Ryan's fastball. The electrons that we know are responsible for holding atoms together need quantum mechanics to explain their behavior. They simply do not obey Newton's laws. Ironically, the forces that hold together salts, composed of ions such as Fe^{3+} and O^{2-} (ions were established by Arrhenius in the late nineteenth century), are almost entirely explained by the classical physics of Coulomb's law. However, negative electrons do *not* collapse into the positive nucleus.

1. J.R. Partington, *A History of Chemistry*, MacMillan, London, 1962, Vol. 2, pp. 478–482.

SECTION VI
CHEMISTRY BEGINS TO SPECIALIZE AND HELPS FARMING AND INDUSTRY

THE ELECTRIC SCALPEL

Count Rumford, whose efforts led to the chartering of the Royal Institution of Great Britain in 1799, took note of the accomplishments and verve of the 23-year-old Humphrey Davy, and had him appointed lecturer in chemistry in 1801.[1,2] The fact that Davy had been critical of Lavoisier's caloric theory probably did not hurt his case.

The handsome, poetic Davy was an immediate hit at the Royal Institution, attracting women as well as men to his lectures. He also worked on practical problems including the chemistry of tanning and agriculture (*Elements of Agricultural Chemistry*, London, 1813). At the time, the scientific world and popular interest had been galvanized by Alessandro Volta's "artificial torpedo (*electric fish*)." It consisted of a pile of alternating circular disks of silver and zinc, each pair separated by a layer of cardboard soaked in brine. Volta (1745–1827) discovered methane in 1776 in Lake Como by stirring up the mud and collecting the bubbles in an inverted bottle filled with water. He described the voltaic pile for the first time in a letter to Sir Joseph Banks, President of the Royal Society, dated March 20, 1800.[3]

During the latter part of the seventeenth century, a variety of chemical experiments had been performed using electricity.[3] Only about a month after Volta's disclosure, Anthony Carlisle and William Nicholson constructed a voltaic pile of 36 pairs of silver and zinc plates (half-crown coins were sometimes used as the silver plates).[2,3] They attached a brass (copper–zinc alloy) wire to a silver plate (Volta's negative pole) and a brass wire to the zinc plate at the other end (positive pole) and dipped the wires into a test tube containing water. Bubbles of hydrogen were produced at the negative pole while the wire attached to the positive pole corroded. When both wires were made of platinum, hydrogen gas was formed at the negative pole and oxygen gas at the positive pole.[3] Davy started to apply himself to electrochemical studies including the electrolysis of water. Figure 106(a) is taken from Davy's *Elements of Chemical Philosophy* (London, 1812; Philadelphia, 1812). It depicts a voltaic pile consisting of 24 pairs of silver and zinc plates, each pair separated by cloth soaked in liquid. By 1806, Davy stated in public that the forces holding compounds together were electrical in nature.[1,2]

In 1807, Davy applied himself to a problem that had vexed Lavoisier, who was convinced that potash (KOH) was a compound even though it resisted "simplification."[1,2] He employed a huge, more powerful, voltaic pile (his "battery of the power of 250 of 6 and 4"—seemingly a pile of 150 pairs of 4-inch square plates connected to a pile of 100 pairs of 6-inch square plates).[4] His attempts to decompose aqueous solutions of potash merely electrolyzed water. However,

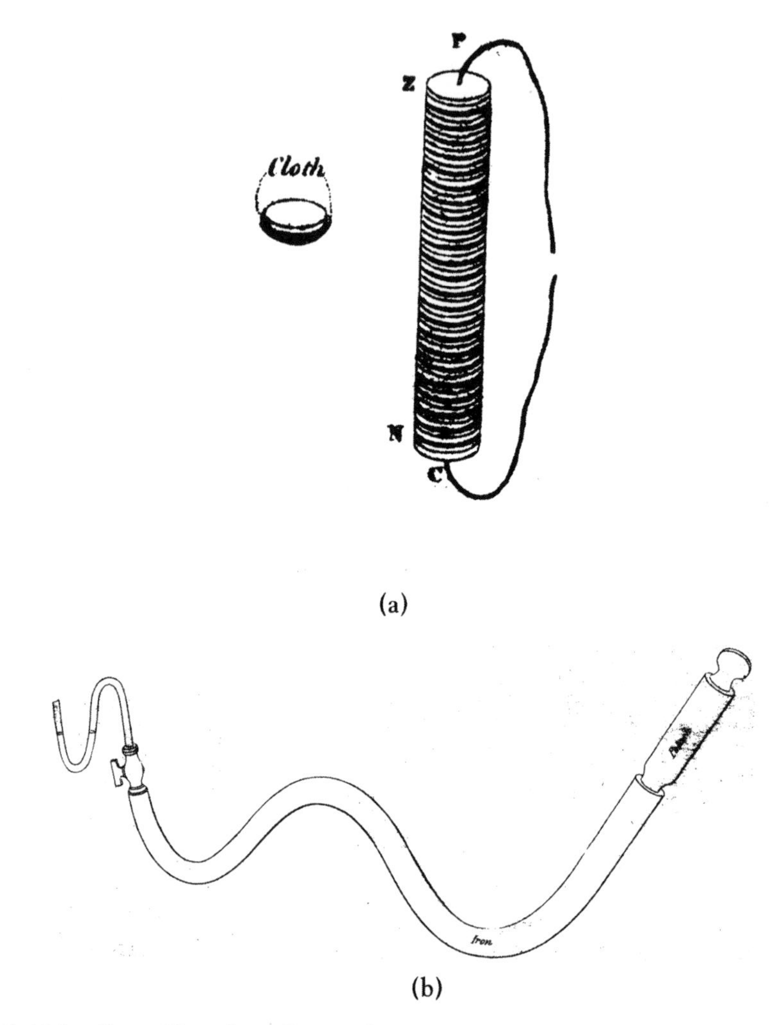

FIGURE 106 ■ From Humphrey Davy, *Elements of Chemical Philosophy* (Philadelphia & New York, 1812; first English, London, 1812): (a) Davy's voltaic pile consisting of alternating zinc and silver plates separated by moistened cloth; (b) another gun-barrel experiment. In the white-hot gun barrel (free of air), potash yields metallic potassium; potassium was first generated using the voltaic pile. When he discovered potassium Davy ". . . actually bounded about the room in ecstatic joy."

when a piece of solid potash was placed on a disk of platinum (connected to the negative pole) and a platinum wire (connected to the positive pole) was touched to the top of the potash, the solid fused at both points of contact. A violent effervescence at the upper surface (positive pole) was due to oxygen gas. At the lower part (platinum plate), beads of a silvery mercurylike liquid appeared, some of which exploded and burned with a bright flame. According to his cousin Edmund Davy, then working as an assistant (report by Humphrey's brother John):[4]

> [When Humphrey Davy] saw the globules of potassium burst through the crust of potash, and take fire as they entered the atmosphere, he could not contain his joy—he actually bounded about the room in ecstatic delight;

> and some little time was required for him to compose himself sufficiently to continue the experiment.

In a few days, Davy successfully isolated sodium. His voltaic pile also gave him the alkaline earths barium, strontium, calcium, and magnesium. (Lavoisier correctly identified their oxides as compounds but could not isolate the metals.) During this period Davy proved that chlorine gas (first isolated by Scheele in 1774) did not contain oxygen. Thus, hydrochloric acid did not contain oxygen, disproving Lavoisier's hypothesis that all acids contained oxygen.

In the year following Davy's electrochemical isolation of potassium, Gay-Lussac and Thenard obtained it chemically. Figure 106(b) (from Davy's *Chemical Philosophy*) shows an experiment performed (once again) in a gun barrel. Iron in an air-free environment is made white hot and the potash in the upper right tube is melted to produce potassium when the melt contacts the iron.

1. J.R. Partington, *A History of Chemistry*, MacMillan, New York, 1964, Vol. 4, pp. 29–75.
2. W.H. Brock, *The Norton History of Chemistry*, Norton, New York, 1993, pp. 147–153.
3. J.R. Partington, op. cit., pp. 4–5; 12–19.
4. J.R. Partington, op. cit., p. 46.

CHEMICAL SCALPELS THROUGH THE AGES

Until the middle of the twentieth century, Humphrey Davy held the record for discovering the most chemical elements: six.[1] He succeeded because he was "the first kid on the block" to apply a new type of scalpel systematically (the voltaic pile or battery) to chemical problems. He very modestly attributed his discoveries to the instruments rather than to his own brilliance:[2]

> The active intellectual powers of man in different times are not so much the cause of the different successes of their labours, as the peculiar nature of the means and artificial resources in their possessions.

Fire is clearly the most ancient chemical scalpel. Indeed, Vulcan's release of Athena from the head of Zeus prior to her chemical marriage [Fig. 36(f)] can be taken as a metaphor for the role of fire in causing chemical change. Prior to 1600, the application of fire ultimately added four new elements (antimony, arsenic, bismuth, and zinc) to the nine elements known to the ancients (carbon, sulfur, and the seven metals: iron, tin, lead, copper, mercury, silver, and gold—one for each day of the week).[1] Flames were themselves dissected by blowpipes and the reducing and oxidizing parts of the flame used as scalpels in Sweden starting in the eighteenth century. Fire powered the stills that produced the new scalpels sulfuric acid (by distillation of green vitriol, $FeSO_4 \cdot 7H_2O$), nitric acid (distill the product produced by adding oil of vitriol to saltpetre), and *aqua regia* (nitric and hydrochloric acids). Oxygen and chlorine (isolated by Scheele) and fluorine (isolated over 100 years later by Moissan) were also potent scalpels.

Radiation, including α-particles and neutrons, eventually led to real transmutation. It is no coincidence that Glenn Seaborg and his associates at Chicago and Berkeley hold the record for discovery of elements since they used these particles to make brand new ones. Expanding the Periodic Table, like expanding the baseball season from 154 to 162 games, almost doesn't seem fair to Davy. Perhaps Seaborg's name should have an asterisk in the record books like the one for Roger Maris when he broke Babe Ruth's home-run record during the first extended season.[3] In the most recent two decades, the laser and the atomic force microscope have been successful in promoting reactions one atom at a time—seemingly the ultimate in chemical dissection.

1. A.J. Ihde, *The Development of Modern Chemistry*, Harper and Row, New York, 1964, pp. 747–749.
2. W.H. Brock, *The Norton History of Chemistry*, Norton, New York, 1993, pp. 187–188.
3. Absolutely no disrespect is meant here. My early adolescent interest in nuclear physics and chemistry made Seaborg the first living chemist whose name I knew. I never met him but I recall the thrill of being at an American Chemical Society lecture when Seaborg quietly and gracefully entered the room.

DAVY RESCUES THE INDUSTRIAL REVOLUTION

"Two great events amazed Britain in 1815: the victory of Wellington over Napolean and the victory of Davy over mine gases."[1] The Industrial Revolution was in danger of stalling in the early nineteenth century due to the dangers in mining with contemporary lamps that used flame and ignited explosions. A disaster near Newcastle in 1812 killed 101 miners, and more than two-thirds of the coal mines in England were considered too dangerous to work because of their levels of coal gas (primarily methane).[1]

In 1815, Humphrey Davy was invited by the Chairman of a "Society for Preventing Accidents in Coal Mines" to invent a solution.[2] His elegant and simple invention is shown (Fig. 107) in the frontispiece of his 1818 book *On The Safety Lamp For Coal Miners; With Some Researches On Flame*. Davy had earlier studied flames and their propagation and noted that flames could not propagate through small holes. Thus, his solution was merely to surround the lamp with a cylinder of wire mesh that still left the flame open to the atmosphere. The mesh conducted away the heat of the flame, thus cooling it so that the temperature methane would encounter at the lamp would be lower than its flash point. The flame itself could not penetrate the mesh.[2]

While on the topic of coal gas, we note that chemist Friedrich Accum (1769–1838) played a key role in support of the introduction of coal-gas lighting in England. It is hard to imagine the change in London nightlife upon its widespread use. "Full moon at night, lovers' delight," but what about the other 27 days? In a London fog on a moonless night, two lovers might hear each other, touch each other, but not see each other. Coal gas, obtained by destructive distillation of coal,[3] consists largely of hydrogen and methane, with smaller

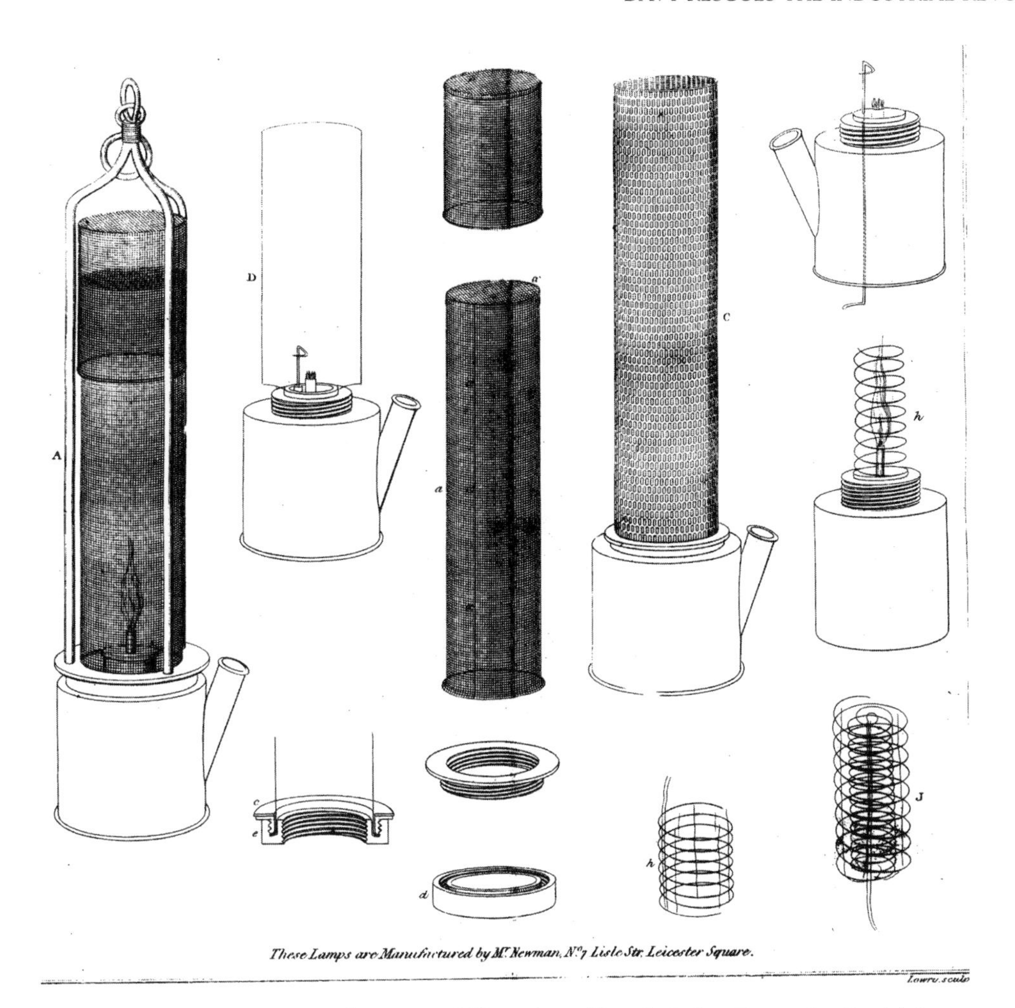

FIGURE 107 ■ Depictions of aspects of Humphrey Davy's Safety Lamp for Coal Miners (London, 1818). His ingenious solution to lamps that would ignite coal gas with deadly results was incredibly simple. The fine metallic mesh would cool the coal gas below its flash point. Thus, although the flame and combustible gas were in open contact, there would be no explosion.

amounts of carbon monoxide, ethylene, and some acetylene as well carbon dioxide, hydrogen sulfide, and ammonia.

Figure 108 is from the third edition of Accum's book *A Practical Treatise on Gas-Light* (3rd ed., London, 1816; first and second eds., 1815). It shows a gas apparatus for exhibit and for testing different coals. At the right is a portable furnace with cast-iron retort for burning the coal; the center unit is a purifier having three chambers (one with water to trap ammonia; the second with aqueous potash (KOH) to trap carbon dioxide and hydrogen sulfide; the third receives other liquid products). The unit at the left of Figure 108 is the gasometer that stores coal gas over water at a moderate pressure. Note the elegant lamps or burners at the top. By 1815, there were already 26 miles of main gas pipes under London streets.

Accum wrote a number of interesting books on chemistry theory and practice and chemical amusements in addition to his *Practical Treatise on Gas-Light*. His book *Death in the Pot: A Treatise of Food, and Culinary Poisons* (London,

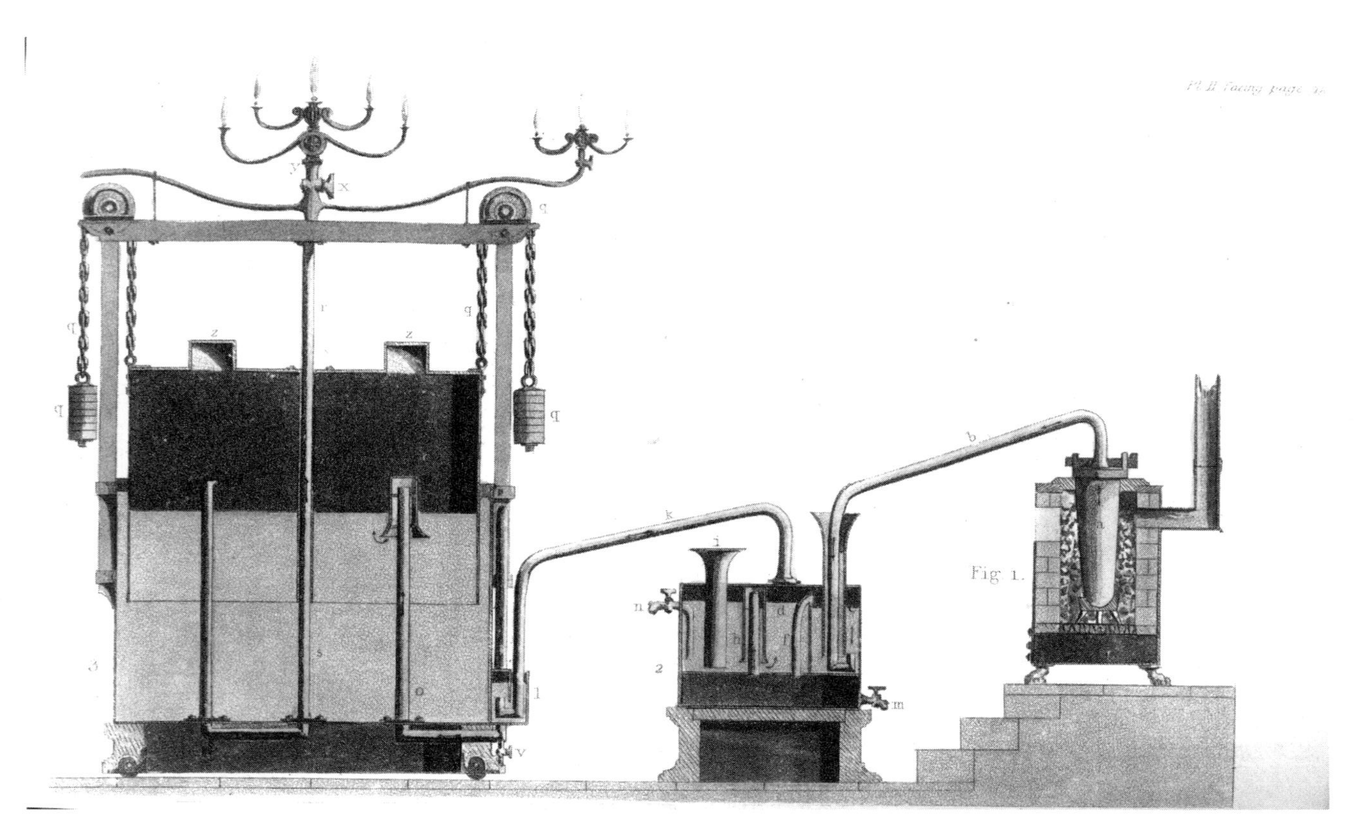

FIGURE 108 ▪ A demonstration model of a coal-gas lighting system (Friedrich Accum, *A Practical Treatise on Gas-Light*, *Third edition*, *London*, *1816*).

1820), made him many enemies. Some of these may have conspired in accusing him of stealing and defacing books in the library of the Royal Institution. He was aquitted but left England in disgrace for a Professor's position in Germany.[4] His excellent 1824 work, *An Explanatory Dictionary of the Apparatus and Instruments Employed in the various Operations of Philosophical and Experimental Chemistry*, was published anonymously.

1. J. Stradins, *Chymia*, No. 9, 125–145, 1964.
2. J.R. Partington, *A Short History of Chemistry*, 3rd ed., Dover, New York, 1989, pp. 189–190.
3. J.R. Partington, *A History of Chemistry*, MacMillan, London, 1962, Vol. 3, pp. 826–827.
4. C.A. Browne, *Chymia*, No. 1, 1–9, 1948.

THE DUALISTIC THEORY OF CHEMISTRY

The early alchemists and natural philosophers believed in the duality of matter —sun and moon; male and female; sulfur (fixed) and mercury (volatile). When Davy electrolyzed pure potash (KOH) and produced a volatile (female) spirit (oxygen) at the positive pole and an explosive, fixed (male) matter (potassium) at the negative pole, this would have been intuitively obvious to them.

Jons Jacob Berzelius[1,2] (1779–1848) was born in Stockholm one year before his great countryman Scheele discovered lactic acid in rotting milk. The title page shown in Figure 109 is from Berzelius' first book. He found lactic acid in muscle ("flesh juice") and this appeared in the second volume of the two-volume set (1806; 1808). Lactic acid was to play a critical role in the development of stereochemistry some 75 years into the future. Like Scheele before him, Berzelius is omnipresent in the chemistry of his day and, indeed, in our modern textbooks. He developed the abbreviations we use for the elements (H, C, and Po, which was subsequently changed to K, Cl, etc.) and wrote versions of our modern-day formulas in which the numbers attached to the elements were superscripted. The modern subscripted formulas such as H_2O were introduced by Liebig and Poggendorff in 1834.[1,2] Berzelius was a great systematizer of chemistry and is credited with the discoveries of selenium and thorium, a share in the discovery of cerium, the first identification of silicon (actually generated earlier by Gay-Lussac and Thenard but not identified), and the first isolation of zirconium and titanium as metals—they had been earlier identified as new elements in their combined states.[1,3] Actually, titanium, currently the "sexiest element in the whole Periodic Table" was really obtained as the pure metal for the first time in 1910.[4] He contributed major work in chemical analysis, including the new and complex realm of organic analysis, and his extraordinarily careful studies (as many as 30 replications) verified Dalton's law of multiple proportions and strengthened atomic theory. He also demonstrated that Dalton's and Berthollet's findings were mutually compatible, differentiated what he termed "empirical formulas" (e.g., C_2H_6O) from what he termed *rational formulas* (e.g., C_2H_4 + H_2O), and defined the terms *isomers* and *allotropes*. In 1827, Berzelius "asserted

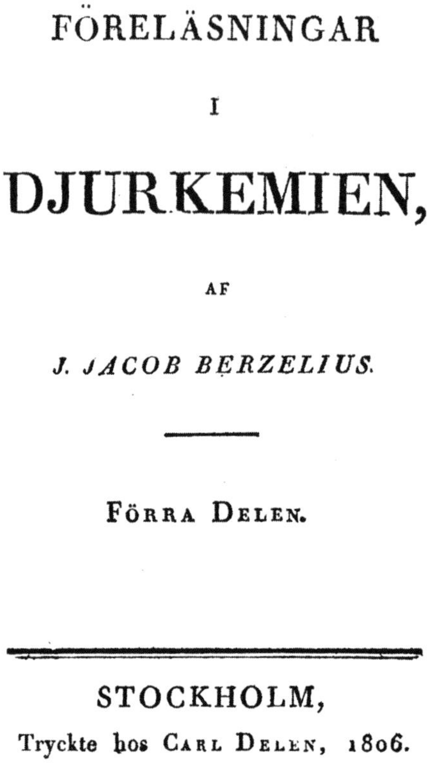

FÖRELÄSNINGAR

I

DJURKEMIEN,

AF

J. JACOB BERZELIUS.

FÖRRA DELEN.

STOCKHOLM,

Tryckte hos CARL DELÉN, 1806.

FIGURE 109 ■ Title page of Jons Jakob Berzelius' first book. He reports the isolation of lactic acid in "flesh juice" (muscle).

that a peculiar vital force intervenes in the formation of organic compounds and their preparation in the laboratory can hardly be expected."[1]

The central tenet of Berzelius' world view was the dualistic theory that still pervades our understanding of chemistry—particularly for ionic compounds such as sodium chloride. Briefly, table salt is composed of a positive part (Na^+) and a negative part (Cl^-). Such dualism was already part of Lavoisier's thinking some 30 years earlier:[1]

Acid = radical + oxygen
Base = metal + oxygen
Salt = base + acid

The term *radical*, introduced by de Morveau and employed by Lavoisier, is defined as "of or from the root"; "foundation or source of something."[5] Berzelius divided ponderable bodies into an electronegative class and an electropositive class. Substances of the electronegative class are attracted to the positive pole (following Davy's convention) and substances of the electropositive class are attracted to the negative pole (Berzelius had initially defined the poles differently but bowed to the widespread acceptance of Davy's definitions).[2] Although organic salts such as sodium acetate fit the dualistic concept, the vast majority of organics did not.

1. J.R. Partington, *A History of Chemistry*, MacMillan, London, 1964, Vol. 4, pp. 142–177.
2. W.H. Brock, *The Norton History of Chemistry*, Norton, New York, 1993, pp. 150–159.
3. A.J. Ihde, *The Development of Modern Chemistry*, Harper and Row, New York, 1964.
4. D. Rabinovich, *Chemical Intelligencer*, October, 1999, pp. 60–62. Professor Rabinovich likes the "iridium" or "palladium" card as successor to the "titanium" card.
5. J.R. Partington, op. cit., pp. 258–262.

ADAMS OPPOSES ATOMS

John Adams and Thomas Jefferson, the second and third Presidents of the United States, came from the two original "power colonies" Massachusetts and Virginia, respectively. They were allies and fundamental forces in the American Revolution, became bitterly estranged later on, but attained a reconciliation in old age.[1] Amazingly, the two men died on July 4, 1826, precisely 50 years after the signing of the Declaration of Independence. Unaware of his friend's fate, Adams' last words were: "Thomas Jefferson survives."

Both of these great leaders were invested in the intellectual life of their young nation. Figure 110 is the dedication page from the book *Plain Discourses on the Laws and properties of Matter: containing the elements or principles of Modern Chemistry, &c*, published in 1806 by Thomas Ewell, M.D. of Virginia, one of the surgeons of the United States Navy. In 1805, he had received from President Jefferson, the following letter:[2]

> Of the importance of turning a knowledge of chemistry to household purposes, I have been long satisfied. The common herd of philosophers seem to write only for one another. The chemists have filled volumes on the composition of a thousand substances of no sort of importance to the purposes of life; while the arts of making bread, butter, cheese, vinegar, soap, beer, cider, &c remain unexplained. Chaptal has lately given the chemistry of wine making; the late Dr. Penington did the same as to bread, and promised to pursue the line of rendering his knowledge useful to common life; but death deprived us of his labors. Good treatises on these subjects should receive general approbation.

When John Gorham assumed the Erving Chair of Chemistry at Harvard in 1817, he received a wonderful congratulatory letter[2] from John Adams. The retired President expressed the view that matter is "a mere metaphysical abstraction" and that he "could not comprehend" atoms and he "could not help laughing" at molecules. Near the end of his delightful letter he exhorts:

> Chymists! Pursue your experiments with indefatigable ardour and perserverance. Give us the best possible Bread, Butter, and Cheese, Wine, Beer and Cider, Houses, Ships and Steamboats, Gardens, Orchards, Fields, not to men-

TO

THOMAS JEFFERSON, *Esq.*

OF VIRGINIA,

THE PRESIDENT OF THE UNITED STATES OF AMERICA.

SIR,

TO inscribe this work to you, I was incited by an impulse given from a view of your station, as well as a sense of favors received. Raised by your own qualities, and the will of a free people, to the first place among them, the legitimacy of your title will be questioned by none.

IN preparing the following plain discourses, I was stimulated by a desire to imitate you in doing good. I was anxious to revolutionize the habits of many of our countrymen; to lessen their difficulties, by acquainting them with important improvements, and to diffuse more widely that genuine happiness derived from the interesting study of the ways of nature.

YOU, sir, have long since enjoyed the luxury of serving your countrymen.

WITHOUT expressing sentiments concerning your services as a statesman, in affairs better suited to my opportunities of observing,

FIGURE 110 ■ The dedication page from Thomas Ewell's *Plain Discourses on the Laws or Properties of Matter* (New York, 1806). President Jefferson had complained that "... chemists have filled volumes on the composition of a thousand substances of no sort of importance to the purposes of life. ..." He asked for a *useful* book and Dr. Ewell delivered.

tion Clothiers or Cooks. If your investigations lead accidentally to any deep discovery, rejoice and cry "Eureka!" But never institute any experiment with a view or a hope of discovering the first and smallest particles of Matter.

1. J.J. Ellis, *American Sphinx*, Knopf, New York, 1997, pp. 12, 290, 292.
2. E.F. Smith, *Old Chemistries*, McGraw-Hill, New York, 1927, pp. 50–52; 60–64.

THE CHEMICAL POWER OF A CURRENT OF ELECTRICITY

The nineteenth century was a period of specialization in the sciences. Organic, inorganic, physical, and analytical chemistries emerged as disciplines. It is noteworthy that Michael Faraday (1791–1867) first saw Humphrey Davy lecture at the Royal Institution in 1812, requested employment in his laboratory, and was appointed as laboratory assistant in 1813.[1] Davy was knighted in 1812 and married a wealthy widow during that year. Although he resigned his Professorship at the Royal Institution in 1813, he continued to visit and perform experiments and maintained his mentor relationship with Faraday. In 1813, Davy started a series of travels to the Continent, packing his chemical apparatus (he performed some experiments in his hotel rooms).[1] Although England and France were at war, there was always an eager audience for Davy. However, the complexities of the war situation had Davy appoint the young Faraday as his "temporary valet" —an appointment the regal and formal Lady Davy apparently took "too literally."[1] In 1815 Faraday received a higher position at the Royal Institution and started presenting public lectures. He started writing his first research papers in 1820, and, at his own request, they were edited by his respected mentor Davy. Although Faraday produced significant research in many areas of chemistry, his most important contributions were in electrochemistry. This was, of course, the field pioneered by Davy as well as Berzelius.

During the period 1831 through 1855 Faraday published a number of series of articles, "Experimental Researches in Electricity," in the *Philosophical Transactions of the Royal Society*. Partington notes that the major studies of electrolysis and the galvanic cell appeared between 1833 and 1840.[1] The most important discovery of these was the electrochemical equivalent:

> The chemical power, like the magnetic force, is in direct proportion to the absolute quantity of electricity which passes.

Figure 111 is from Faraday's Seventh Series of Lectures, presented to the Royal Society on January 9, 1834 and read on January 23, February 6 and 13, 1834. *Figs*. 64 to 66 (in Figure 111) are variants of the apparatus invented by Faraday to measure the quantity of gases generated by *electrolysis* (Faraday's term) of water. The gases were sometimes collected separately or together. He showed that the amount of water decomposed was directly proportional to the quantity of electricity employed, and he briefly defined "a degree of electricity" as that quantity that released 0.01 cubic inch of dry, mixed gas (corrected for temper-

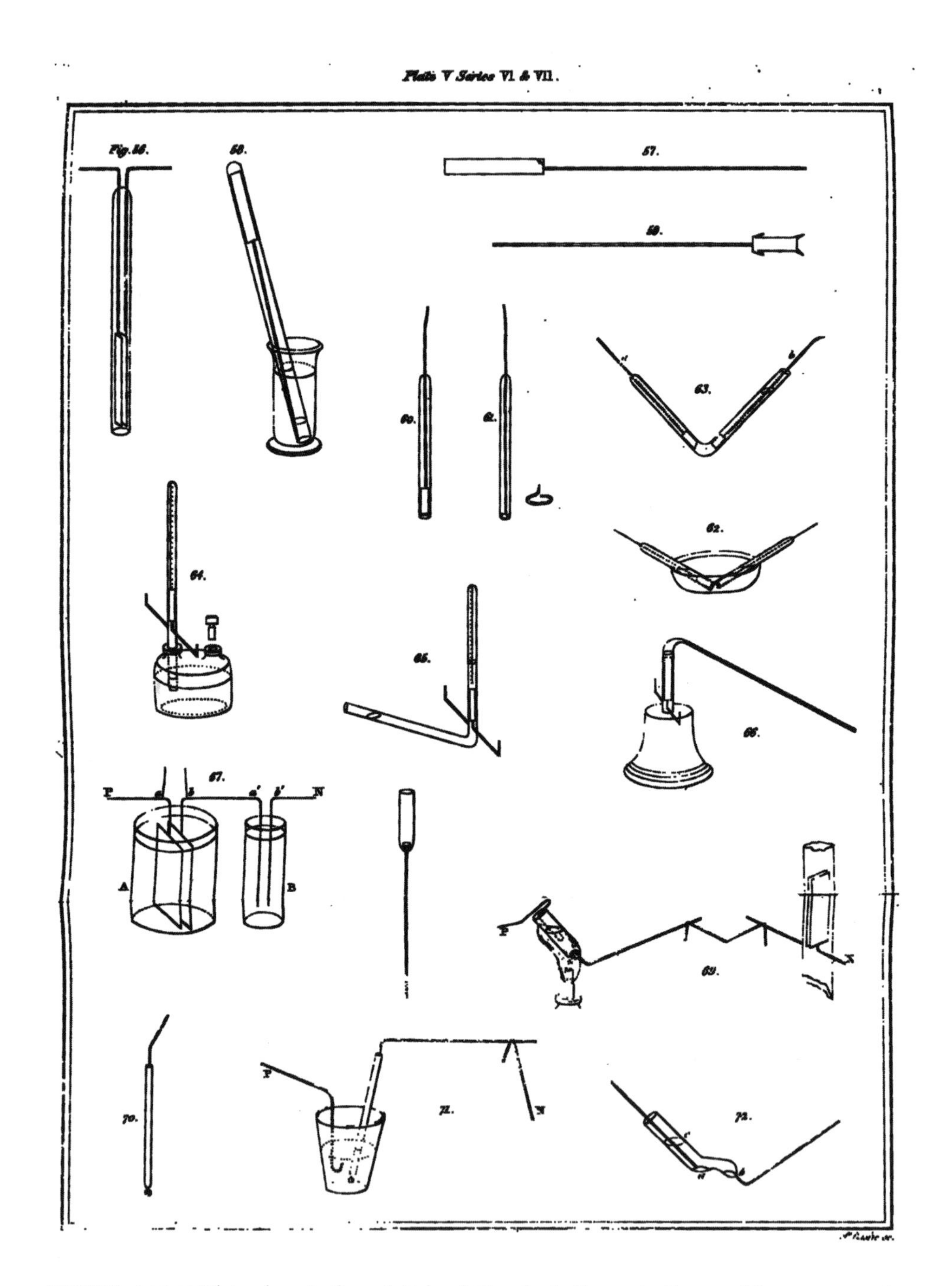

FIGURE 111 ■ This plate is from Michael Faraday's Seventh Series of Lectures to the Royal Society (January–February, 1834) and depict his Volta electrometer (coulometer since 1902) in which electrical current is measured by the volume of gas produced by electrolysis (Faraday's term) of water (see *64* to *66*). In *69* to *72* we see apparatus for electrolysis of melts. Faraday discovered that the mass of matter produced by electrolysis was proportional to the current and demonstrated electrical equivalents of matter.

ature and pressure).[1] He realized that this apparatus was useful in determining quantity of electricity and called it the "Volta-electrometer" (later termed by him the *voltameter*; since 1902 it has been termed the *coulometer*).[1]

Faraday discovered (accidentally) that ice is an insulator and also that, while salts are insulators, their melts are good conductors of electricity. *Figures* 69, *71*, and *72* depict three versions of Faraday's apparatus for electrolysis of molten salts. *Figure* 69 includes a glass tube into which a platinum wire with a bulb at one end is fused. The other platinum wire *P* is dipped into the molten salt. The apparatus was connected to a battery through a voltameter. Starting with molten stannous chloride ($SnCl_2$), chlorine is released, combines with stannous chloride, and forms hot gaseous stannic chloride ($SnCl_4$, boiling point 114°C), which is collected. Metallic tin deposits on the preweighed platinum wire. Once the apparatus is allowed to cool, fused $SnCl_2$ is scraped off of the wire, and the increase in the wire's mass due to tin plating is determined. Faraday found that 3.2 grains of tin were collected and this coincided with collection of 3.85 cubic inches of gas. On a scale of $H = 1$, the equivalent mass of tin [Sn(II) in stannous chloride] was found to be 58.53 (four determinations).[1] This is quite close to the modern value of 118.7/2 = 59.35). Although, like his mentor Davy, Faraday was uncomfortable with the reality of atoms, he was forced to conclude that:[1]

> The equivalent weights of bodies are simply those quantities of them which contain equal quantities of electricity. Or, if we adopt the atomic theory . . . the atoms of bodies which are equivalent to each other in their ordinary chemical actions, have equal quantities of electricity associated with them. But I must confess I am jealous of the term "atom"; for though it is very easy to talk of atoms, it is very difficult to form a clear idea of their nature, especially when compound bodies are under consideration.

In addition to the term *electrolysis*, with the collaboration of William Whewell, a broadly trained scholar, Faraday developed the terms *electrode*, *anode*, *ion*, *cathode*, *anion*, *cation*, and *electrolyte*.[2] He is considered to be the inventor of the test tube.[3] He made early studies of the liquefaction of gases. For example, when a syringe was used to compress chlorine gas into a tube, a small amount of oily, green liquid was formed. He also used the newly discovered solid carbon dioxide in a bath of acetone (1835, by Thilourier; compression of CO_2 into a liquid and rapid expansion of the liquid rapidly cools the substance and forms dry ice[1]) to achieve a temperature of −78°C. This permitted Faraday to liquefy ethylene and other low-boiling-point gases using high pressure and cooling and led him to conclude that certain gases, such as hydrogen, were "permanent gases." For example, the critical temperature of methane (T_c) is equal to −82.6°C. At this temperature, the critical pressure (P_c) of 45.4 atm (4.60 MPa) will condense it to a liquid. However, at −78°C, no amount of pressure will condense methane, and hence it is a "permanent gas" at this and higher pressures.

1. J.R. Partington, *A History of Chemistry*, MacMillan, London, 1964, Vol. 4, pp. 99–128.
2. A.J. Ihde, *The Development of Modern Chemistry*, Harper and Row, New York, 1964, pp. 133–138.
3. W.H. Brock, *The Norton History of Chemistry*, Norton, New York, 1993, p. 191.

A PRIMEVAL FOREST OF THE TROPICS

"Organic chemistry appears to me like a primeval forest of the tropics, full of the most remarkable things" wrote Friedrich Wöhler to Berzelius in 1835.[1] I remember receiving from my father his personal copy of Karrar's *Organic Chemistry* (3rd ed., 1947) on the eve of taking my first organic chemistry course along with the admonition that I needed to learn everything in it (almost 1000 large pages). Thirty-five years later I realize that he may have been pulling my leg a bit, but it took me almost half the semester to gain my footing in the course. I thus have a great deal of empathy with students in my own organic chemistry course.

Organic chemistry is the chemistry of carbon compounds. We note that minerals such as carbonates are not considered to be organic, nor are certain gases such as carbon dioxide (or carbon monoxide) that may be derived from them. Although Lavoisier did not make this differentiation, organic chemistry was regarded as different from the remainder of chemistry and, through the early nineteenth century, relegated to descriptive sections on "Animal Chemistry" and "Vegetable Chemistry" in chemistry texts. The sheer complexity of the mixtures, the complexities of the formulas, and the fact that organic compounds did not obey the dualism seen for inorganics such as water or sodium chloride added to these conceptual problems.

I visited the web page of Chemical Abstracts Service (CAS) on May 24, 1999. At precisely 11:17:11 A.M. EDT, there were 19,632,211 registered substances of which some 12 million were organic (68%). The remaining substances were biosequences (17%), coordination compounds (6%), polymers (4%), alloys (3%), and tabular inorganics (2%). Of these, about 160,000 substances are of sufficient practical importance to be on national or international chemical inventories and registry lists. As of 1997, 1.3 million new substances were being added to the list each year. The cause of these daunting numbers and incredible diversity is, primarily, the carbon atom, which can form bonds with almost all other elements including other carbons. It forms four bonds in combinations of single, double, and triple bonds, as well as chains, rings, and cages. Two minutes later (11:19:12 A.M. EDT), there were 19,632,221 substances (10 new ones!) in the CAS registry.

Figure 112, from the Youmans 1857 edition of the *Chemical Atlas* depicts the atmospheric part of the carbon cycle involving plants and animals. On the right we see animals that inhale (arrows down) oxygen (note it is written as monoatomic rather than O_2) with the food that nourishes them to produce carbon dioxide and water. In the Youmans text, confusion reigning for 50 years persists in the formula for water (HO) and atomic weights for oxygen (8), carbon (6), sulfur (16), and others. The formula for ammonia (NH_3) is correct and the atomic weight of nitrogen is correctly 14. On the left are plants that incorporate carbon dioxide and water (arrows down) to produce oxygen (arrows up).

The confusion so evident in Youmans's book over formulas, atomic weights, isomers, and valence will all clear up within the following ten years or so.

1. J.R. Partington, *A History of Chemistry*, MacMillan, London, 1964, Vol. 4, p. 233.

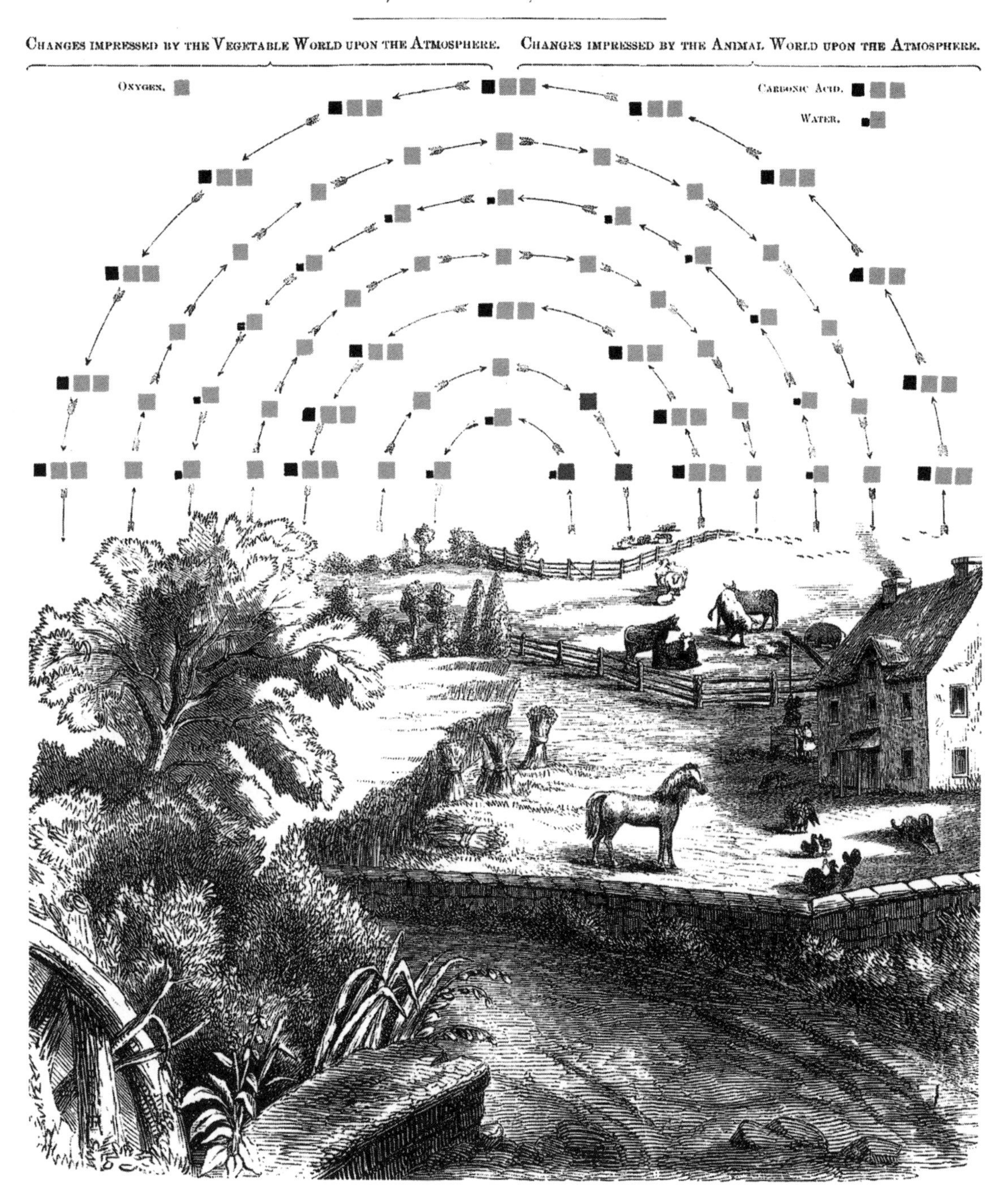

FIGURE 112 ■ This beautiful hand-colored figure (see color plates) is from the 1857 edition of Edward Youmans' *Chemical Atlas* (New York, first published in 1854).

TAMING THE PRIMEVAL FOREST

In nature, organic compounds are usually found in incredibly complex mixtures. Destructive distillation of a sample of coal produces hundreds of compounds in easily measurable quantities and thousands of compounds if one wishes to measure trace levels. If a chemist wishes to determine the formula of a compound, it must first be separated from other compounds and rigorously purified. Even today, absolute separation of compounds cannot always be achieved using *fin-de-millennium* techniques.

Lavoisier did not believe that organic compounds were outside the normal realm of chemistry and analyzed the amount of oxygen consumed and carbon dioxide formed in the combustion of charcoal using the usual apparatus (e.g., Fig. 90, *Fig. 1*). He also burned alcohols, fats, and waxes.[1] However, his data on the composition of H_2O and CO_2 were inaccurate:[1]

	Lavoisier	Correct
CO_2	28% C; 72% O	27.2% C; 72.8% O
H_2O	13.1% H; 86% O	11.1% H; 88.9% O

These errors may appear to be negligible. While they would be for determining simple formulas such as CH_4, the errors would be significant for formulas such as $C_{18}H_{38}O$ and would interfere with the understanding of carbon's valence.

Gay-Lussac and Thenard made the first accurate determinations of carbon content of organic compounds by using potassium chlorate ($KClO_3$) as the oxidizing agent.[1] The sample for analysis and potassium chlorate were pressed together into a pellet, which was dropped carefully into a vessel heated by charcoal. The resulting CO_2 was absorbed by potash. They eventually replaced $KClO_3$ with cupric oxide (CuO), which was safer and did not oxidize organic nitrogen. Apparatus continued to evolve notably due to improvements by Berzelius.[1]

Justus Liebig (1803–1873) developed the method for C, H, and O analysis essentially in use today.[1–3] His first book on organic analysis was published in 1837 and is quite rare. Figure 113(a) is from the first English edition, *Hand-Book of Organic Analysis* (London, 1853). Liebig notes that organic substances often absorb water and that they must first be free of water prior to analysis. Figure 113(a) depicts a drying apparatus. The siphon attached to the three-necked flask on the right draws off water, creating a slight vacuum that pulls air through drying tube C (filled with calcium chloride) on the left. The sample itself is in a tube A, which is not seen here because it is in the hot bath above the furnace. It is connected by glass tubing to C as well as to tube *D*, which condenses any water released from the sample. Tube A is periodically removed from the heat and weighed until no further change occurs. Tube *D* can also be weighed if necessary to determine water content.

Figure 113(b) depicts Liebig's *kaliapparat* (A, *kalia* refers to potassium and the potash solutions that occupy the three lower bulbs of this five-bulbed piece

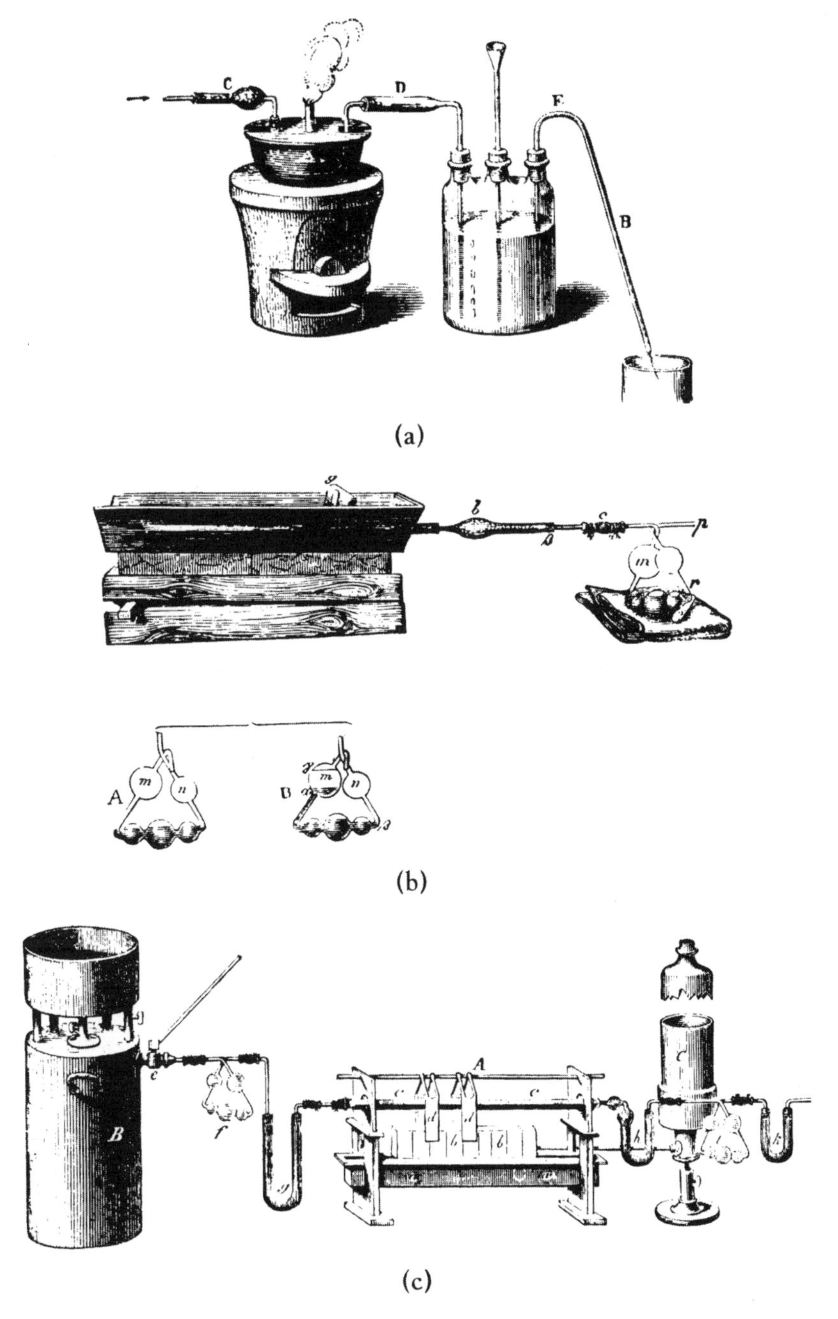

FIGURE 113 ■ Figures from Justus von Liebig's *Handbook of Organic Analysis* (first English Edition, London, 1853; Liebig's first book on organic analysis was published in 1837 in Braunschweig and is extremely rare): (a) Apparatus for quantitative drying of organic substance to be analyzed; (b) Apparatus for carbon and hydrogen determination; Liebig's ingenious *kaliapparat*, the five-bulbed glassware containing potash solution in the lower three bulbs for quantitation of CO_2 is shown as part of the apparatus and separately; (c) Apparatus capable of using pure oxygen as well as air for carbon/hydrogen analysis.

of glassware). The sample for analysis is placed in a hard-glass tube sitting in an iron trough heated over a flame.[1,2] Water derived from combustion is trapped in the preweighed drying tube to the right of the iron trough. Carbon dioxide is trapped in the potash solutions present in the lower three bulbs of the *kaliapparat*. The upper two bulbs at both ends of the *kaliapparat* serve two functions: as depicted in Figure 113(b), prior to the start of the experiment, some air is drawn out by mouth from the apparatus and potash solution climbs into bulb *m*. If the vacuum is maintained and the liquid level in *m* does not drop, then there are no leaks in the apparatus. These bulbs also prevent loss of potash solution due to splashing. The preweighed *kaliapparat* is followed by a preweighed drying tube that traps any water vapors lost from the *apparat*.[1,2] The apparatus depicted in Figure 113(c) employs pure oxygen for combustion—the oxygen is generated in *B*, passed through the *kaliapparat f* containing concentrated sulfuric acid, and then tube *g* is filled with calcium chloride. The dry oxygen is introduced to combustion tube *cc* in a bed of magnesium oxide in an iron trough. The combustion tube has a thick plug of copper turnings at the left and is two-thirds filled with copper oxide. Dried atmospheric air, free of carbon dioxide can be introduced using the apparatus on the right.

Here are the results of a state-of-the-art analysis reported by Adolph Strecker in Liebig's laboratory at the University of Giessen in 1848:[3] the formula for cholic acid was found to be $C_{48}H_{39}O_9$ (with the atomic weights C = 6; O = 8). The present day formula is $C_{24}H_{40}O_5$ (C = 12; O = 16).[3] It is obvious that the results are accurate but not enough to "hit the formula on the head." Yet that is precisely what is needed to make sense of carbon's valence.

Liebig was born and raised in rather poor circumstances. He was an intense, irascible man who, as a student, was arrested for his political activities. He was sponsored by Karl Wilhelm Kastner (1783–1857) at the University of Bonn and later Erlangen. Kastner persuaded the Erlangen faculty to award Liebig an honorary doctorate *in absentia* in 1822. As Brock states:[3] "It is one of the ironies of Liebig's teaching career that he himself never presented a thesis for his doctorate." He engaged in acrimonious debates throughout his career, was unkind in his later criticisms of his kind patron Kastner, and did not hesitate to attack Friedrich Wöhler when the two found the same formula for Liebig's silver fulminate and Wöhler's silver cyanate. The controversy was settled when the two performed their analyses together, discovered the first example of isomerism, and began one of the greatest friendships in the history of chemistry. The gentle and wise Wöhler counselled the "Type-A" Liebig in 1843 thusly:[1]

> To make war against Marchand, or indeed against anyone else, brings no contentment with it and is of little use to science. You merely consume yourself, get angry, and ruin your liver and your nerves—finally with Morrison's pills. Imagine yourself in the year 1900, when we are both dissolved into carbonic acid, water, and ammonia, and our ashes, it may be, are part of the bones of some dog that has despoiled our graves. Who cares then whether we have lived in peace or anger; who thinks then of the polemics, of the sacrifice of thy health and peace of mind for science? Nobody. But thy good ideas, the new facts which thou hast discovered—these, sifted from all that is immaterial, will be known and remembered, to all time. But how comes it that I should advise the lion to eat sugar?

Liebig had joined the Giessen faculty in 1824 and became Co-Editor of the *Magazin fur Pharmazie*. In 1832, he assumed sole editorship, changed the title to *Annalen der Chemie und Pharmazie* and the tough, caustic "lion" made it a vital chemical journal. He built a renowned research and teaching school at Giessen and by 1852 he had influenced about 700 students of chemistry and pharmacy.[3] In that year, he moved to the University of Munich but his health no longer permitted him to work in the laboratory. His intensity probably contributed to his poor health and he spent his final 20 years in bitter chemical controversies.[3] His friend Wöhler, whose sense of humor was reflected in an 1843 paper he published in the *Annalen* under the pseudonym "S.C.H. Windler,"[4] lived to reach the age of 82. He trained over 20 American chemists. Among these were Ira Remsen, who started at Johns Hopkins University the first American Ph.D. program in chemistry as well as Edgar Fahs Smith at the University of Pennsylvania.[5] It is delightful to realize that Wöhler synthesized urea in 1828 and E.F. Smith, his student, published the delightful book *Old Chemistries* in 1927.

1. A.J. Ihde, *The Development of Modern Chemistry*, Harper and Row, New York, 1964, pp. 173–183.
2. J.R. Partington, *A History of Chemistry*, MacMillan, London, 1964, Vol. 4, pp. 234–239.
3. W.H. Brock, *The Norton History of Chemistry*, Norton, New York, 1993, pp. 194–207.
4. W.H. Brock op. cit., p. 218.
5. A.J. Ihde, op. cit., p. 264.

THE ATOMIC WEIGHT OF CARBON AND RELATED CONFUSIONS

Confusion over molecular formulas and atomic weights was an unfortunate by-product of the early atomic theory. Dalton's Rule of Greatest Simplicity provided incorrect formulas such as HO for water and NH for ammonia. Although Gay-Lussac's Law of Combining Volumes (1808), Avogadro's hypothesis (1811), the Law of Dulong and Petit (1819), and other studies began to clear up the confusion, it was not until Cannizzaro's 1858 paper and the 1860 Karlsruhe Conference that atomic formulas, equivalents, and atomic weights were really clarified.

Figure 114 is from Edmund Youmans' *Chemical Atlas* (New York, 1854; 1857 printing). It exemplifies the continuing confusion over the atomic weights of carbon and oxygen relative to hydrogen (assumed to be 1). Thus, Dulong determined that the ratio of densities of CO_2/O_2 = 1.38218.[1] Therefore, the same volume of oxygen gas containing 100 g would contain 138.218 g of carbon dioxide. If one accepts Avogadro's hypothesis, then the mass ratio of oxygen to carbon is 100.00/38.218. Using the assumption of Gay-Lussac and Dumas, the formula of "fixed air" being CO not CO_2, the atomic weight of carbon would be 6.12 if oxygen is 16.0. Berzelius determined that fixed air is CO_2 and assigned an atomic of 12.24 to carbon. However, in 1840 Dumas and Stas published their very precise studies of the combustion of purified graphite in a stream of pure

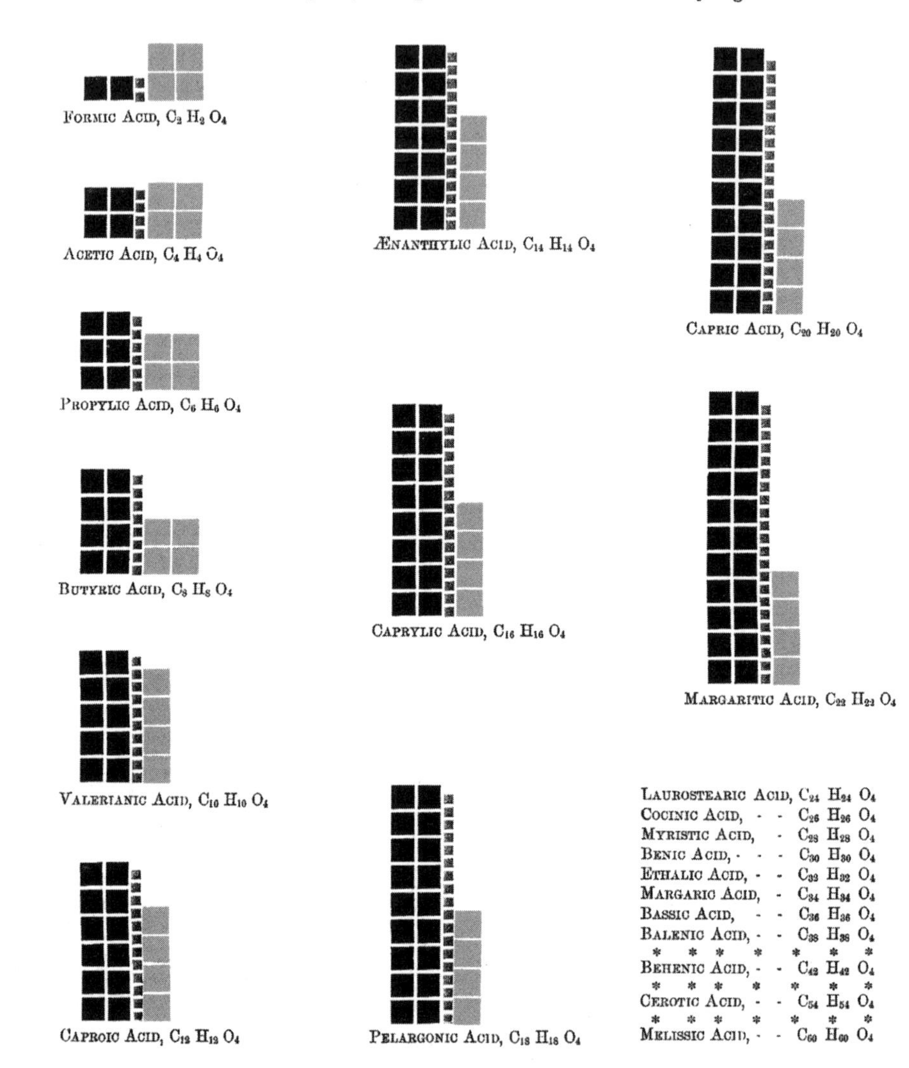

FIGURE 114 ■ Plate from Youmans' *Chemical Atlas* (Fig. 112). The organization of the "Primeval Forest" of organic chemistry by Laurent and Gerhardt included the concept homology. The units of homology are CH_2 rather than C_2H_2 as shown. The confusion was the result of discrepancies in the atomic weight of elements and assumed formulas. These would be cleared up very shortly in the Karlsruhe Congress of 1860.

oxygen. Weighing any unburned ash, they determined carbon's atomic weight at 12.0 (if oxygen = 16.0 and fixed air is CO_2). Nevertheless, it is clear from Figure 114 that confusion continued for about 20 more years. The problems of formulas and atomic weights would only be settled in Karlsruhe. In Youmans' *Chemical Atlas* the atomic weights of carbon (6) and oxygen (8) are half of the accepted (post-Karlsruhe) values while nitrogen is correct at 14. Thus, acetic acid, given as $C_4H_4O_4$ in Figure 114 is really $C_2H_4O_2$, butyric acid is really $C_4H_8O_2$, and the common difference in a homologous series is CH_2, not C_2H_2.

1. A.J. Ihde, *The Development of Modern Chemistry*, Harper and Row, New York, 1964, pp. 183–184.

WHY'S THE NITROGEN ATOM BLUE, MOMMY?

The beautiful hand-colored plates in Youmans' *Chemical Atlas* (e.g., Figure 114) have unique colors representing individual atoms. These are explained by Youmans: oxygen changes the color of blood to bright red in the lungs, hence it is represented as red; the sky, which is 80% nitrogen, is blue, thus nitrogen atoms are colored blue; carbon, sulfur, and chlorine are depicted in their natural elemental colors of black (sort of), yellow, and green. It is "cool" that most molecular models, made of wood, metal, or plastic, keep the same color scheme. Moreover, most modern computer programs that model molecules represent oxygen as blood red and retain the other traditional colors. I am further reminded of such traditions when I recall a professor's quip at a chemical meeting that "everybody knows that *p* orbitals are blue and green"—a reference to the influence of and colors used in the work on orbital symmetry by Robert Burns Woodward and Roald Hoffmann.[1]

1. R.B. Woodward and R. Hoffmann, *The Conservation of Orbital Symmetry*, Verlag Chemie, Weinheim, 1970.

I CANNOT HOLD MY CHEMICAL WATER—I CAN MAKE UREA!

Figure 115 is from Edward L. Youmans' *Chemical Atlas*[1] and depicts concepts of isomerism from the mid-nineteenth century. The field of organic chemistry is vast. As of the year 2000, there should be over 13 million known organic compounds. This enormous diversity is due, in large part, to the occurrence of isomers—molecules with the same formula but different arrangements of atoms.

In the 1820s two great organic chemists, Justus Liebig and Friedrich Wöhler, discovered that two very different substances, silver fulminate and silver cyanate, respectively, had the same composition. Liebig initially attacked (after fulminating?) Wöhler's results but after meeting and comparing results, they agreed that they were consistent—and thus *very* confusing.[2,3] The quandary was essentially resolved by Berzelius in 1830, who came up with the concept of

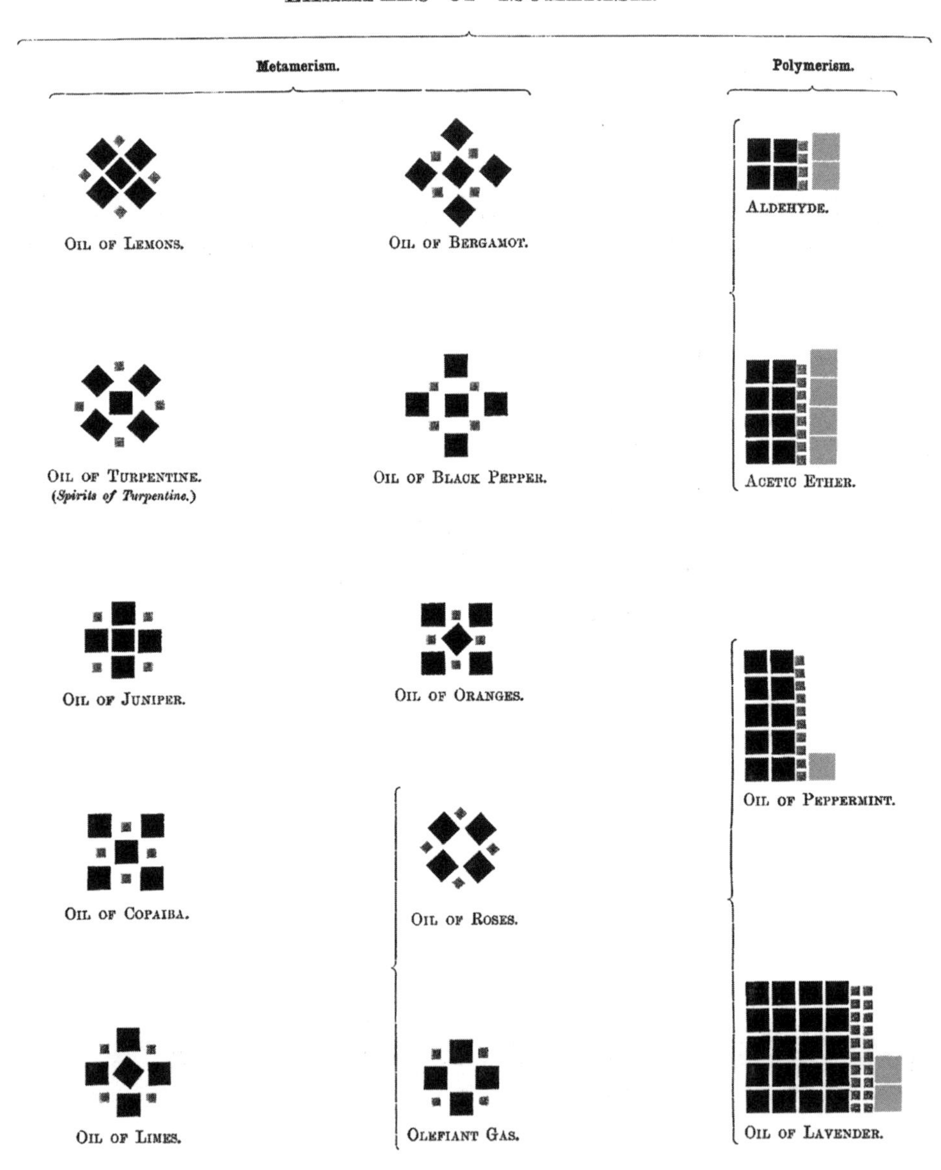

FIGURE 115 ■ Plate from Youmans' *Chemical Atlas*. Although Wöhler and Liebig discovered that silver fulminate and silver cyanate were isomers (term coined by Berzelius in 1830) and it was suspected that the origin was the different arrangement of atoms, the concept of valence remained to be discovered. In this plate Youmans depicts isomers as different arrangements of atoms. But he postulates that their chemical history plays a role. For example, since the atomic arrangements in the carbon allotropes graphite and diamond are different (presumably different charcoals are also allotropes), then hydrocarbon isomers (butane and isobutane, for example) maintain the different carbon arrangements of the allotropes from which they were (presumably) derived.

isomers.[2] However, he differentiated isomers that were *metamers*, basically similar to our modern concept of isomers, and *polymers*, which had the same formula but different densities. Thus, the density of gaseous butylene (C_4H_8) is double that of gaseous ethylene (C_2H_4) even though the two have the same composition (85.7% C; 14.3% H or CH_2).[2] In Youmans' text, isomers are considered to arise from different allotropes of carbon.

It is interesting that one is more likely to find information on fulminates in a history of chemistry book than in a chemistry text. There are two readily accessible fulminates,[4] mercury fulminate, which has been used as a "primer" in percussion caps, and silver fulminate, which is considered to be too dangerous for use (kudos, therefore, to the famous Liebig!). Solid silver fulminate has the structure Ag-CNO (for some time in the twentieth century, it was *thought* to be CNO-Ag).[5] The structure of solid silver cyanate is Ag-NCO.[5,6] And here is some interesting irony. The cyanato ligand (NCO^-) could, in principle, combine with metals at N or O. Although the solid Ag-NCO should be named silver *isocyanate*, Brittin and Dunitz[6] bowed to history and indicated tolerance or at least resigned acceptance for retaining Liebig's nomenclature since the NCO-Ag isomer (the *real* silver isocyanate) is unknown.

In 1828, Wöhler attempted to synthesize ammonium cyanate (NH_4OCN) and found instead a substance having the same formula as the target compound but identical in all of its properties to urea (H_2NCONH_2). Urea is a component of mammalian urine and Wöhler wrote to his mentor Berzelius: "... I cannot, so to say, hold my chemical water, and must tell you that I can make urea, without thereby needing to have kidneys, or anyhow, an animal, be it human or dog ...."[7] This was the beginning of the end for the theory of Vitalism, which held that "organic" substances have a kind of vital force, since they had always been isolated from or at least related to living organisms. Thus, they could not be synthesized from nonorganic (really, elemental) substances.

Actually, Wöhler apparently *may* have first made ammonium cyanate as he had intended.[8] However, upon heating and evaporating off water, ammonium cyanate isomerizes *in solution* to urea.[8] Two years later, in 1830, Liebig and Wöhler actually synthesized solid ammonium cyanate by reaction of dry ammonia and cyanic acid.[9] In a sealed vessel under ammonia the crystals are now known to be stable, but if the vessel is opened, conversion to urea is complete in two days.[9] Dunitz and colleagues[9] also determined the structure of ammonium cyanate by x-ray crystallography. They note the difficulty in differentiating the N versus O end of cyanate by x-rays even using end-of-twentieth-century technology. The structure is found to be NH_4NCO. Just as in the silver case, in a formal sense, the solid could be called ammonium isocyanate but it is not.

Wöhler is said to have remained a believer in Vitalism.[8] It has been noted that just as Priestley, the phlogistonist, discovered oxygen, which was the downfall of phlogiston theory, Wöhler, the Vitalist, synthesized urea, which was the beginning of the downfall of Vitalism.[8] The true end for Vitalism came in the 1840s when the German chemist Hermann Kolbe effectively demonstrated the synthesis of acetic acid (the active component of vinegar, which is related to wine, which is related to sugar—therefore ORGANIC and ALL-NATURAL) from its constituent chemical elements through the following sequence:[10]

$$H_2 + O_2 \rightarrow H_2O$$
$$FeS_2 + C \rightarrow CS_2 + Fe$$
$$CS_2 + 2Cl_2 \rightarrow CCl_4 + 2S$$
$$2CCl_4 \rightarrow C_2Cl_4 + 2Cl_2$$
$$C_2Cl_4 + 2H_2O + Cl_2 \rightarrow CCl_3CO_2H + 3HCl$$
$$CCl_3CO_2H + 3H_2 \rightarrow CH_3CO_2H + 3HCl$$

where CH_3CO_2H is acetic acid.

There remained other issues to clarify in structural chemistry. In 1841, Berzelius developed the concept of allotropism—different arrangements of atoms in pure elements.[2] Modern examples include oxygen (O_2) and its allotrope ozone (O_3), sulfur, which commonly has octagons of sulfur atoms but can be heated to form "plastic sulfur" (long chains of sulfur atoms), and carbon, which is considered to have three allotropes (graphite, "infinite" sheets of carbon atoms; diamond, an "infinite" three-dimensional network of carbon atoms; and fullerenes such as C_{60}, buckminsterfullerene, a soccer-ball arrangement of carbon atoms). One could imagine each significant fullerene (C_{60}, C_{70}, C_{84}) as an allotrope. The confusing issue of different crystalline arrangements of the same substance was also solved by Berzelius who termed them *polymorphs*.[2]

We end this essay on an amusing note. John Darby, Professor of Chemistry and Natural Sciences in East Alabama College, describes isomeric bodies in his 1861 *Text Book of Chemistry*.[11] He correctly notes that ethyl formate and methyl acetate are "isomeric bodies" (their formulas are both $C_3H_6O_2$). He then goes on to say: "An explanation of these phenomena that attributes them to a different arrangement of atoms, is not satisfactory, as elementary bodies assume different states in inorganic chemistry, which is called allotropism, when such a cause is evidently impossible. We can only refer it, at present, to the will of the Creator."

1. E.L. Youmans, *Chemical Atlas*, Appleton, New York, 1857.
2. A.J. Ihde, *The Development of Modern Chemistry*, Harper and Row, New York, 1965, pp. 170–173.
3. W.H. Brock, *The Norton History of Chemistry*, Norton, New York, 1993, pp. 201–202; 214.
4. A.G. Sharpe, *Comprehensive Coordination Chemistry*, G. Wilkinson, R.D. Gillard, and J.A. McCleverty (eds.), Pergamon, Oxford, 1987, Vol. 2, pp. 12–14.
5. K. Vrieze and G. Van Koten, *Comprehensive Coordination Chemistry*, G. Wilkinson, R.D. Gillard, and J.A. McCleverty (eds.), Pergamon, Oxford, 1987, Vol. 2, pp. 227–236.
6. D. Brittin and J.D. Dunitz, *Acta Crystallographica*, **18**: 424–428, 1965.
7. P.S. Cohen and S.M. Cohen, *Journal of Chemical Education*, **73**: 883–886, 1996. I am grateful to Dr. Daniel Rabinovich for bringing this paper and the one in Reference 7 to my attention.
8. G.B. Kauffman and S.H. Chooljian, *Journal of Chemical Education*, **56**: 197–200, 1979.
9. J.D. Dunitz, K.D.M. Harris, R.L. Johnston, B.M. Kariuki, E.J. MacLean, K. Psalidas, W.B. Schweitzer, and R.R. Tykwinski, *Journal of the American Chemical Society*, Vol. **120**: 13 274–13 275, 1998.
10. W.H. Brock, op. cit., pp. 620–621.
11. J. Darby, *Text Book of Chemistry—Theoretical and Practical*, Cooper, Savannah and Barnes & Burr, New York, 1861, pp. 275–276.

TWO STREAMS IN THE PRIMEVAL FOREST

The darkness in Wöhler's "primeval forest" only deepened during the 1840s and early 1850s as the complexity of organic chemistry became ever more apparent.[1,2] Ironically, the vast majority of organic compounds are composed of only four elements: carbon, oxygen, hydrogen, and nitrogen.

For starters, there remained three sets of atomic masses for H, C, and O:[1,2] 1, 12, 16, Berzelius; 1, 6, 8, Liebig; 1, 6, 16, Dumas. Berzelius' Dualistic Theory of chemistry remained an important organizing principle. Today we recognize that the vast majority of organic compounds, such as ethyl alcohol (C_2H_6O), are held together by covalent (electron-sharing) bonds; simple inorganic salts, such as sodium chloride, are composed of ions held together by electrostatic forces. However, it was not until 1884 that Svante August Arrhenius recognized ions as real entities independent of electrochemistry.[3]

The work of Davy and Berzelius clearly established the importance of electrical forces in holding compounds such as sodium chloride and water together. Electropositive elements substituted for other electropositive elements (e.g., HCl, KCl, $MgCl_2$); electronegative elements substituted for electronegative elements (e.g., Na_2O, NaCl, NaBr). In 1815, Gay-Lussac's studies of prussic acid (HCN) led him to discover cyanogen, $(CN)_2$, and a series of other compounds such as potassium cyanide (KCN) and silver cyanide (AgCN), which kept the CN radical intact as if it were an "atom." Indeed, cyanogen seemed to be as "elementary" as Cl_2.[1] Even more complex radicals were soon discovered: in studies of benzaldehyde and derivatives published in 1832, Liebig and Wöhler discovered the benzoyl radical (modern formula C_7H_5O)—exciting because it appeared to be an "intact" unit of three elements. But many vexing problems were coming to the fore, for example:

1. Gay-Lussac found that reaction of prussic acid (HCN) with chlorine gas produced cyanogen chloride (ClCN). How could an electronegative element replace an electropositive one?
2. Similarly, how is it that the hydrogen in chloroform ($CHCl_3$) can be replaced by chlorine to produce CCl_4?
3. If benzoyl radical combines with chlorine to form benzoyl chloride, it should be electropositive. How can it include the electronegative element oxygen?
4. We understand that $SO_3 + H_2O \rightarrow H_2SO_4$. We know that $C_2H_4 + H_2O \rightarrow C_2H_6O$ (ethyl alcohol) and with HCl, C_2H_4 (the radical "etherin") forms C_2H_6Cl (ethyl chloride). But why does ethyl alcohol appear to release C_2H_5 radicals when reacted with sulfuric acid to form "sulfuric ether" ($C_4H_{10}O$)? Do the molecules break into different radicals when they please?

Figure 116 is from Youmans' *Chemical Atlas*. The year of this printing, 1857 (first printing 1854), occurs near the end of this chaotic period. The figure illustrates the prevailing confusion in theories as well as atomic weights.[1,2,4]

Brock's book has very nicely captured August Kekulé's metaphor of two streams of thought in the organization of organic chemistry.[2] The top illustration in Figure 116 shows the theory of compound radicals largely developed by Ber-

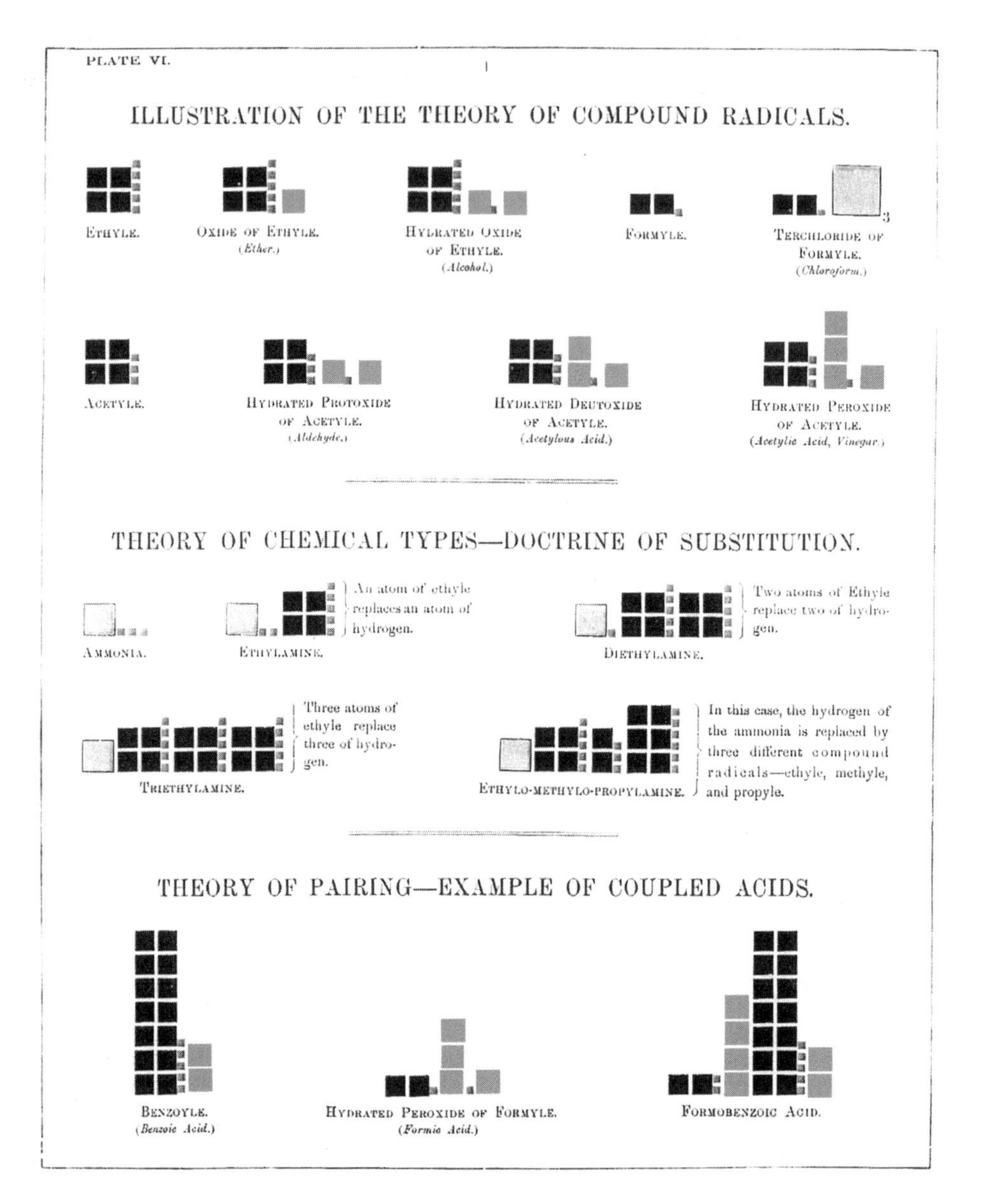

FIGURE 116 ■ Plate from Youmans' *Chemical Atlas* (original in color) depicting the three prevailing theories of organic chemistry structure in reactivity prior to Karlsruhe.

zelius and favored by the German and English schools. In this stream of thought, organic radicals that exist independently join each other to form organic compounds. These radicals can be ethyl (C_2H_5, represented as C_4H_5 using atomic weight of C = 6 and the double carbon atom favored in Germany and used in Youmans' book) and hydroxyl (OH, represented as HO_2, using the atomic weight of oxygen as 8). The "formyl" radical, here described as C_2H–meaning CH–can combine with three chlorine radicals to form chloroform.

The middle illustration depicts the other stream of thought, the "theory of types," advanced initially by Dumas, in which compounds are related to a Chemical Type (or Class). Classes make sense. Acids, which share H in common, comprise a class—HCl, HBr, HCN. Similarly, salts such as NaCl, KCl,

$MgCl_2$, NaBr, NaCN, and Na_2O also form a class. Here things get a bit "dicey" in light of being able to replace H by Cl (see above). Indeed, in an 1840 issue of *Annalen der Pharmacie und Chemie*, a certain S.C.H. Windler (also known as Friedrich Wöhler) published a satirical paper in which he reported replacing, in logical stages, all of the atoms in manganous acetate ("MnO + $C_4H_6O_3$") by chlorine, thus "demonstrating" that manganous acetate, a salt, and chlorine, a gas, were of the same "type."[1] Students saddled with the "Amines I" chapter of a typical modern-day organic text will have no trouble recognizing this figure: ethylamine, diethylamine, and triethylamine (and countless other amines) are in the "ammonia-type class" because they can be directly derived from ammonia. The "water-type class" was much more complicated. It included alcohols, ethers, carboxylic acids, esters, and acid anhydrides (at least five chapters in the modern Organic Chemistry text) and even extended to acids such as sulfuric and phosphoric. Gerhardt ultimately recognized four fundamental types: the nitrogen-type (amines and amides); water-type (see above); hydrogen-type (hydrocarbons, ketones, aldehydes); and hydrogen chloride-type (alkyl chlorides, acid chlorides, and related bromides).

The Theory of Pairing (Fig. 116, bottom) was introduced by Berzelius to modify his radical theory. The idea was that one radical in the original compound retains its character in the new compound while the other changes its character (is "copulated") in the new compound, through substitution or rearrangement. This was used to explain, for example, Dumas' discovery in 1838 that chlorination of acetic acid produced trichloroacetic acid having essentially the same properties.[2] The acidic radical remained essentially constant, while the other associated radical changed.

Actually, the organization of organic chemistry by Auguste Laurent and Charles Frederic Gerhardt, *Les Enfants Terribles*,[2] ultimately more or less brought together aspects of the radical and type theories into the ten-pound organic texts lugged by the pre-med students of today and the multi-ounce CD ROMs of next year.

1. A.J. Ihde, *The Development of Modern Chemistry*, Harper and Row, New York, 1964, pp. 203–216.
2. W.H. Brock, *The Norton History of Chemistry*, Norton, New York, pp. 210–240.
3. A.J. Ihde, op. cit., pp. 413–415.
4. J.R. Partington, *A History of Chemistry*, MacMillan, London, 1964, Vol. 4, pp. 352–464.

WANT A GREAT CHEMICAL THEORY? JUST LET KEKULÉ SLEEP ON IT

Toward the end of the 1850s, two major advances occurred to begin the taming of the "primeval forest." In 1858, Stanislao Cannizzaro emphasized the importance of Avogadro's Hypothesis, first published in 1811, that equal volumes of gas (same temperature and pressure) had equal numbers of ultimate units (molecules of Cl_2, O_2, P_4; atoms of Hg vapor). Using the ideal gas law ($PV = nRT$) and the Dumas technique, the atomic masses of atoms not volatile in the elemental state could also be measured in molecules if the other atoms' weights were known. Thus, at the end of the decade and at the Karlsruhe Conference in 1860, the atomic mass problem was largely solved. Not only did this set the table for the periodic law, but it also brought coherence to chemical formulas.

The other major advance was the realization that carbon is tetravalent, first enunciated narrowly by Friedrich August Kekulé (1829–1896) in 1857, and then for all carbon-containing compounds in 1858.[1] Kekulé started at the University of Giessen as an architecture student before he was drawn into chemistry by Liebig. The concept of valence, sometimes credited to Kekulé for work published in 1854, appears to be due to William Odling and others a bit earlier.[1] Equally important was his idea that carbons are directly linked—equally sharing their "affinities." This idea was in direct conflict with the electrochemical-dualism theory.[1] Kekulé recalled how he had integrated pictures with the extant data to arrive at the tetravalence theory in 1854 while riding a London omnibus:[2]

> I fell into a reverie. The atoms were gambolling before my eyes. I had always seen them in motion, these small beings, but I had never succeeded in discerning the nature of their motion. Now, however, I saw how, frequently, two smaller atoms united to form a pair; how a larger one embraced two smaller ones; how a still larger one kept hold of three or even four of the smaller; whilst the whole kept whirling in a giddy dance. I saw how the larger ones formed a chain and the smaller ones hung on only at the end of the chain.

History indicates that Archibald Scott Couper (1831–1892) discovered the tetravalence of carbon (and carbon–carbon bonding) simultaneously with and independently of Kekulé. His publication was delayed for technical reasons by his Director Adolph Wurtz. When he was "scooped" by Kekulé, he complained bitterly and was promptly fired by Wurtz. A physical breakdown before he was 30 effectively ended this promising scientist's career.[3]

The recognition that carbon is tetravalent established the foundation for structural organic chemistry. In 1861 Aleksandr Butlerov first stated that the particular arrangement of atoms in a molecule is responsible for the substance's physical and chemical properties:[4]

> Only one rational formula is possible for each compound, and when the general laws governing the dependence of chemical properties on chemical structure have been derived, this formula will express all of these properties.

This sentence still belongs in the first lecture of any modern course in organic chemistry. Figure 117 is taken from the 1868 Leipzig edition of Butlerov's *Lehrbuch Der Organischen Chemie* (original Russian edition, 1864). The formulas show the tetravalence of carbon and clearly express the differences in structures between isomers.

Benzene, first obtained from compressed oil gas in 1825 by Faraday, was an interesting enigma. With a formula of C_6H_6, it was highly "unsaturated" and would have been expected to undergo addition reactions like ethylene and other olefins. Its chemistry was remarkably different from that expected. Once again Kekulé claims to have dreamed up a solution:[5]

Normaler Butylalkohol

$\left\{\begin{array}{l} CH_2[CH_2(CH_3)] \\ CH_2 \\ \quad H \end{array}\right. \Big\} O$

Normale Buttersäure

$\left\{\begin{array}{l} CH_2[CH_2(CH_3)] \\ CO \\ \quad H \end{array}\right. \Big\} O$

Primärer Pseudobutyl- (dimethylirter Aethyl-) Alkohol

$\left\{\begin{array}{l} CH(CH_3)(CH_3) \\ CH_2 \\ \quad H \end{array}\right. \Big\} O$

Iso- oder Pseudobutter- säure (vgl. § 170)

$\left\{\begin{array}{l} CH(CH_3)(CH_3) \\ CO \\ \quad H \end{array}\right. \Big\} O$

FIGURE 117 ■ Figure from Butlerov's 1868 Leipzig edition of *Lehrbuch Der Organischen Chemie* (original Russian Edition, 1864). Butlerov first enunciated the modern structural basis for organic chemistry.

> I was sitting, writing at my text-book; but the work did not progress; my thoughts were elsewhere. I turned my chair to the fire and dozed. Again the atoms were gambolling before my eyes. This time the smaller groups kept modestly in the background. My mental eye, rendered more acute by repeated visions of the kind, could now distinguish larger structures, of manifold conformation: long rows, sometimes more closely fitted together; all twining and twisting in snake-like motion. But look! What was that? One of the snakes had seized hold of its own tail, and the form whirled mockingly before my eyes. As if by a flash of lightning I awoke; and this time also I spent the rest of the night working out the consequences of the hypothesis.

The sausagelike model of benzene is at the top right of the group of structures in Figure 118 from Kekulé's 1865 paper in the *Bulletin Société Chimiques*.[5] Benzene, chlorobenzene, and a dichlorobenzene are shown later. Later work by Ladenburg and Körner caused Kekulé to postulate two equivalent alternating structures of benzene in 1872. What was Kekulé reading before he dozed off? Perhaps Libavius' *Alchymia* (Figure 52) inspired his serpent dreams. Perhaps Porta's tortoise [Fig. 26(b)] was the clue he needed for benzene's structure. I am convinced, however, that (1) Kekulé accomplished more asleep than I have awake; (2) I am going to seek a lighter course load to add napping to my yearly activity report.

Figure 119 is from a rare pamphlet distributed at an 1886 meeting of the German Chemical Society that celebrated Kekulé's structure work.[6] The monkeys adopt two rapidly alternating structures (tails entwined and not entwined). The modern representations for benzene [dotted or solid circle in the hexagon —see Fig. 26(b)] were contributed by Johannes Thiele in 1899 and Sir Robert Robinson in 1925.[7]

The ability to explain the complex substitution chemistry of benzene and other related aromatic derivatives as the result of benzene's structure signaled the triumph of structural chemistry. When Hermann Kolbe, who hated Kekulé, ces principes permettra-t-elle de prévoir de nouvelles métamorphoses et de nouveaux cas d'isomérie.

Qu'il me soit permis, en terminant, de faire une observation sur les formules rationnelles par lesquelles on pourrait représenter la composition des substances aromatiques et sur la nomenclature qu'il conviendrait de leur appliquer.

Il est vrai que les substances aromatiques présentent sous plusieurs rapports une grande analogie avec les substances grasses, mais on ne peut pas manquer d'être frappé du fait que sous beaucoup d'autres rapports elles en diffèrent notablement. Jusqu'à présent, les chimistes ont insisté surtout sur ces analogies; ce sont elles qu'on s'est efforcé d'exprimer par les noms et par les formules rationnelles. La théorie que je viens d'exposer insiste plutôt sur les différences, sans toutefois négliger les analogies qu'elle fait découler, au contraire, là où elles existent réellement, du principe même.

Peut-être serait-il bon d'appliquer les mêmes principes à la notation des formules, et, quand on a de nouveaux noms à créer, aux principes de la nomenclature.

Dans les formules on pourrait écrire, comme substitution, toutes les métamorphoses qui se font dans la chaîne principale (noyau); on pourrait se servir du principe de la notation typique pour les métamorphoses qui se font dans la chaîne latérale, lorsque celle-ci contient du carbone. C'est ce que l'on a tenté dans ce Mémoire pour plusieurs formules, en supprimant toutefois des formules typiques la forme triangulaire que la plupart des chimistes ont acceptée, en suivant l'exemple de Gerhardt, et que l'on ferait bien, selon moi, d'abandonner complétement à cause des nombreux inconvénients qu'elle entraîne.

Je ne dirai rien sur les principes que l'on pourrait suivre en formant des noms. Il est toujours aisé de trouver des noms qui expriment une idée donnée, mais tant qu'on n'est pas d'accord sur les idées, il serait prématuré d'insister sur les noms.

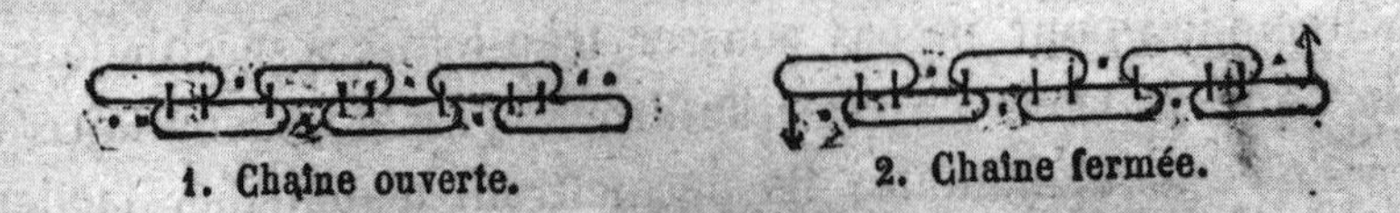

1. Chaîne ouverte. 2. Chaîne fermée.

3. Benzine.

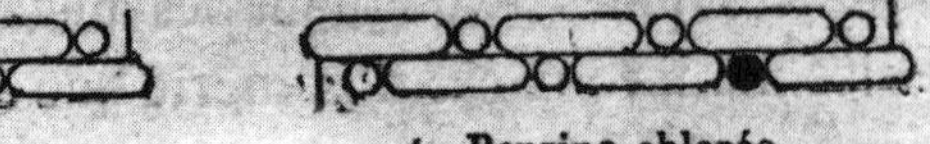

4. Benzine chlorée.

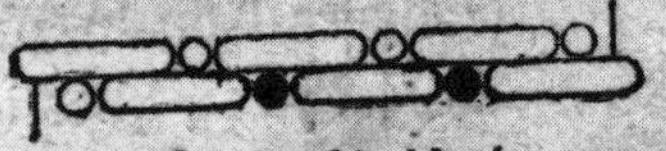

5. Benzine bi-chlorée.

FIGURE 119 ▪ Satirical celebration of Kekulé's benzene structures by the German Chemical Society in 1886 (see E. Heilbronner and J.D. Dunitz, *Reflections on Symmetry*, VCH, Weinheim, 1993, p. 52; courtesy of John Wiley–VCH).

died in 1884, the last real resistance to structural chemistry died with him.[8] In the eloquent words of Brock:[8]

> Just as Picasso had transformed art by allowing the viewer to see within and behind things, so Kekulé had transformed chemistry. Chemical properties arose from the internal structures of molecules, which could now be "seen" and "read" through the experienced optic of the analytical and synthetic chemist.

1. J.R. Partington, *A History of Chemistry*, MacMillan, London, 1964, Vol. 4, pp. 533–565.
2. J.R. Partington, op. cit., p. 537.
3. A.J. Ihde, *The Development of Modern Chemistry*, Harper and Row, New York, 1964, pp. 222–225.
4. W.H. Brock, *The Norton History of Chemistry*, Norton, New York, 1993, p. 256.
5. A.J. Ihde, op. cit., pp. 310–319.
6. E. Heilbronner and J.D. Dunitz, *Relections On Symmetry*, VCH, Weinheim, 1993, p. 52.
7. W.H. Brock, op. cit., p. 555.
8. W.H. Brock, op. cit., pp. 263–269.

←

FIGURE 118 ▪ Kekulé's "sausage formulas" for benzene and two benzene derivatives appearing in *Bulletin de la Société Chimique* (Paris), Vol. 3, p. 98, 1865) (courtesy Edgar Fahs Smith Collection, Rare Book & Manuscript Library, University of Pennsylvania).

"MY PARENTS WENT TO KARLSRUHE AND ALL I GOT WAS THIS LOUSY TEE-SHIRT!"

I apologize, gentle reader, for this tacky take-off on the archetypal All-American souvenir tee-shirt. In all likelihood, the 140 chemists who came to the pleasant Rhineland town of Karlsruhe in 1860 did little boating and souvenir shopping. Human endeavors, such as science, are collective efforts. We need human interaction to jostle, needle, annoy, and inspire us—the sum *is* greater than the parts. Are we different from the termites that must first assemble in great number before they can develop a "collective idea" and construct a mound? Well, frankly yes—we're big, they're small; we get coffee breaks and TV, they don't. Still, we *do* touch "antenna," use the phone, send e-mail, send snail mail, give talks, attend seminars and symposia, read articles and books, chat at breakfast, and hide from assessment reports at professional meetings—the remoter, the better.[1]

Attempts to classify the chemical elements began in the early nineteenth century. Johann Wolfgang Dobereiner (1780–1849) noted during 1816 and 1817 that strontium, which is chemically similar to calcium and barium, had an

	Simboli delle molecole dei corpi semplici e formule dei loro composti fatte con questi simboli, ossia simb. e form. rappresentanti i pesi di volumi eguali allo stato gassoso		Simboli degli atomi de'corpi semplici, e formule dei composti fatte con questi simboli		Numeri esprimenti pesi corrispondent
Atomo dell'idrogeno . . .	$\mathfrak{H}^{1}/_{2}$	=	H	=	1
Molecola dell'idrogeno . .	$\mathfrak{H}$	=	H^2	=	2
Atomo del cloro	$\mathfrak{Cl}^{1}/_{2}$	=	Cl	=	35,5
Molecola del cloro. . .	$\mathfrak{Cl}$	=	Cl^2	=	71
Atomo del bromo	$\mathfrak{Br}^{1}/_{2}$	=	Ar	=	80
Molecola del bromo . . .	$\mathfrak{Br}$	=	Br^2	=	160
Atomo dell' iodo	$\mathfrak{J}^{1}/_{2}$	=	I	=	127
Molecola dell'iodo. . . .	$\mathfrak{J}$	=	I^2	=	254
Atomo del mercurio . . .	$\mathfrak{Hg}$	=	Hg	=	200
Molecola del mercurio . .	$\mathfrak{Hg}$	=	Hg	=	200
Molec. dell'acido cloridrico	$\mathfrak{H}^{1}/_{2}\mathfrak{Cl}^{1}/_{2}$	=	HCl	=	36,5
Mol. dell'acido bromidrico.	$\mathfrak{H}^{1}/_{2}\mathfrak{Br}^{1}/_{2}$	=	HBr	=	81
Mol. dell'acido iodidrico .	$\mathfrak{H}^{1}/_{2}\mathfrak{J}^{1}/_{2}$	=	HI	=	128
Mol. del protocl. di merc.	$\mathfrak{Hg}\mathfrak{Cl}^{1}/_{2}$	=	HgCl	=	235,5
Mol. del protobr. di merc.	$\mathfrak{Hg}\mathfrak{Br}^{1}/_{2}$	=	HgBr	=	280
Mol. del protoiod. di merc.	$\mathfrak{Hg}\mathfrak{J}^{1}/_{2}$	=	HgI	=	327
Mol. del deutoclor. di merc.	$\mathfrak{Hg}\mathfrak{Cl}$	=	$HgCl^2$	=	271
Mol. del deutobr. di merc.	$\mathfrak{Hg}\mathfrak{Br}$	=	$HgBr^2$	=	360
Mol. del deutoiod. di merc.	$\mathfrak{Hg}\mathfrak{J}$	=	HgI^2	=	454

FIGURE 120 ■ Cannizzaro's system of atomic weights based upon Avogadro's Hypothesis (of 1811) and recalled in his 1858 paper and presentation at the 1860 Karlsruhe Congress. This figure is from S. Cannizzaro, *Scritti Intorno Alla Teoria Molecolare Ed Atomica ed alla Notazione Chimica Di S. Cannizzaro* (Palermo, 1896).

atomic weight that was the arithmetic average of the other two.[2] By 1829, he had noted other such "triads" and claimed to have correctly predicted the atomic weight of the newly discovered bromine by averaging chlorine and iodine.[2]

On September 3, 1860 the Karlsruhe Conference convened in order to attempt to settle vexing issues pertaining to atoms, molecules, equivalents, nomenclature, and atomic weights.[3] The clarity on atomic weights provided by Cannizzaro's 1858 pamphlet and presentations at the conference moved Julius Lothar Meyer to comment:[2]

> The scales seemed to fall from my eyes. Doubts disappeared and a feeling of quiet certainty took their place. If some years later I was able myself to contribute something toward clearing the situation and calming heated spirits no small part of the credit is due to this pamphlet of Cannizzaro.

Stanislao Cannizzaro (1826–1910), born in Palermo, was the star of Karlsruhe. He recalled for all assembled the importance of Avogadro's hypothesis, combined it with the Law of Dulong and Petit and other findings and clarified the atomic weights that became the "*y* axis" for the future Periodic Table even as the chemical properties were to become the "*x* axis." In Figure 120, we see Cannizzaro's delineation of atoms, molecules, and atomic weights. The "half-molecule" concept derives from Avogadro's 1811 nomenclature. Figure 121 is a very straightforward exposition of the Law of Dulong and Petit. In this table, all triatomic solids have almost the same specific heat per atom, regardless of the identity of the atom. It is a powerful validation of atomic theory as well as the atomic weights employed.

Formule dei composti	Pesi delle loro molecole $=p$	Calorici specifici dell'unità di peso $=c$	Calorici specifici delle molecole $=p\times c$	Numeri di atomi nelle molecole $=n$	Calorici specifici di ciascun atomo $=\frac{p\times c}{n}$
$HgCl^2$	271	0,06889	18,66919	3	6,22306
$ZnCl^2$	134	0,13618	18,65666	3	6,21888
$SnCl^2$	188,6	0,10161	19,163646	3	6,387882
$MnCl^2$	126	0,14255	17,96130	3	5,98710
$PbCl^2$	278	0,06641	18,46198	3	6,15399
$MgCl^2$	95	0,1946	18,4870	3	6,1623
$CaCl^2$	111	0,1642	18,2262	3	6,0754
$BaCl^2$	208	0,08957	18,63056	3	6,21018
HgI^2	454	0,04197	19,05438	3	6,35146
PbI^2	461	0,04267	19,67087	3	6,55695

FIGURE 121 ■ Cannizzaro's use of the Law of DuLong and Petit to strengthen his system of atomic weights (see Fig. 120).

1. I gratefully acknowledge Lewis Thomas' book, *Lives of a Cell*, Viking Press, New York, 1974, for its influence on this essay.
2. A.J. Ihde, *The Development of Modern Chemistry*, Harper and Row, New York, 1964, pp. 236–237.
3. A.J. Ihde, op. cit., pp. 228–229.

THE ICON ON THE WALL

In an observant Moslem household, a page of verses from the Koran handwritten in beautiful calligraphy may grace the wall. In a Catholic household, one might see a crucifix; in an observant Jewish household there will be a mezuzah affixed to the doorway; a Bodhisattva in a Buddhist household; an image of the family deity in a Hindhu household. And in every house of chemistry, every classroom, lecture hall, and laboratory, hangs our icon—the Periodic Table.

Figure 122(a) is from *Grundlagen der Chemie* (St. Petersburg, 1891), the first German edition of Mendeleev's textbook on chemistry, and shows a Periodic Table from this time. It lacks the rare gases and the "islands" of inner transition metals (lanthanides and actinides) but in other ways looks similar to modern Periodic Tables.

Shortly after the Karlsruhe Conference, John Alexander Newlands (1837–1898) published some papers on regularities in atomic weights.[1] In 1864 he published a version of a table of the elements and noted his law of octaves: ". . . the eighth element starting from a given one is a kind of repetition of the first, like the eighth note of an octave in music."[1] Newlands published a modified table in 1865 and further improved it in 1866. William Odling (1829–1921) published a table of the elements in order of atomic weights in 1865. Lothar Meyer made a table (unpublished) in 1868 that placed carbon, nitrogen, oxygen, fluorine, and lithium at the top of their respective groups. A modified version was first published in 1869.

Credit for the Periodic Table is accorded to Dmitry Ivanovich Mendeleyev (Mendeleeff or Mendeleev, 1834–1907).[2] Mendeleev's mother was a hero in every sense of the word.[2] Following her husband's physical collapse and the destruction by fire in 1848 of the glass factory she had restored and managed, she took her scientifically gifted son, the youngest of 14 children, to Moscow. Unsuccessful in enrolling him in the University because he was Siberian, she moved Dmitry Ivanovich to St. Petersburg and managed to enroll him in the Pedagogical School in 1850, the year she died. In a dedication to a paper published in 1887, Mendeleev wrote: "She instructed by example, corrected with love, and in order to devote him to science, left Siberia with him, spending her last resources and strength."[2]

Subsequent to his training as a teacher in St. Petersburg, Mendeleev wrote a Masters thesis at the University of St. Petersburg and was given a position there. He later studied in Paris and Heidelberg, returned to St. Petersburg in 1861, became Professor at the Technological Institute and later Professor at the University. In 1868, while writing his *Principles of Chemistry*, Mendeleev is said to have started thinking about the periodic law, having been to Karlsruhe but

GRUPPE:	I.	II.	III.	IV.	V.	VI.	VII.	VIII.
Reihe: 1	. **H**			RH^4	RH^5	RH^2	RH	Wasserstoffverbindungen.
» 2	**Li**	**Be** .	**B**	**C** .	**N** .	**O** .	**F** .	
» 3	. **Na**	Mg	. Al	. Si	. P	. S	Cl	
» 4	K .	Ca .	Sc .	Ti	V .	Cr :	Mn .	Fe. Co. Ni. Cu.
» 5	. (Cu)	. Zn	. Ga	Ge	. As	. Se	. Br	
» 6	Rb. .	Sr .	Y .	Zr .	Nb	Mo .	— .	Ru. Rh. Pd. Ag.
» 7	. (Ag)	Cd	. In	. Sn	. Sb	. Te	. J	
» 8	Cs .	Ba .	La	Ce .	Di? .	— .	— .	— — — —
» 9	. —	. —	. —	. —	. —	. —	. —	
» 10	— .	— .	Yb .	— .	Ta .	W .	— .	Os. Ir. Pt. Au.
» 11	. (Au)	. Hg	. Tl	. Pb	. Bi	. —	. —	
» 12	— .	— .	— .	Th .	— .	U .	— .	
	R^2O	R^2O^2 RO	R^2O^3	R^2O^4 RO^2	R^2O^5	R^2O^6 RO^3	R^2O^7	Höchste salzbildende Oxyde RO^4

(a)

Зная атомность радикаловъ, легко предугадать ихъ обыкновеннѣйшія соединенія, наблюдая всегда чтобы сумма атомностей всѣхъ радикаловъ была четное количество.

Простѣйшіе виды соединеній будутъ:

$$R'R', \quad R'^2 R'', \quad R'^3 R'''.$$

Потому водородъ образуетъ слѣдующія типическія соединенія:

Главные типы: $\left.\begin{matrix}H\\H\end{matrix}\right\}$, $\left.\begin{matrix}H\\H\end{matrix}\right\}O$ и $N\left\{\begin{matrix}H\\H\\H\end{matrix}\right.$

Производные типы: $\left.\begin{matrix}H\\Cl\end{matrix}\right\}$ $\left.\begin{matrix}H\\H\end{matrix}\right\}S$ $P\left\{\begin{matrix}H\\H\\H\end{matrix}\right.$

$\left.\begin{matrix}H\\Br\end{matrix}\right\}$ $As\left\{\begin{matrix}H\\H.\\H\end{matrix}\right.$

Орг. химія, Менделѣева. 2

(b)

FIGURE 122 ■ (a) Mendeleev's first Periodic Table was published in 1869. The one in this figure appeared in his 1891 book *Grundlagen der Chemie* (St. Petersburg, 1891) (see text). (b) This intriguing figure comes from the second Russian Edition of Mendeleev's text on organic chemistry (*Organicheskaja Khimia*, St. Petersburg, 1863). Although Partington indicates that Mendeleev began thinking about the Periodic Table in 1868, ideas seem to be occurring earlier. Dr. Roy G. Neville, who is cataloguing his book collection, feels there is evidence for organizing the elements in Mendeleev's 1856 Masters' thesis (personal conversation with A. Greenberg).

unaware of Newlands' work.[2] Mendeleev's first Periodic Table was printed in 1869, the same year as Lothar Meyer's and a modified version published in 1871.[2]

With Partington's above-noted historical summary in mind, it is still fascinating to find what might be considered an early idea for the Periodic Table. Figure 122(b) is taken from the second edition of Mendeleev's *Organicheskaja Khimia* (St. Petersburg, 1863). It certainly has a prescient look about it.

The brilliance and primacy of Mendeleev's Periodic Table rest upon his audacious act of leaving gaps in it, where he predicted that elements, as yet unknown, were missing.[3] In Figure 122(a), we see below aluminum (Al) the element gallium (Ga). This element was unknown in 1871, but Mendeleev predicted the existence of a new element he termed *eka*-aluminum as well as its atomic weight, density, melting point, and the formula of its oxide. (The term eka means "something added.") In 1875, it was discovered by Paul Emile Lecoq de Boisbaudran and named after Gaul to soothe his countrymen's egos after their defeat in the Franco-Prussian War. In 1879, Lars Frederik Nilson of Sweden discovered *eka*-boron, well matching the properties predicted by Mendeleev, and named it scandium (Sc). In 1886, Clemens Winkler discovered *eka*-silicon, again matching Mendeleev's predictions, and named it germanium (Ge)—payback time for French chauvinism.

Mendeleev's predictions were not always correct. He courageously placed iodine *after* the heavier tellurium, incorrectly predicting that new experiments would correctly reverse their masses. He also predicted new elements that were never to be. Unbeknownst to Mendeleev, the source of order for the Periodic Table was not the atomic weight, but the atomic number, and this would be discovered by Henry Moseley just before the First World War.

In 1999, the penultimate year of the millenium, it has been a sheer delight to discover a beautiful article by physician-writer (of *Awakenings* fame) Oliver Sacks who confesses his lifelong fascination with the Periodic Table[4]:

> My kitchen is papered with periodic tables of every size and sort—oblongs, spirals, pyramids, weather vanes—and on the kitchen table, a very favorite one, a round periodic table made of wood that I can spin like a prayer wheel.

Clearly, "chemistry is spoken" in the Sacks household and he even keeps two small periodic icons in his wallet. Perhaps at an appointed hour each day, Dr. Sacks faces St. Petersburg and meditates, contemplating the bearded, prophetlike Mendeleev.

1. J.R. Partington, *A History of Chemistry*, MacMillan, London, 1964, Vol. 4, pp. 886–891.
2. J.R. Partington, op. cit., pp. 891–899.
3. A.J. Ihde, *The Development of Modern Chemistry*, Harper & Row, New York, 1964, pp. 231–256.
4. O. Sacks, "Everything in its Place—One Man's Love Affair with the Periodic Table," *New York Times Magazine*, April 18, 1999, pp. 126–130.

THE PEOPLE'S CHEMISTRY

A Muck Manual for Farmers (Fig. 123) and *600 Receipts Worth Their Weight in Gold* (Fig. 124) are nineteenth-century American books that continue a tradition dating back to the early sixteenth century. Books of secrets,[1] such as Porta's *Natural Magick*, gave recipes for cosmetics, wines, and other concoctions. Books

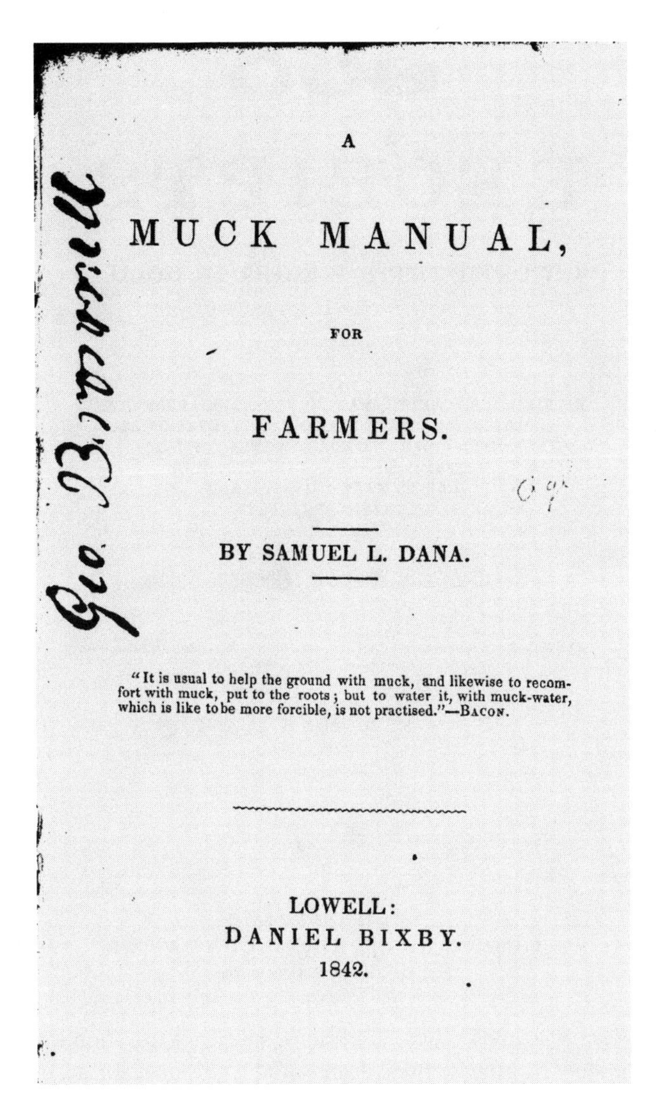
A

MUCK MANUAL,

FOR

FARMERS.

BY SAMUEL L. DANA.

"It is usual to help the ground with muck, and likewise to recomfort with muck, put to the roots; but to water it, with muck-water, which is like to be more forcible, is not practised."—BACON.

LOWELL:
DANIEL BIXBY.
1842.

FIGURE 123 ■ Here is a nice title for a practical American book. Samuel L. Dana was a respected American chemist who authored a practical and never patronizing book for farmers on soils and compost (a kind of Agricultural Extension Service).

of recipes[2] and "household" books also provided practical home remedies, information on preserving food, and thousands of other bits of technical assistance for daily living.

A Muck Manual, by American chemist Samuel L. Dana, provides an intelligent, accessible and never-patronizing introduction to minerals, rocks, soils, manures, and composts. Dana (1795–1868) was an esteemed technical chemist and inventor of the "American System" of bleaching.[3] His book follows a tradition exemplified by works like *A Treatise Shewing the Intimate Connection that Subsists between Agriculture and Chemistry* (Archibald Cochrane, 9th Earl of Dundonald, London, 1795) and Humphrey Davy's *Elements of Agricultural Chemistry* (London, 1813). An interesting aspect of Dana's book is his introduction of a new term, *urets*, for minerals such as metal sulfides. He carefully accounts to his practical audience for his need to introduce a new term to the chemical lexicon.

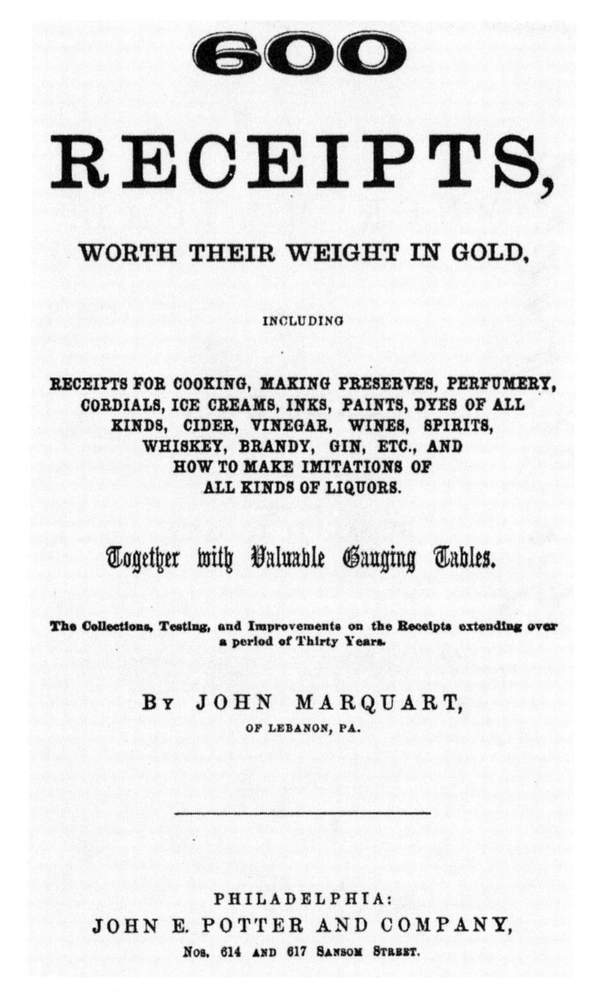
600

RECEIPTS,

WORTH THEIR WEIGHT IN GOLD,

INCLUDING

RECEIPTS FOR COOKING, MAKING PRESERVES, PERFUMERY, CORDIALS, ICE CREAMS, INKS, PAINTS, DYES OF ALL KINDS, CIDER, VINEGAR, WINES, SPIRITS, WHISKEY, BRANDY, GIN, ETC., AND HOW TO MAKE IMITATIONS OF ALL KINDS OF LIQUORS.

Together with Valuable Gauging Tables.

The Collections, Testing, and Improvements on the Receipts extending over a period of Thirty Years.

BY JOHN MARQUART,
OF LEBANON, PA.

PHILADELPHIA:
JOHN E. POTTER AND COMPANY,
NOS. 614 AND 617 SANSOM STREET.

FIGURE 124 ■ Here is a book whose traditions date back to "home manuals" of the sixteenth century: how to prepare inks, drinks, and a catharsis for an overblown cow.

Muck Manual also discusses the chemical nature of geine—humus—whose complexity remains daunting. Today, state land-grant universities run Agricultural Extension Service programs to fulfill the practical teaching role for farmers assumed by Dana's useful manual.

I am a transplanted Brooklyn Yankee[4] living in Charlotte, North Carolina as this book is being written. It is interesting to speak with people whose families have lived for long periods in this region. One friend[5] tells me of the desperate importance of salt, for food preservation and refrigeration, as the Confederacy was losing the Civil War and in dire straits. Unsalted meat would often rot in transit. Heavily salted meat would arrive preserved but have to be repeatedly boiled in water in order to make it barely edible. Destroy the sources of salt and you have dealt the Confederate soldiers a crippling blow. Strategically, a major battle was fought in 1864 for control of Saltville, Virginia, a location of natural salt-licks.[6] Southern families were reduced to desperately digging up the soil from dirt floors under smokehouses and "boiling" it in water in order to recover the salt from previous seasons. This was a very sad life-or-death people's chemistry.

We end this essay on lighter, though not uplifting, notes. In Marquart's 1867 book, we find Recipe No. 83, "A remedy for Black Teeth": pulverized cream of tartar and salt; wash your teeth in the morning then rub them with this powder; Recipe No. 479, "To cure Hoven or Blown in Cattle [cattle over-eating rich food, bloating due to "overcharged" first stomach and incapable of expelling its contents—a life-threatening situation]—see Recipe No. 480": 1 pound of Glauber's salt ($Na_2SO_4 \cdot 10H_2O$—a cathartic); 2 ounces ginger powder; 4 ounces of molasses, mix, and then pour 3 pints of boiling water on the mass. When "new-milk warm" (i.e., fresh from the udder or "udderly" fresh), give the entire dose (cover your ears and hold your nose).

1. J.R. Partington, *A History of Chemistry*, MacMillan, London, 1961, Vol. 2, pp. 27–31.
2. J.R. Partington, op. cit., pp. 68–69.
3. E.F. Smith, *Chemistry In America*, Appleton, New York, 1914, p. 222.
4. A Brooklyn Yankee is an oxymoron—ask anybody who ever went to a baseball game at Ebbets Field. We hated "Yankees" as much as any Carolinian.
5. I learned of the history of boiling soil from floors under smokehouses from retired North Carolina state trooper, Harold Eaker, whose family has lived in the vicinity of King's Mountain, NC since the American Revolution. See also: Charles Frazier, *Cold Mountain*, Vintage, New York, 1997, p. 103.
6. G.G. Walker, *The War in Southwest Virginia: 1861–1865*, 6th ed., A & W Enterprise, Roanoke, 1985, pp. 10, 71–106. I thank Mr. R. Stewart Lillard for enlightening discussion and for making me aware of this book.

INK FROM PEANUTS AND THE FINEST SUGAR IN THE SOUTH

In 1947, Nobel laureate Roald Hoffmann was ten years old, in a displaced persons (DP) camp in Germany, when he was fascinated by the biographies (in German translation) of Marie Curie and George Washington Carver.[1] Carver (ca. 1860–1943) was born to slaves belonging to Moses Carver just prior to the start of the Civil War.[2] At the end of the War, Moses Carver discovered that his only living former slave was the five-year-old George who was seriously ill with whooping cough. Returned to his former Master's house, George remained almost ten years before traveling and developing his interests in plants and animals and talents in art and music. He obtained a high-school degree in his late 20s and became the first Afro-American graduate of what is now Iowa State University (1894), soon attaining the Masters degree there in 1896. His agricultural knowledge led him to act as a kind of extension service for African-American farmers. This dedication led Carver to Tuskegee Institute in Alabama, then under the Presidency of Booker T. Washington.

As a university administrator running the newly-organized Agriculture Department at Tuskegee, Carver was not strong on bureaucratic practice or budget balancing. However, applied research was his real calling and Carver pioneered crop rotation and planting of legume products such as soybeans and peanuts to replenish the soil's nutrients. He and his collaborators at Tuskegee developed about 300 products derived from peanuts (e.g., inks, plastics, dyes, coffee) and over 100 others from sweet potatoes. Peanuts evolved from being a "noncrop" to the second leading cash crop (following "king cotton") in the South. In 1990, Carver and organic chemist Dr. Percy L. Julian, were the first African Americans to be inducted into the National Inventors Hall of Fame. Dr. Julian (1899–1975) pioneered the synthesis of physostigmine, used to treat glaucoma, perfected an economical route to the steroid cortisone, so effective for treating rheumatoid arthritis, and became the first African-American Director of Research for a major company (Glidden Company in Chicago).[3]

Norbert Rillieux (1806–1894) was the son of inventor Vincent Rillieux (the great uncle of artist Edgar Degas) and a free woman of color, Constance Vivant, with whom he had a long-standing relationship.[4] The younger Rillieux, a chemical engineer educated in Paris, developed the triple-effect-evaporator for sugar refining in the 1830s. In partnership with Jewish plantation owner Judah P. Benjamin (later Jefferson Davis's Secretary of War), the sugar produced by Rillieux's apparatus won awards and recognition and the apparatus was widely adopted. It is thought that Degas may have used the two men as models for one of his double portraits.[4]

1. R. Hoffmann and V. Torrence, *Chemistry Imagined—Reflections On Science*, Smithsonian Institution Press, Washington, D.C., 1993, pp. 30–32.
2. R. Holt, *George Washington Carver: An American Biography*, rev. ed., Doubleday, Doran and Co., Garden City, 1963.
3. E.J. McMurray (ed.), *Notable Twentieth-Century Scientists*, Gale Research Inc., Detroit, MI, 1995, Vol. 2, pp. 1045–1047.
4. *Chemical Heritage*, **16** (1):10, Summer, 1998.

SECTION VII
TEACHING CHEMISTRY TO THE MASSES

MICHAEL FARADAY'S FIRST CHEMISTRY TEACHER

Mrs. Jane (Haldimand) Marcet (1769–1858) was born in England and married a prominent Swiss physician and respected amateur chemist Alexander Marcet.[1,2] Influenced by Humphrey Davy's public lectures she tried some experiments and decided to write a book to explain the science:

> In venturing to offer to the public, and more particularly to the female sex, an Introduction to Chemistry, the author, herself a woman, conceives that some explanation may be required: and she feels it the more necessary to apologize for the present undertaking, as her knowledge of the subject is but recent, and as she can have no real claims to the title of chemist.

(Compare this strategically diplomatic *Apologia* with the one cited earlier from Mrs. Fulhame's 1794 book [p. 156]. Mrs. Fulhame is openly contemptuous of narrow and ignorant people who would limit a woman's role). The first London edition of *Conversations* (Fig. 125) is said to have appeared in 1805[1] (another opinion is 1806[2]). Edgar Fahs Smith avers that about 160,000 copies of its numerous editions were sold before 1853.[1]

The most careful perusal of the title page and the rest of the text of the early editions will not provide a hint of the author's identity. Part of the reason was Mrs. Marcet's own modesty about her lack of formal training. However, the etiquette of the day is also a likely cause. Most outrageously, later editions (e.g., 1822, 1826, 1829, and 1831, edited by Dr. J.L. Comstock) were published by men who, while crediting the "authoress," were quick to add their own criticisms. One defender of Mrs. Marcet wrote[1]:

> We are informed by one of the American editors of this work that his reason for not placing the name of Jane Marcet on the title-page, was because scientific men believed it fictitious!

Conversations on Chemistry is a delightful interplay between a Mrs. B. (sometimes referred to as Mrs. Bryan[2]) and Caroline and Emily[2] (ages 13 to 15). Its coverage of chemical principles, while accessible, is not at all superficial, and Mrs. Marcet updated her own editions by including the latest work of her correspondent Davy and other prominent chemists. Here is a selection found on pages 198–199 of the 1814 American edition.

Mrs. B.: From its own powerful properties, and from the various combinations into which it enters, sulphuric acid is of great importance in many

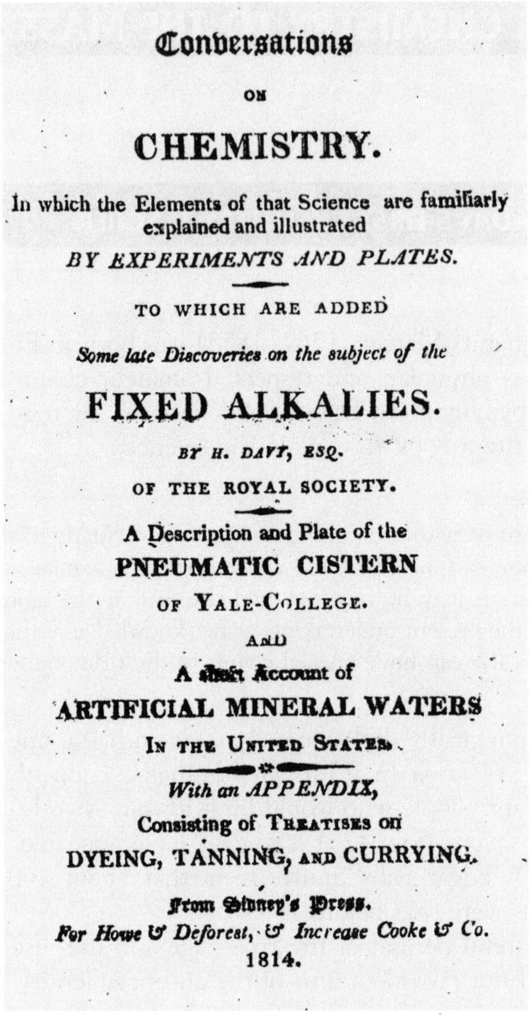

Conversations

ON

CHEMISTRY.

In which the Elements of that Science are familiarly explained and illustrated

BY EXPERIMENTS AND PLATES.

TO WHICH ARE ADDED

Some late Discoveries on the subject of the

FIXED ALKALIES.

BY H. DAVY, ESQ.

OF THE ROYAL SOCIETY.

A Description and Plate of the

PNEUMATIC CISTERN

OF YALE-COLLEGE.

AND

A [illegible] Account of

ARTIFICIAL MINERAL WATERS

IN THE UNITED STATES.

With an APPENDIX,

Consisting of TREATISES on

DYEING, TANNING, AND CURRYING.

From Sidney's Press.

For Howe & Deforest, & Increase Cooke & Co.

1814.

FIGURE 125 ■ *Conversations on Chemistry* was actually authored by Mrs. Jane Marcet. It is a beautiful teaching text that uses Socratic dialogue involving a Mrs. B. and two adolescent girls, Caroline and Emily. It inspired the young Michael Faraday's interest in chemistry and appeared in a number of editions over almost 50 years and sold over 160,000 copies.

of the arts. It is also used as a medicine in a state of great dilution; for were it taken internally, in a concentrated state, it would prove a most dangerous poison.

Caroline: I am sure it would burn the throat and stomach.

Mrs. B.: Can you think of any thing that would prove an antidote to this poison?

Caroline: A large draught of water to dilute it.

Mrs. B.: That would certainly weaken the power of the acid, but it would increase the heat to an intolerable degree. Do you recollect nothing that would destroy its deleterious properties more effectually?

Emily: An alkali might, by combining with it; but then, a pure alkali is itself a poison, on account of its causticity.

Mrs. B.:	There is no necessity that the alkali should be caustic. Soap, in which it is combined with oil, or magnesia, either in a state of carbonat, or mixed with water, would prove the best antidotes.
Emily:	In those cases, then, I suppose, the potash and the magnesia would quit their combinations to form salts with the sulphuric acid?
Mrs. B.:	Precisely.

It appears that the novelist Maria Edgeworth read Mrs. Marcet's book and may have saved the life of her younger sister, who had swallowed acid, by administering milk of magnesia.[2] It is intriguing that in her 1998 novel of suspense,[3] historian Barbara Hambly provides a schoolteacher, a free woman of color, with a book titled *Conversations in Chemistry More Especially for the Female Sex* that is authored by a (presumably Mrs.) Mercer.

The great nineteenth-century scientist Michael Faraday came from a family of very modest means and worked as a bookbinder starting in 1804 at the age of 13. He was first introduced to chemistry by Mrs. Marcet's book[1]:

> So when I questioned Mrs. Marcet's book by such little experiments as I could find to perform, and found it true to the facts as I could understand them, I felt that I had got hold of an anchor in chemical knowledge, and clung fast to it. Hence my deep veneration for Mrs. Marcet: first, as one who had conferred great personal good and pleasure on me, and then as one able to convey the truth and principle of those boundless fields of knowledge which concern natural things, to the young, untaught, and inquiring mind.
>
> You may imagine my delight when I came to know Mrs. Marcet personally; how often I cast my thoughts backward, delighting to connect the past and the present; how often, when sending a paper to her as a thank-offering, I thought of my first instructress, and such like thoughts will remain with me.

Mrs. Marcet's influence on Faraday is probably doubly profound. In addition to his fundamental contributions to science, Michael Faraday was renowned for his public lectures to lay audiences and his book *A Course of Six Lectures on the Chemical History of a Candle* (1861) became a classic for popularizing chemistry.

1. E.F. Smith, *Old Chemistries*, McGraw-Hill, New York, 1927, pp. 64–71.
2. M. Rayner-Canham and G. Rayner-Canham, *Women in Chemistry: Their Changing Roles from Alchemical Times to the Mid-Twentieth Century*, American Chemical Society and the Chemical Heritage Foundation, Washington, D.C. and Philadelphia, 1998, pp. 32–35.
3. B. Hambly, *Fever Season*, Bantam, New York, 1998, p. 292. I thank Professor Susan Gardner for bringing this to my attention.

"CHEMISTRY NO MYSTERY"

At the very start of Marcet's *Conversations in Chemistry*, Caroline says: "To confess the truth, Mrs. B., I am not disposed to form a very favorable idea of chemistry, nor do I expect to derive much entertainment from it. I prefer those sciences that exhibit nature on a grand scale, to those which are confined to the minutiae of petty details." Four years after Dalton's Atomic Theory and already "I'm bored" from teen-age students!

John Scoffern, a surgeon and occasional chemical assistant at the London Hospital,[1] wrote a book titled *Chemistry No Mystery* (London, 1839) that offered excitement to young and old alike:

FIGURE 126 ■ "Step right up, ladies and gentlemen, and get your nice hot tootsie-frootsie chemistry!" (homage to Marx—Chico, not Karl). This figure, drawn by George Cruikshank (who illustrated *Oliver Twist*), appears in Dr. John Scoffern's *Chemistry No Mystery* (London, 1839). A practical application of chemistry (a stinkbomb) has been released in a circus tent.

FIGURE 127 ■ More chemical mischief in *Chemistry No Mystery*: the Old Philosopher ("O.P.") has allowed his class to participate in the nitrous oxide experiment. I imagine the following dialogue afterward: "Lucky you are tenured," sayeth O.P.'s Department Chair; "Academic freedom," responds O.P.; "Don't press your luck," responds the Chair who sees no excuse for laughter in a lecture hall.

> If I were to present myself before you with an offer to teach you some new game:—if I were to tell you an improved Plan of throwing a ball, of flying a kite, or of playing leap-frog, oh, with what attention you would listen to me. Well, I am going to teach you many new games. I intend to instruct you in a science full of interest, wonder, and beauty; a science that will afford you amusement in your youth, and riches in your mature years. In short, I am going to teach you the science of chemistry.

How wonderfully fitting that the title page (Fig. 126) depicts a scene outside a show-caravan wherein the imaginary narrator ("The Old Philosopher" or "O.P.") recalls a scene from his misspent youth. He enjoyed practical jokes and released hydrogen sulfide gas (rotten-egg odor) under the flooring of the stage driving out the show's giant and its dwarf. He apparently soon felt the giant's wrath and spent two days in the hospital afterward.[1]

Figure 127 depicts one of O.P.'s hypothetical lectures in which he makes and then foolishly distributes laughing gas to the students in his lecture hall.[1] The illustrator and caricaturist George Cruikshank (1792–1878), who produced these drawings, was probably the first to provide lively, humorous pictures for children's books, and he illustrated Charles Dickens' *Oliver Twist* (1838).[2]

1. J. Read, *Humour and Humanism in Chemistry*, Bell, London, 1947, pp. 208–214.
2. *Encyclopedia Brittanica*, Chicago, 1986, Vol. 3, p. 763.

THE CHEMICAL HISTORY OF A CANDLE

A Course of Six Lectures on the Chemical History of a Candle [London, 1861; New York, 1861—see Fig. 128(a)], derived from notes at Faraday's public lectures, is the culmination of a wonderful 60-year heritage of popularizing chemistry involving three individuals: Humphrey Davy, Jane Marcet, and Michael Faraday. We have already met Count Rumford, whose "boring experiment" (Fig. 97) disproved Lavoisier's caloric theory. He married the widowed Madame Lavoisier in 1805 and they effectively separated within two months (a boring husband?). In 1799 Rumford's ideas for improving the education of the middle classes and improving arts and manufacturing led to the chartering of the Royal Institution of Great Britain. He brought in young Humphrey Davy as Assistant Lecturer in Chemistry, Director of the Laboratory, and Editor of the Institution's chemical journal. Davy's public lectures were popular and well attended. One of those

FIGURE 128 ■ (a) Title page from Michael Faraday's *Chemical History Of A Candle* (the London edition was also published in 1861). The book was not written by Faraday but derived using notes from his public lectures at the Royal Institution. Faraday's interest in teaching chemistry to the public follows a 60-year strand through Mrs. Marcet from Humphrey Davy. (b) Collecting the invisible vapors of a candle. *Illustration continued on following pages*

A

COURSE OF SIX LECTURES

ON THE

CHEMICAL HISTORY OF A CANDLE:

TO WHICH IS ADDED

A LECTURE ON PLATINUM.

BY

MICHAEL FARADAY, D.C.L., F.R.S.,

FULLERIAN PROFESSOR OF CHEMISTRY, ROYAL INSTITUTION; FOREIGN ASSOCIATE OF THE ACADEMY OF SCIENCES, ETC.

Delivered before a JUVENILE AUDITORY *at the* ROYAL INSTITUTION *of* GREAT BRITAIN *during the Christmas Holidays of* 1860–1.

EDITED BY WILLIAM CROOKES, F.C.S.

WITH NUMEROUS ILLUSTRATIONS.

NEW YORK:
HARPER & BROTHERS, PUBLISHERS,
FRANKLIN SQUARE.
1861.

(a)

Fig. 7.

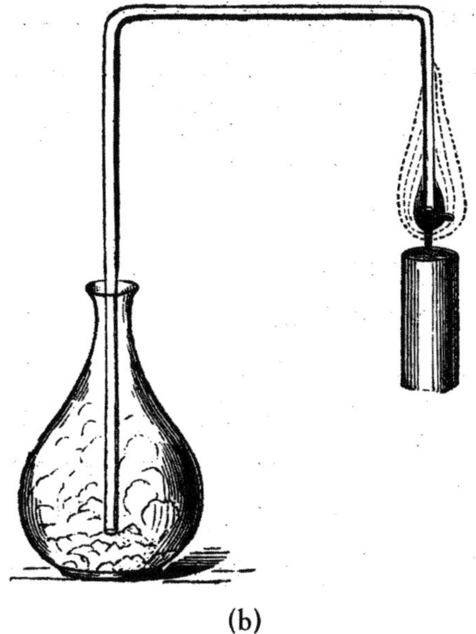

(b)

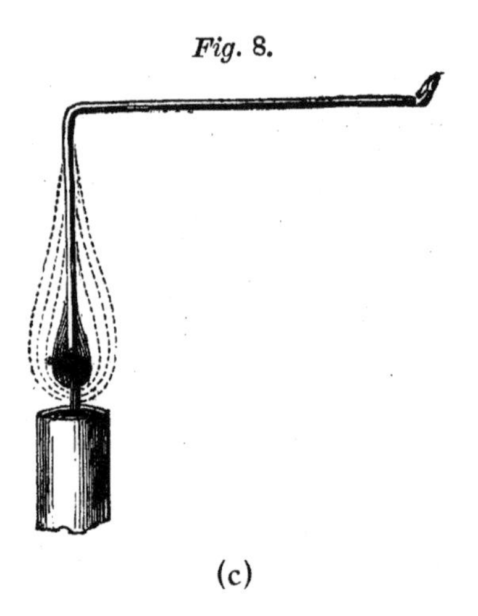

FIGURE 128 ■ *Continued* (c) An articulated candle.

attending was Mrs. Jane Marcet. Davy's lectures inspired Mrs. Marcet's interest in chemistry and ultimately stimulated her to write *Conversations on Chemistry*, which went through numerous editions and sold over 160,000 copies. Mrs. Marcet included some of Davy's latest work in her editions and maintained scientific correspondence with him.

Michael Faraday (1791–1867) was born to the family of a poor blacksmith.[1] At the age of 13 he was apprenticed to a bookbinder. With the owner's permission, he read and was inspired by Mrs. Marcet's book. Faraday started to attend public chemical lectures and, in 1812, a customer rewarded him with a ticket to Davy's lecture at the Royal Institution. Shortly afterward, Faraday sent a copy of the lecture notes he wrote out to Davy and requested to be employed as his assistant. Davy hired the young man and by 1820 Faraday had published his first paper. Throughout his career, Faraday joyfully acknowledged his debt to Mrs. Marcet and remained her correspondent and friend. Faraday took a course in elocution in 1818 and was "a splendid lecturer."[1]

The *Chemical History of the Candle* was derived from Faraday's public lectures. The book was reprinted throughout the nineteenth century in many languages. In fact, the most recent reissue appears to be in 1993 (Cherokee Press, Atlanta). Here is Faraday's rationale presented in Lecture 1:

> I propose to bring before you, in the course of these lectures The Chemical History of a Candle. There is no better, there is no more open door by which you can enter the study of natural philosophy than by considering the physical phenomena of a candle. There is not a law under which any part of this universe is governed which does not come into play, and is not touched upon, in these phenomena. I trust, therefore, I shall not disappoint you in choosing this for my subject rather than any newer topic, which could not be better, were it even so good.

Figure 128(b) is from Lecture 2. The glass tube opens at one end into the dark middle part of a candle flame. At the other end, the invisible wax vapors from this part of the flame are seen to condense. Faraday then differentiates vapors from gases for his audience. He proceeds to heat some candle wax in another flask and pours the vapors into a basin and sets them on fire. In another demonstration [Fig. 128(c)] he uses a piece of glass tubing in communication with the middle of the flame and lights the other end of the glass tubing to form a kind of articulated candle. He notes further that if the glass tubing communicated with the top, rather than the middle, of the flame, there would be no vapor to carry through since it is burned in the upper region. He thus demonstrated the presence of invisible, flammable vapors present in the center of the flame but not at the top. Faraday quips: "Talk about laying on gas—why we can actually lay on a candle."

1. J.R. Partington, A *History of Chemistry*, MacMillan, London, 1964, Vol. 4, pp. 99–140.

INTO THE HEART OF THE FLAME

Figure 129 is the wild and wonderfully stylized illustration of a candle's flame in the 1857 edition of Edward Youmans' *Chemical Atlas*. The formula of carbon dioxide is shown correctly, but the author errs (see discussion of Fig. 116) in describing water as HO, fuel as CH, and in depicting gaseous oxygen as atoms rather than as O_2 molecules. The lower interior region of the flame (which we see as blue) is shown as fuel rich and lacking oxygen. We now know that this part of the flame and the incandescent regions immediately above and around it are full of short-lived, exotic, and ultrareactive carbon-rich molecules, molecular fragments and particles.[1] These regions are reducing in nature since the carbon-rich species hungrily grab oxygen atoms from calxes, such as tin oxide, to produce the metals. (The carbon-oxygen bond in CO is the strongest bond in any neutral compound.[2]) In contrast, the outer blue edge of the flame is oxidizing—rich in the super-reactive hydroxyl radical (truly $HO^{\bullet}$) as well as oxygen, carbon dioxide, and water.[1] In this region, tin would be immediately oxidized to its calx. These details have been known for almost 200 years through the application of a kind of "flame scalpel" called the blowpipe.

In his book *The Use of the Blowpipe in Chemical Analysis, and in the Examination of Minerals* (Stockholm, 1820; London, 1822), Jons Jacob Berzelius traces the history of the blowpipe to traditional applications by jewelers. He dates its earliest application to "dry" chemistry at about 1733. An ideal blowpipe is made of a brass tube with an ivory tip (having an opening of roughly $^3/_8$-inch diameter) attached at one end, to facilitate the chemist's exhalation, with a fused platinum tip (about $^1/_{16}$-inch opening diameter) following a 90° bend at the other end. The platinum tip is inserted into the flame and blowing is performed in a forceful but steady manner to excise the reducing or oxidizing parts for contact with the matter of interest. The author notes that inexperienced users seem to require exhausting bursts of lung power—"they might as well have proposed to play on a wind instrument with a bladder."[3] He details a technique in which the cheeks are filled with air and continuously replenished and used to generate a steady but forceful airstream. The blowpipe was a sensitive instru-

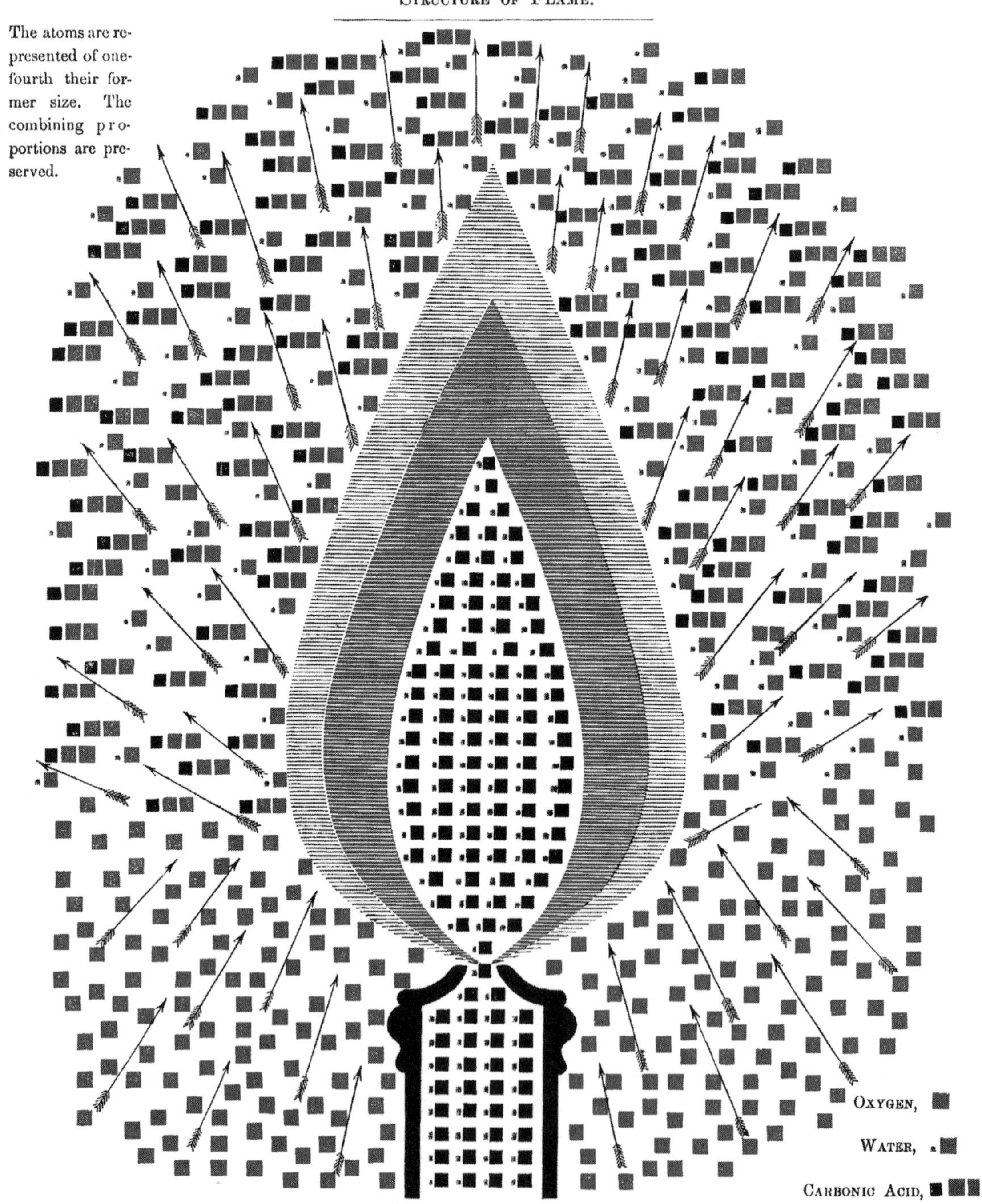

FIGURE 129 ▪ An ebullient flame from the 1857 edition of Youmans' *Chemical Atlas* (see Fig. 112; the errors in formulas such as HO for water are discussed in the text).

ment for analysis of mineral samples and could provide evidence for metallic impurities at levels too low to weigh. For example, the ashes of a piece of paper, subjected to the reducing flame from a blowpipe, yielded microscopic particles of metallic copper.[3]

1. P.W. Atkins, *Atoms, Electrons and Change*, Scientific American Library, Freeman, New York, 1991, pp. 105–109.
2. I thank Professor Joel F. Liebman for calling this to my attention.
3. J.J. Berzelius, *The Use of the Blowpipe in Chemical Analysis, and in the Examination of Minerals* (translated by J.G. Children), Baldwin, Cradock and Joy, London, 1822, pp. 5, 8.

POOF! NOW YOU SMELL IT! NOW YOU DON'T!

Here is an imaginative way to teach chemistry from a highly imaginative person. In his 1823 book *Diagrammes Chimiques* Henri Decremps puts ideas into flow diagrams that dissect substances into parts and reassemble them following chemical reactions. Decremps was a lawyer and amateur magician.[1] In 1784, he published a book titled *La Magia Blanche Devoilee* ("White Magic Revealed"). A rival conjuror of great fame, Pinetti, who claimed to be a Knight, Professor of Mathematics and Natural Science, etc., borrowed liberally from *La Magia* without sharing credit. Decremps published many books attempting to debunk Pinetti but they only increased Pinetti's fame.[1] Finally, long in the tooth and gray in the beard, Decremps tried his hand at writing a chemistry text.

Figure 130 describes how "two odorless bodies placed in contact produce a very sharp odor and two other bodies form by their reunion a visible, palpable body." At the top left, we see sulfuric acid join with the components of limestone ($CaCO_3$), which are lime (CaO) and carbon dioxide (CO_2) liberated by addition of heat ("caloric"). Calcium sulfate ($CaSO_4$) and water remain in the top retort while carbon dioxide and "caloric" travel to the bottom-most flask. In the middle left, we add lime (CaO) and "caloric" to ammonium chloride (NH_4Cl), which will react under these conditions to release ammonia gas (NH_3)—also known as the "piquant odor," leaving calcium chloride ($CaCl_2$) and water in the retort in the middle left. Ammonia and "caloric" join carbon dioxide and "caloric" (undoubtedly in the presence of some water) to form ammonium carbonate $(NH_4)_2CO_3$—a visible, palpable body.

More magic in Figure 131: infusion of violets is actually an acid–base indicator (the first was discovered by Boyle in 1675). When vinegar, an acid, is added to the neutral blue infusion of violet, the solution turns red. When excess aqueous ammonia base is added, the solution goes from red to blue to green. The first human blow-hard neutralizes the solution back to blue by blowing in carbon dioxide, which forms carbonic acid in water. The second blow-hard returns the color to red by adding more carbonic acid.

Figure 132 shows the reader how to picture the molecular structure of copper sulfate (one oxygen short). Ionic compounds were not understood until the work of Arrhenius late in the nineteenth century. The third figure in this

Diagrammes Chimiques. *Pl. 2.*

Deux corps inodores *mis en contact produisent une* odeur très piquante *et deux autres* corps invisibles *forment par leur réunion un* corps visible et palpable; *préparation du gaz ammoniacal et du carbonate d'ammoniaque,* sel volatil d'angleterre.

Ingrédients | Appareil | Produits ou résultats.

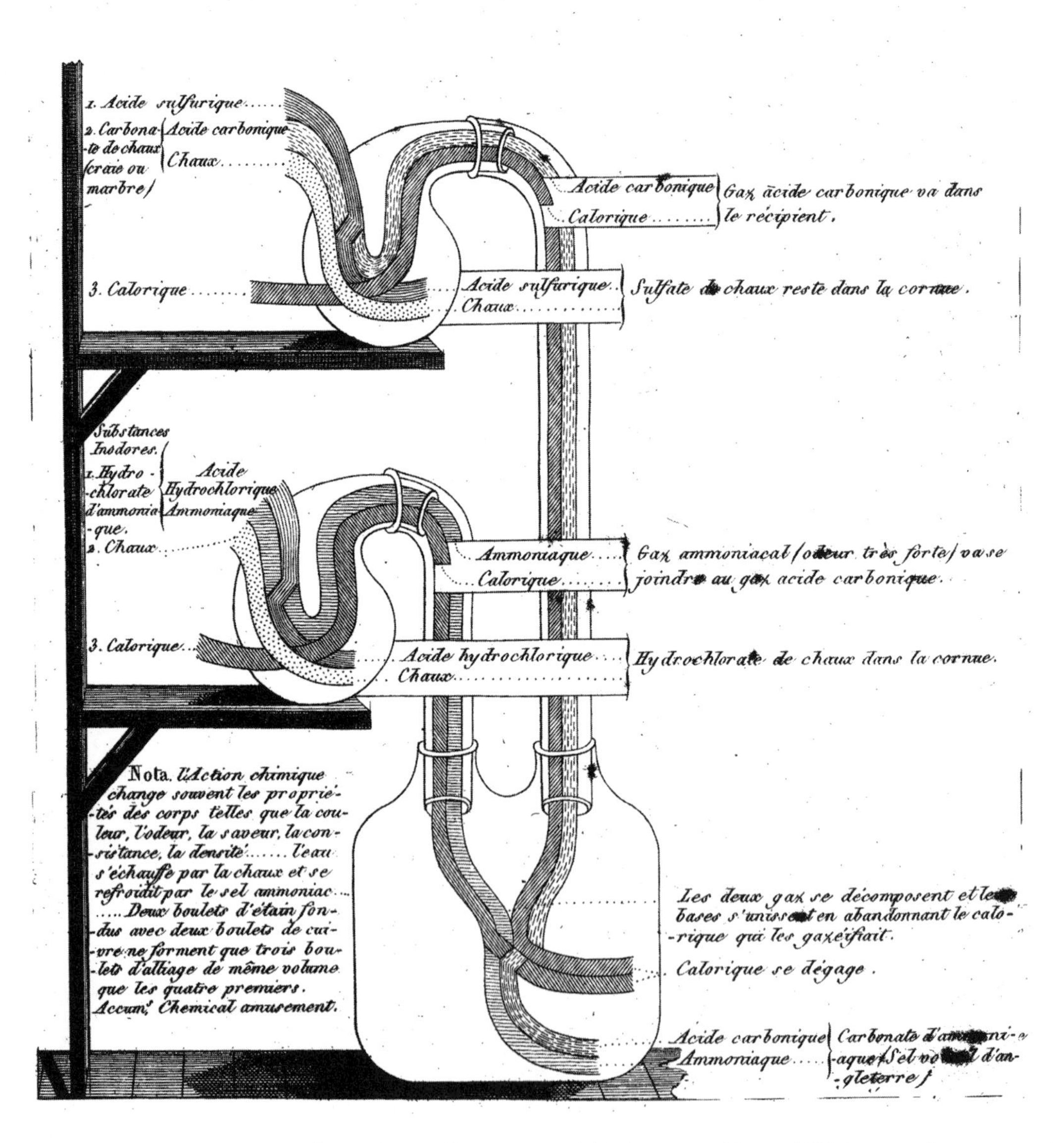

FIGURE 130 ■ Henri Decremps, the author of *Diagrammes Chimiques . . .* (Paris, 1823), was a famous magician for most of his life. The fascinating diagrams in this book seem to invent imaginary apparatus to conduct conceptual streams of chemicals, their "dissection," and subsequent reactions. In this figure, two odorless substances cause the formation of a "piquant" substance—ammonia.

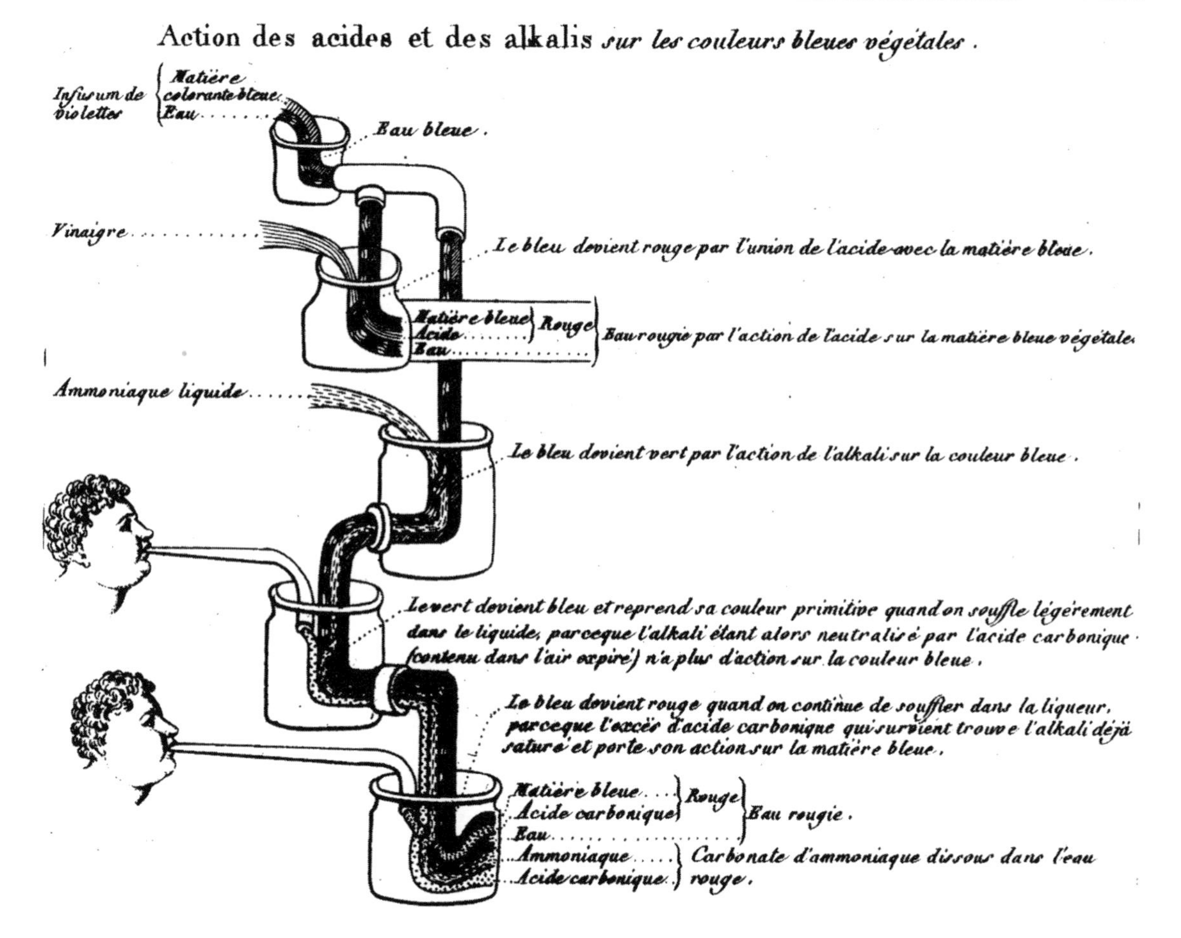

FIGURE 131 ■ Color changes in *Diagrammes Chimiques* brought about by adding vinegar to a neutral solution (colored blue by the indicator) and observing the solution turn red; ammonia is added and the solution goes back to blue then green (basic). Blowing carbon dioxide into the solution then neutralizes to blue. The second blowhard makes it acidic again (red).

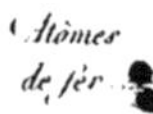

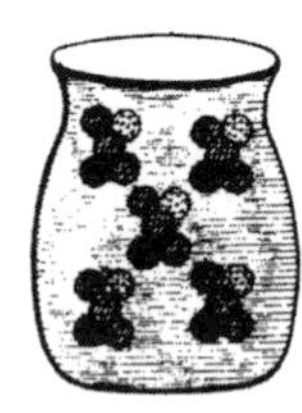

FIGURE 132 ■ Decremps did not know about ions. Also copper sulfate is $CuSO_4$. However, his diagram nicely shows that atomic iron will oxidize and reduce copper to the metal.

drawing shows the addition of metallic iron (iron atoms) to copper sulfate. The iron atoms lose their electrons (are oxidized) to copper ions, which are reduced to atoms and precipitate out.

Figure 133 reminds the world that it was the French who defeated phlogiston. The top diagram shows metallic lead composed of "earth of lead" (lead calx or oxide) plus phlogiston. Heating of metallic lead causes loss of phlogiston,

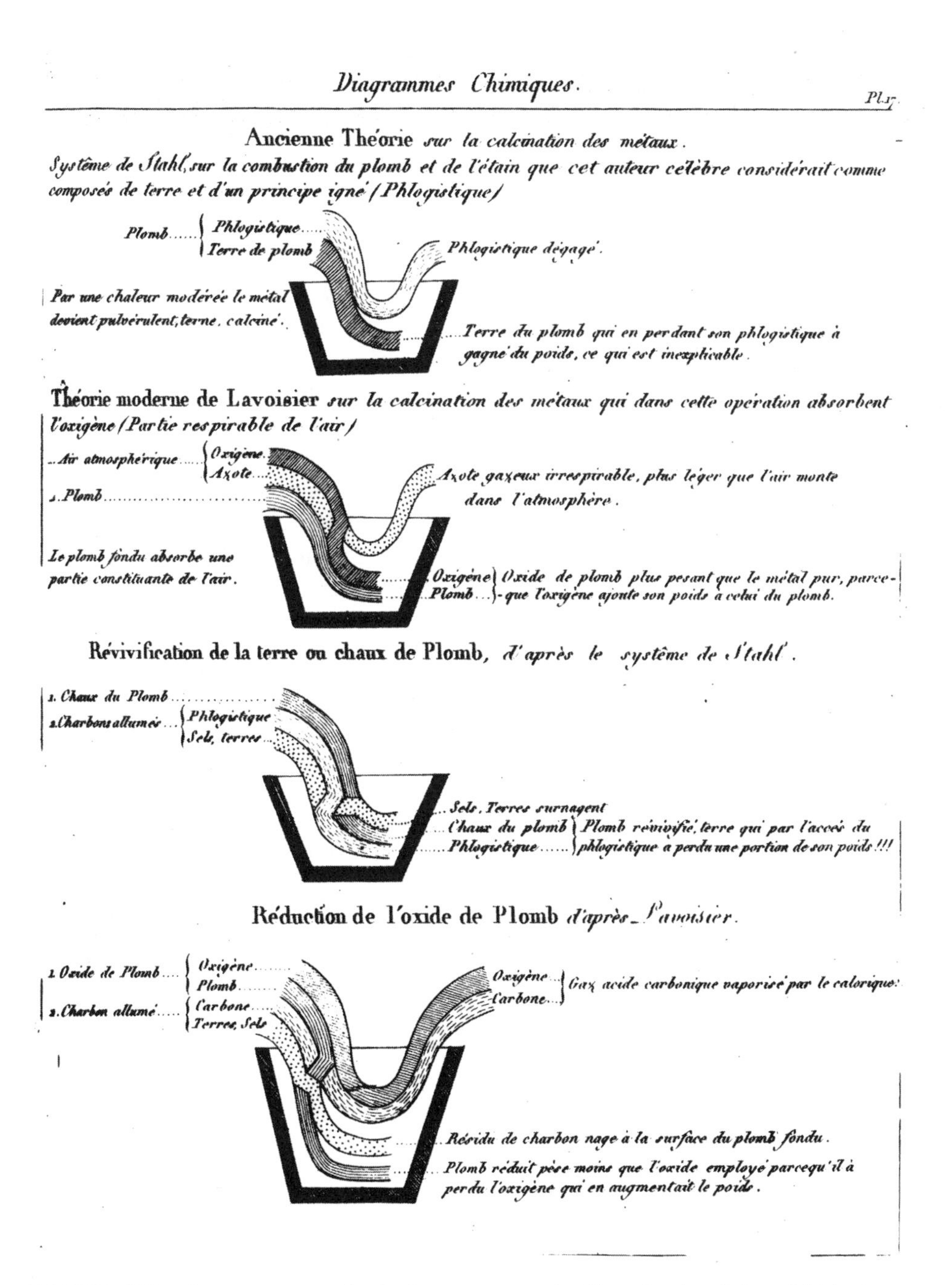

FIGURE 133 ■ Just in case anybody forgot, it was the French who defeated Phlogiston. This figure demonstrates that the gain in weight upon calcination of lead involves a gain in mass not a loss as would occur if the metal lost Phlogiston.

leaving behind the calx. The diagram notes that the calx is heavier than the metal and this is impossible (unless phlogiston has negative mass). Thus "calorique," developed in France, receives kinder treatment than phlogiston, developed in Germany and championed in England.

1. C. Milbourne, *Panorama of Magic*, Dover, New York, 1962, pp. 27–31.

CHLORINE FAIRIES?

Real Fairy Folks or The Fairy Land of Chemistry (Lucy Rider Meyer, Boston, 1887) was a rather too precious take on Jane Marcet's marvelous *Conversations on Chemistry*, first published 80 years earlier. Twins (Joseph and Josephine or Joey and Jessie—sentimental descendants of Sol and Luna?) learn chemistry from their uncle Richard James, a chemist also known as "The Professor."

Chlorine fairies [Fig. 134(a)] are the molecules in chlorine gas. The chlorine atoms each have one arm (monovalent); they wear green dresses; the fully spread wings signal volatility [remember the winged dragon in Basil Valentine's Third Key—Fig. 17(c)?]. Bromine, a liquid, has one-armed fairies in red dresses with their wings folded; mild heating causes the bromine fairies to spread their wings and fly. The one-armed fairies in solid iodine wear purple dresses and have their wings folded and their legs tucked up. "My, my!" exclaimed Jessie "They must be just the *teentiest-weentiest* kind of people." Sodium and chlorine fairies wed to form salt [Fig. 134(b)] and their dress is now white (what else?) and their wings folded and legs tucked up. Hydrogen fairies and hydrochloric acid (really gaseous HCl) fairies are shown in Figures 134(c) and 134(d). Figure 135(a) correctly depicts the atmosphere, which is 80% nitrogen fairies and 20% oxygen fairies. The oxygen fairies correctly have two arms [see water fairies in Fig. 135(b)] but the nitrogen fairies should have three arms each rather than one—perhaps a bit too monstrous? Come to think of it, how would Ms. Meyer know how many arms were linking the fairy atoms? The octet rule remained some thirty years into the future.

Uncle Richard has his niece and nephew and their neighborhood friends sniff chlorine, bromine, and hydrogen sulfide. He also keeps a bottle of strychnine in the house to show to the children. He composes poetry: "Hg, Mercuree, What a poet, I be!." I wouldn't want him near my children. Michael Faraday was inspired by Jane Marcet's book to become a chemist. Had he read *Fairy Land of Chemistry* he might have become a CPA.

FIGURE 134 ■ From *Real Fairy Folks or The Fairy Land of Chemistry* (Lucy Rider Meyer, Boston, 1887): (a) Two chlorine fairies linked with one arm since chlorine's valence is one; (b) a chlorine fairy and a sodium fairy marry—their wings and legs are folded since salt is a solid; (c) hydrogen fairies each have one arm; (d) hydrochloric acid fairies form a sharp couple.

FAIRIES OF THE AIR.

(a)

FAIRY PICTURE OF WATER.

(b)

FIGURE 135 ■ (a) Note that Ms. Meyer's fairies of the air are in the correct proportion of 4:1 N_2 to O_2; (b) wouldn't mermaids have been better for water than fairies?

"RASCALLY" FLUORINE: A FAIRY WITH FANGS?

Uncle Richard finishes his lesson about the halogens by talking about fluorine:

> Fluorine is the last of the cousins. Its fairies are very wilful [sic], harder to catch, and harder still to keep. it is supposed that they have very active feet and wings, and wear the invisible cloak, but they are such little rascals that no one is quite sure of ever having caught them, separate from everything else.

Did Ms. Meyer know that Henri Moissan isolated fluorine gas in 1886, the year before her book was published? Perhaps. Fluorine is the most reactive element—the bonds in F_2 are quite weak, those between carbon and fluorine and in HF are incredibly strong. The molecule will grab electrons from almost anybody. It does not react with argon but does react with xenon. XeF_2 is stable relative to Xe and F_2 while Kr and F_2 are stable relative to KrF_2.[1] The mineral fluorspar (CaF_2) had been employed for hundreds of years and the presence of a fourth halogen that could not be separated from its compounds was understood by 1830.[2] It was known by 1670 that addition of oil of vitriol (sulfuric acid) produced a gas (HF) that could etch glass.[2] At least two early nineteenth-century chemists died exploring the chemistry of gaseous fluorine compounds and many others became seriously ill. Although we might think of fluorine as a fairy with fangs, it has been called the *Tyrannosaurus rex* of the elements, although I prefer to call it the Tasmanian devil of the elements.[3,4] Finally, Moissan obtained fluorine gas from potassium acid fluoride ($KF \cdot HF$ or KHF_2) in liquid HF (−50°C),[4] using electrolysis with inert platinum–iridium alloy in an inert platinum vessel.[2,4] Moissan's efforts in fluorine chemistry took a toll on his health as well. He received the Nobel Prize in Chemistry in 1906, months before he died at the age of 55. He won the Prize by one vote over Mendeleev, who died the following year and thus would never win it.[4]

1. I thank Professor Joel F. Liebman for this insight.
2. A.J. Ihde, *The Development of Modern Chemistry*, Harper and Row, New York, 1964, pp. 366–369.
3. G. Rayner-Canham, *Descriptive Inorganic Chemistry*, Freeman, New York, 1996, pp. 349–352.
4. D. Rabinovich, *The Chemical Intelligencer*, **3**: 64–65, October 1997.

A MID-SEMESTER NIGHT'S DREAM

The laboratory fairies at Haverford College recognized a talented artist in the 19-year-old Maxfield Parrish (1870–1966).[1] They did their best to whisk him through his chemistry course to his true calling as a painter, illustrator, and designer.[2] Indeed, he later placed these fairies on retainer and often used them as a leitmotif in his woodland scenes.[2]

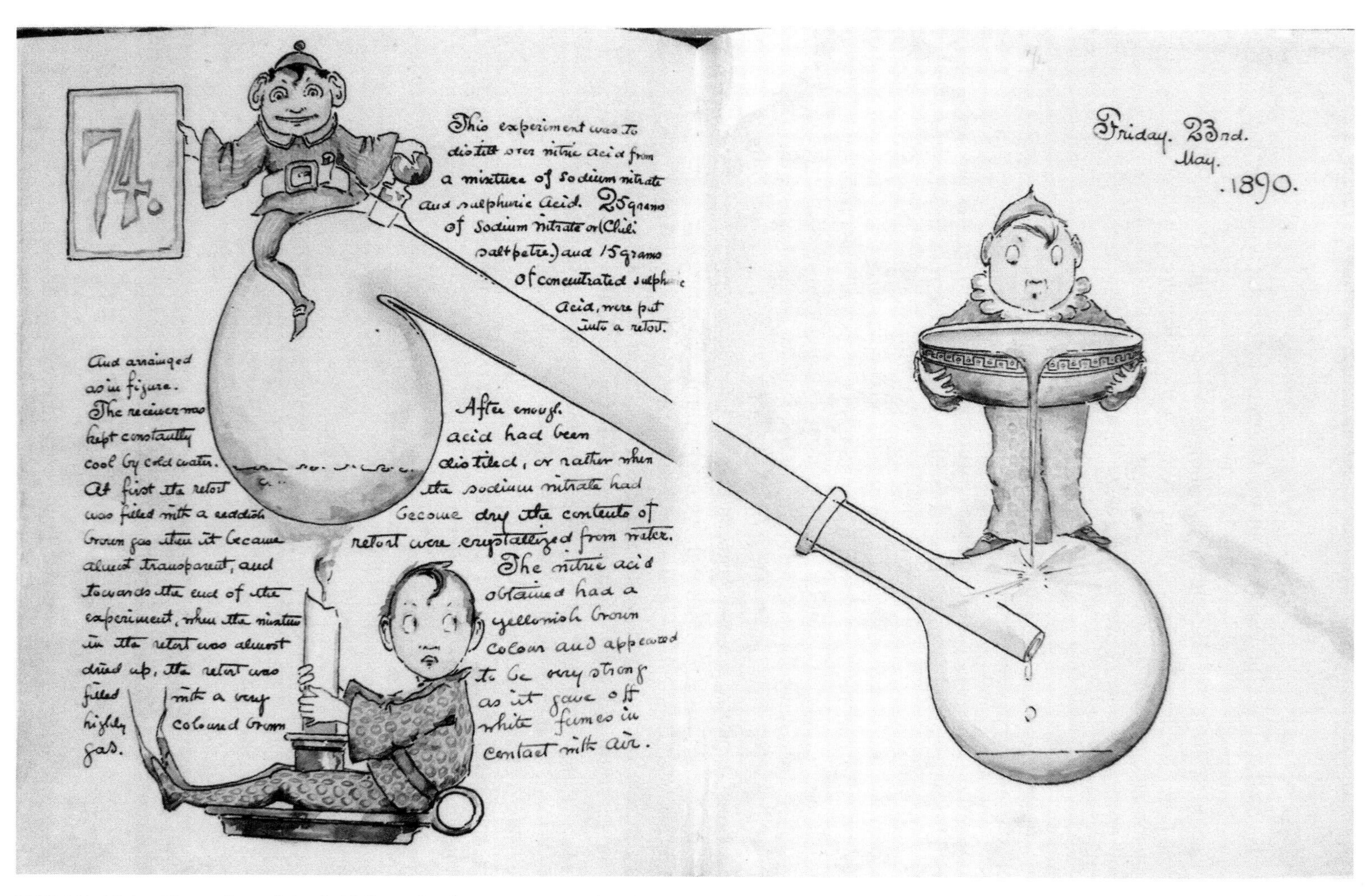

FIGURE 136 ■ Pages from Maxfield Parrish's beginning notebook. Courtesy of the Quaker Collection, Haverford College Library.

Figure 136 is from Parrish's laboratory notebook, presently part of the Quaker Collection of the Haverford College Library.[3] Now, how should a professor respond to such a notebook? On the one hand, we ask for scrupulous accuracy in the description of an experiment. However, it is highly unlikely that more than one fairy at a time assisted an individual student at Haverford. Indeed, would the fairies have consented to their portraiture? On the other hand, his professor, Lyman Beecher Hall, duly noted that Parrish's "observations and experimental summaries are concise and carefully written."[1] Since Professor Hall made very few notations in the book (and these in light pencil) and since Parrish presented him with the book some 20 years later (in 1910), we can safely assume that the course ended amiably for the young artist.

1. J. Chesick, *Chemical Heritage*, Vol. 17, No. 2, p. 42 (1999). The original figure (and drawing) is in color.
2. J. Turner (ed), *The Dictionary of Art*, Vol. 24, 1996, p. 210, New York: Macmillan.
3. Although Parrish's family were Quakers, he married a non-Quaker and, although a declaration of sincere interest would have allowed him to remain a Quaker, it may be supposed that he elected not to rejoin the Quakers. I am grateful for discussions with Diana F. Peterson, Haverford College Library, and Barbara Katus, Pennsylvania Academy of Fine Arts, which ran the first-ever critical retrospective of Maxfield Parrish in 1999.

AND NOW TURN TO PAGE 3 OF OUR CHEMICAL PSALM BOOK

The illuminated title page of the gentle 1873 English Christian psalm book *Chemistianity* is depicted in Figure 137.

> This work may prove a memory burnisher.
> To teen-youth or octagenarian,
> And act as match or chemistian torch
> For needed light to order Ignorance.

Its clarion call to study chemistry is a bit "forced":

> Chemistry lore should be
> Well known on land and sea
> To sow the seed of *Chemistry*, so heigh, so ho, so hee

Our Service begins on page 3 (ALL RISE):

> MATTER, is the body of the universe,
> That, by the aid of Chemical Science,
> With the best of all known appliances,
> Has been resolved into *Sixty-three bodies*
> (Or conditions of free, real essence)
> Term'd ELEMENTS, or Simple Substances;
> These, we have been unable to split up.

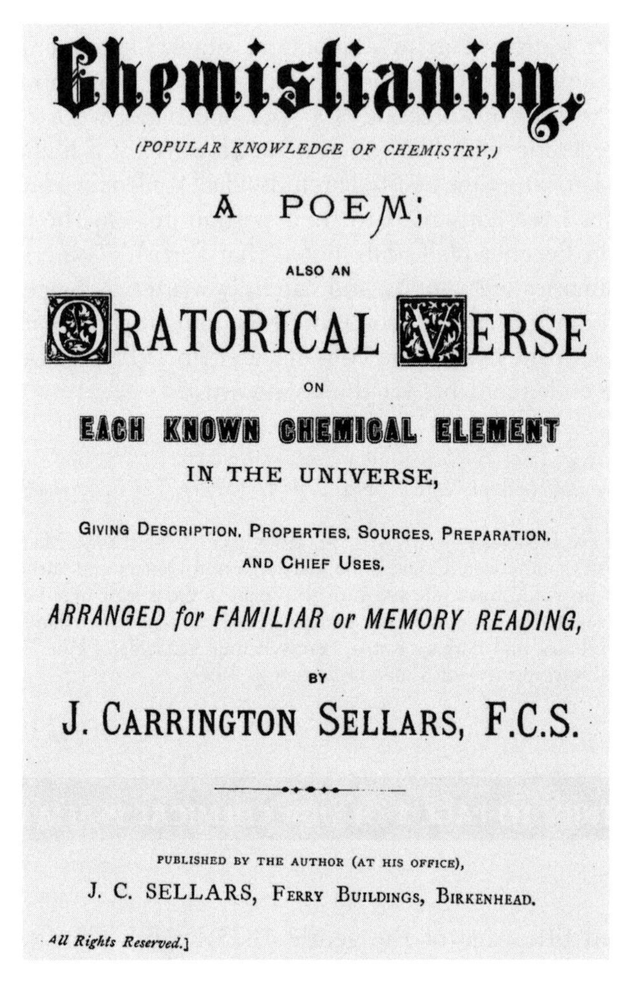

Chemistianity,

(POPULAR KNOWLEDGE OF CHEMISTRY,)

A POEM;

ALSO AN

ORATORICAL VERSE

ON

EACH KNOWN CHEMICAL ELEMENT

IN THE UNIVERSE,

GIVING DESCRIPTION, PROPERTIES, SOURCES, PREPARATION, AND CHIEF USES,

ARRANGED for FAMILIAR or MEMORY READING,

BY

J. CARRINGTON SELLARS, F.C.S.

PUBLISHED BY THE AUTHOR (AT HIS OFFICE),

J. C. SELLARS, FERRY BUILDINGS, BIRKENHEAD.

All Rights Reserved.]

FIGURE 137 ▪ The idea of this chemical Psalm book is to teach teens and octagenarians, who both supposedly have short memories, chemistry by reciting psalms. The poetry in this book is among the worst published and if you prefer calling glass "die-bee-day," then this is your book!

> Or subdivide, into more Primal being.
> Named in order of their combining weights,
> And forty-three known, proved, real *Metals*,
> Arranged under Chemist Roscoe's system,
> By classing in *ten families* or *Klans*;
> The bodies appertaining to each Klan
> Are writ in order of their *combining weight*
> Or type of their Chemical energy.

Please turn now to page 61:

> OXYGEN, the Queen of Body Affection;
> The supporter of man's Earthual life;
> The needed Air-puff for all common forms
> Of combustion in term'd live Animals,

In ordinary burning Wood or Coal;
And the prime mover in most heat-felt goceptions,
Is a colorless gaseous metalloid,
Tasteless and devoid of odour.

PLEASE BE SEATED

(The author has coined the term *goception* for chemical action and God is called *The Great Goceptor.*)

Sellars writes that: "In reading the names of chemical compounds, many persons are disappointed at their length and unmeaningness to them." (This remains a common complaint among students in Freshman Chemistry courses.) He, thus, develops a simpler alphabetical nomenclature which will be very briefly illustrated. For the five lightest elements known to the author we have:

Alphabetical Name	Composition Name in Brief		Pronounced Present Name
ABGEN	Ab	Abb	Hydrogen
AMYAN	Am	Amm	Boron
ATYAN	At	Att	Carbon
BAGEN	Ba	Bay	Nitrogen
BEGEN	Be	Bee	Oxygen

Using this nomenclature, water (H_2O, which we *could* call today dihydrogen oxide but don't) would be pronounced "die-abb-bee." Common glass (silicon dioxide) would have the pleasing sound "die-bee-day" and P_2O_3 the jolly "try-bee-die-dee." However, nitrous oxide or laughing gas (N_2O) is "die-bay-bee," not likely to encourage a dental patient, but fortunately it is not N_3O, pronounced "try-bay-bee."

This gentle and heartfelt effort, doomed by its doggerel and nomenclature, is a compelling argument for separation of Church and Oxidation State.

MOLECULAR MECHANICS IN THE YEAR 1866

The desire to calculate mathematically the shapes of molecules and the forces that hold them together has been with us since the development of Newtonian physics. John Freind's 1704 lectures published in his *Chymical Lectures* (London, 1712; see Figs. 104 and 105) were an early attempt to apply Newtonian physics to the problem. At the start of the twentieth century it became abundantly clear that quantum mechanics (ultimately, the solution of the Schrödinger equation developed in 1926) was required to solve these problems. Classical Newtonian physics was simply inappropriate for calculating the properties of electrons in atoms. However, accurate quantum calculations of really interesting molecules (say, more than five atoms but not huge) had to await the computing power of the late twentieth century. The value of such calculations was explicitly recognized in 1998 by the award of the Nobel Prize in Chemistry to John A. Pople and Joachim Kohn.

But what of the large unsymmetrical molecules found in nature—alkaloids such as morphine, the plethora of complex proteins? Here, a technique suggested initially by Frank Westheimer and now widely termed *molecular mechanics* is employed.[1] It is based essentially on classical physics. Bonds are treated as springs subject to Hooke's Law. The parameters used are derived from experiment—there are lots of them. The technique is both theoretically indefensible and incredibly useful. It forms the basis of the programs widely used in the pharmaceutical industry to design new drugs.

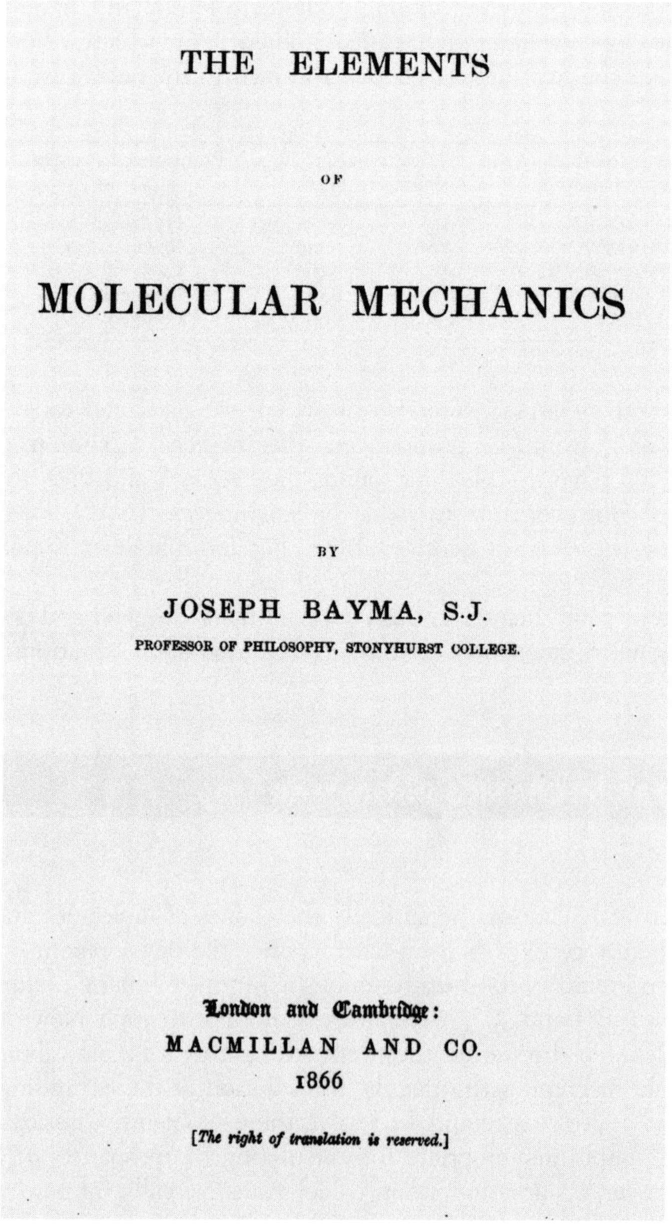

THE ELEMENTS

OF

MOLECULAR MECHANICS

BY

JOSEPH BAYMA, S.J.

PROFESSOR OF PHILOSOPHY, STONYHURST COLLEGE.

London and Cambridge:

MACMILLAN AND CO.

1866

[*The right of translation is reserved.*]

FIGURE 138 ■ Molecular mechanics in the year 1866?

Now we wish to make it clear that Joseph Bayma did not "scoop" Allinger and Burkert[1] and there are no copyright or patent infringements. His book (Fig. 138) contains lots of math and physics and virtually no chemistry. Apparently, the author did present his work before the Royal Society and it met a rather skeptical audience. An English book dealer friend of mine called it a "Nutter." Nonetheless, it is interesting. Book IV, *Dynamical Constitution of Primitive Polyhedric Systems of Elements*, offers as "Problem I. Four repulsive elements having equal powers are so arranged as to form a regular tetrahedron around an attractive centre. Find the dynamical formula of this system." Do you think van't Hoff and Le Bel were reading this? What about Nyholm and Gillespie and the VSEPR (valence shell electron pair repulsion) theory of the mid-twentieth century?

1. N.L. Allinger and I. Burkert, *Molecular Mechanics*, American Chemical Society, Washington, D.C., 1977.

SECTION VIII
THE APPROACH TO MODERN VIEWS OF CHEMICAL BONDING

RIDING PEGASUS TO VISIT CHEMISTRY IN SPACE

Optical activity was a fundamental mystery of matter during most of the nineteenth century. Jean Baptist Biot discovered that certain minerals were optically active—they rotated the plane of polarized light. In 1815 he found that certain liquids, oil of turpentine and camphor in alcohol solution for example, were also optically active.[1] However, it was Louis Pasteur's genius that perceived the molecular connection in 1848 even though rational structural chemistry remained some fifteen years or so in the future.

Pasteur first stated the oft-quoted: "Chance favors only the prepared mind."[2] Indeed, serendipity was working in his favor in a (fortunately) cold laboratory in Dijon when he crystallized sodium ammonium tartrate. A close look at the large hemihedral crystals indicated that they were "right-handed" and "left-handed" in the sense of being mirror images (like our hands or feet) that cannot be superimposed point-for-point on each other. (Structures VIII and IX in Figure 139 are flat pictures of right-handed and left-handed hemihedral crystals of ammonium bimalate—the three-dimensional structures are not superimposable.) Meticulously separating the two sets of crystals by hand and dissolving each set in separate solutions, Pasteur discovered that each solution was optically active—but in an equal, yet opposite sense. One solution rotated the plane of polarized light clockwise (called *dextrorotatory*); the other solution was *levorotatory*. Pasteur had affected the first resolution of an equal mixture of enantiomers termed the racemate.

Pasteur's observations began to connect with others.[1] For example, in 1770 Scheele had isolated lactic acid [$CH_3CH(OH)COOH$] from fermented milk. In 1807, Berzelius isolated lactic acid from muscles. Subsequently, lactic acid from fermented milk was found to be optically inactive while that from muscle was found to be optically active. What was the origin of this dichotomy?

The solution to the problem was discovered in 1874 by Jacobus Henricus van't Hoff, 22 years old, and Joseph Achille Le Bel, age 27. Although they both worked in the laboratory of Adolph Wurtz in Paris in 1874, their discoveries were completely independent.[1,2] Van't Hoff would continue to make major contributions to physical chemistry and won its first Nobel Prize (1901) for his discovery of laws of osmotic pressure of solutions.

In Figure 139 we see the plate printed in the first English edition[3] of van't Hoff's work, translated from the second French edition. The two young chemists postulated that a carbon atom at the center of a tetrahedron with four different atoms or groups attached to it (at the corners of the tetrahedron) would be asymmetric, existing as nonsuperimposable mirror images. These were the enantiomers earlier described. Structures I and II in Figure 139 show flat formulas

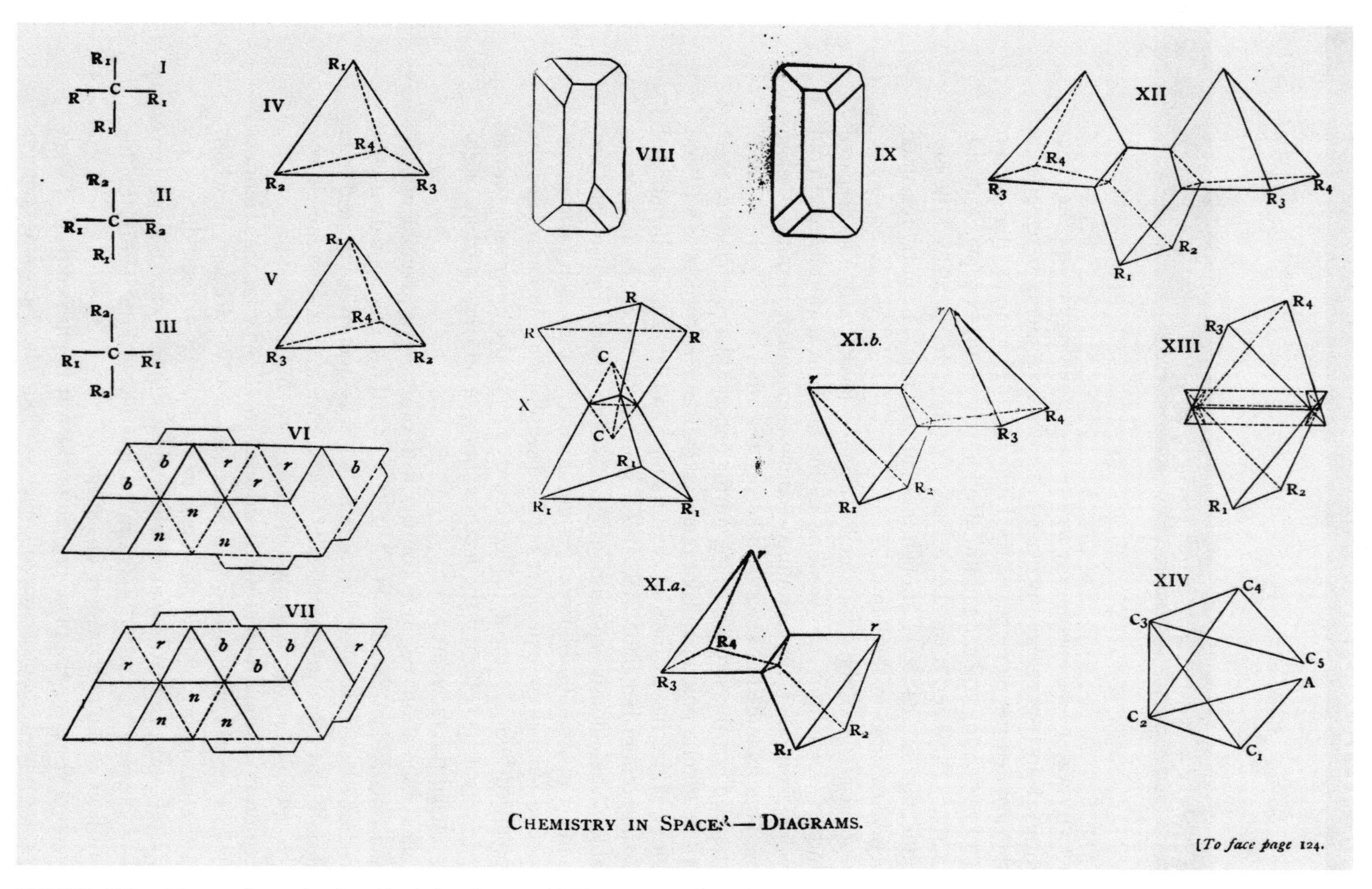

FIGURE 139 ▪ Figures from the first English edition of J.H. van't Hoff's *Chemistry in Space* (Oxford, 1891). His discovery simultaneously with (independently of) LeBel may have played some role in his receipt of the first Nobel Prize in Chemistry (1901). However, the prize was awarded specifically for his discovery of laws of osmotic pressure of solutions.

of generalized enantiomers with four different groups (R_1 to R_4) attached to the asymmetric carbon. Structures III and IV are the corresponding three-dimensional tetrahedral representations that are not superimposable. (Structure V depicts a nonasymmetric carbon since it is attached to two identical groups —no enantiomers are possible.) Structures VI and VII are cut-outs for making models of structures IV and V (*n* is black; *r* is red; *b* is blue; unmarked parts are white; see the beautiful book by Heilbronner and Dunitz[4] for a photo of van't Hoff's personal set of handmade models).

The solution to the lactic acid dichotomy was now clear. Lactic acid has an asymmetric carbon atom. The four different groups (R_1 to R_4 in structures I and II or III and IV of Fig. 139) are H, CH_3, OH, and COOH. Scheele's lactic acid from fermented milk had both enantiomers in equal quantity (the racemate) and was optically inactive, while Berzelius' lactic acid from muscle was optically active because only one enantiomer was present.

In 1876, van't Hoff was appointed to a junior-faculty position at the Veterinary College of the University of Utrecht in Holland. In 1877, the 1875 French translation of his work was translated into German. He received very strong support from Johann Wislicenus at Wurzberg but a far different reception from Herr Professor Doktor Hermann Kolbe at Leipzig:

> A Dr. J.H. Van't Hoff, of the Veterinary College, Utrecht, appears to have no taste for exact chemical research. He finds it a less arduous task to mount his Pegasus (evidently borrowed from the Veterinary College) and to soar to his Chemical Parnassus, there to reveal in his "La Chimie dans L'Espace" how he finds the atoms situate in the world's space.[5]

Nasty stuff! Sadly, this may be Kolbe's most quoted passage although he was an accomplished scientist and over 30 years earlier hammered the nails into Vitalism's coffin (see pp. 203–204). Ironically, it was Wislicenus who succeeded Kolbe in the Chair at Leipzig in 1885.

Structure X in Figure 139 shows interpenetrating tetrahedra with carbon centers and a single bond between these carbons. Van't Hoff correctly postulated that there is free rotation about such single bonds. Structures XIa, XIb, XII, and XIII rationalize *cis* and *trans* isomerism (e.g., the difference between maleic acid and fumaric acid). Structure XIV explains the widespread occurrence of six-membered rings in chemistry and aspects of Baeyer's strain theory.[6]

We conclude this tour of the molecular third dimension with a bit of verse by occasional poet, full-time theoretician and long-time friend Joel F. Liebman[7]:

Owed to van't Hoff and Le Bel

Lacking magnifying scopes and magic wands
We cannot see molecules and their bonds.
No need though, for it's plain to see
That tetracoordination means planar C*
(What else can it be?)

Enter van't Hoff, Le Bel and their dissensions:
Molecules, invisible, but in three dimensions.

How so? It's really plain to see;
Four-bonded carbon links tetrahedrally**
(What else can it be?)

*Now we clearly don't mean as square,
For that, of course, would be unfair.
Should not the groups with bigger heft
Get more room, small ones get what's left.
Four groups form a quadrilateral;
Any disputation is caterwaul.

**Now clearly we don't mean T_d
Bigger groups remain greedy.
Should not the groups with bigger heft
Get more room, small ones get what's left.
All angles not arccosine minus one third.
That's it, no need for another word.

Joel F. Liebman

1. J.R. Partington, *A History of Chemistry*, MacMillan, London, 1964, Vol. 4, pp. 749–759.
2. W.H. Brock, *The Norton History of Chemistry*, Norton, New York, 1993, pp. 257–264.
3. J.E. Marsh, *Chemistry In Space, from Professor J.H. van't Hoff's "Dix Annees Dans L'Histoire D'Une Theorie,"* Clarendon, Oxford, 1891.
4. E. Heilbronner and J.D. Dunitz, *Reflections on Symmetry*, VCH, Weinheim, 1993.
5. J.E. Marsh, op. cit., p. 16.
6. For a discussion of strain theory see A. Greenberg and J.F. Liebman, *Strained Organic Molecules*, Academic, New York, 1978. For a masterful treatment of stereochemistry, see E.L. Eliel and S. Wilen, *Stereochemistry*, Wiley, New York, 1996.
7. *Journal of Molecular Structure* (*Theochem*), Vol. 338 frontis matter (1995). Courtesy Professor Joel F. Liebman.

IS THE *ARCHEUS* A SOUTHPAW?

Pasteur's brilliant work with enantiomers and racemates included his realization that a single enantiomer of one substance (e.g., an optically active base) could be used to separate the enantiomers of another (e.g., a racemic acid). The analogy with hands is simple—we refer to molecules that have nonsuperimposable mirror images as chiral (handed). A right glove can differentiate ("separate") a right hand from a left hand. Pasteur and others soon realized that all of their optically active compounds had come from living organisms. Lactic acid from muscle was optically active; synthetic lactic acid was not. Moreover, living organisms readily resolved racemates by selectively metabolizing one enantiomer. Thus, Pasteur incubated 3 g of optically inactive secondary amyl alcohol (the four groups attached to the asymmetric carbon are H, OH, CH_3, and C_3H_7) with a suspension containing yeast mold. After one month, the alcohol distilled from the mixture was found to be dextrorotatory.[1] The yeast had selectively metabolized the levorotatory enantiomer.

Was there a Vital Force in living organisms that allowed them to be the only source of optically active substances? Was this Vital Force ultimately the only means for resolution of racemates? Do you remember our earlier discussion of the *Archeus*, the Spiritual Alchemist, a Vital Spirit, thought by Paracelsus to reside near our stomachs (see Fig. 64)? The *Archeus* was thought to have a head and two hands only and to function by separating the nutritional from the poisonous parts of food and air. Now, if the *Archeus* were left-handed, for example, we might have understood earlier how the body separates left from right. Happily, there was no return to Vitalism by serious scientists and today billions of dollars are earned by companies that have learned to make the pure enantiomer of a drug without contamination from the other enantiomer and without paying the salaries of any *Archei*.

1. J.E. Marsh, *Chemistry In Space From Professor J.H. van't Hoff's "Dix Annees Dans L'Histoire D'Une Theorie,"* Clarendon, Oxford, 1891, p. 41.

JOHN READ: STEREOCHEMIST

I have expressed my admiration at other points in this book for the prodigious intelligence, scholarship, and wit of Dr. John Read who wrote the wonderful trilogy on alchemy and chemistry. Read was an early stereochemist "present at the creation" of at least two very important discoveries in the field.

Interestingly enough, while both van't Hoff and Le Bel postulated a single asymmetric carbon as necessary for optical activity, by the end of the nineteenth century no optically active compounds had been isolated having fewer than three carbon atoms in a chain.[1] This prompted Norwegian chemist F. Peckel Möller to postulate his "Screw-Theory" in a section of his book *Cod-Liver Oil and Chemistry*[2] titled "Position of Atoms In Space."[3] A believer in the Universal Ether, disproven by the Michelson–Morley experiment published in 1887 but still adhered to by famous scientists including Mendeleev, Moller postulated that a three-carbon chain is the minimum requirement for optical activity. The idea was that three carbons in a zigzag chain were the minimum for a chiral corkscrew capable of creating right-handed or left-handed vortices through the ether thus accounting for dextro- or levorotatory properties. Ironically, the second English edition of van't Hoff's work,[1] published in 1898, three years after Moller's work, adds the new axiom of the three-carbon requirement. I would have liked to have seen van't Hoff's face when he learned of this piece of editing.

The first optically active compound containing only one carbon atom [$CHClI(SO_3H)$] was reported in 1914.[4] It was synthesized and optically resolved by the great stereochemist William Jackson Pope and our own John Read at Cambridge University.[3] Pope and his co-workers extended stereochemical concepts from carbon only to nitrogen, phosphorus, sulfur, selenium, silicon, and tin.[3]

Read had earlier obtained his Ph.D. with Alfred Werner at the University of Zurich. Werner received the 1913 Nobel Prize in Chemistry for his extension of chirality to metallic compounds. Strictly speaking, there were six carbons in the cobalt compound whose resolution was reported in 1911.[5] However, the molecule's chirality was due to the spatial relationship about the hexacoordinate cobalt not to the carbons. Werner's collaborator in this revolutionary resolution —John Read.[6]

However, the history of this discovery is not so simple. It appears that Edith Humphrey, one of the very few women engaged in doctoral research 100 years ago, is likely to have actually made the original resolution 10 years before John Read, although it was not realized at the time.[6] Dr. Humphrey died in 1977 at the age of 102. At her 100th birthday she is quoted as saying "There were very few women students in Zurich, but fairly soon I was made assistant to the professor. I think being English helped—and also I knew more physical chemistry than most people there."[6]

1. J.H. van't Hoff, *The Arrangement of Atoms in Space*, 2nd ed., A. Eiloart (translator), Longmans, Green, London, 1898.
2. F.P. Möller, *Cod-Liver Oil and Chemistry*, Peter Möller, London, 1895.
3. A. Greenberg, *Journal of Chemical Education*, **70**: 284–286, 1993.
4. W.J. Pope and J. Read, *Journal of the Chemical Society (London)*, **105-I**: 811, 1914.
5. A. Werner, *Berichte*, **44**: 2447, 1911.
6. I. Bernal, *The Chemical Intelligencer*, **5**, (1):28–31, January, 1999.

FINDING AN INVISIBLE NEEDLE IN AN INVISIBLE HAYSTACK

Atmospheric air should be colorless and odorless although over certain segments of the New Jersey Turnpike it can be seen and even tasted. During the 1770s Scheele and Priestley demonstrated that the atmosphere is roughly 80% phlogisticated air (nitrogen) and 20% dephlogisticated air (oxygen). [Figure 135(a) depicts four nitrogen fairy couples and one oxygen fairy couple.]

During the 1890s, Lord Rayleigh (John William Strutt, Third Baron), a physicist, and chemist William Ramsay noted inconsistencies between the densities of "chemical nitrogen" and "atmospheric nitrogen." The density, at 0°C and 760 mm of "atmospheric nitrogen" (1.2572 g per liter) was apparently about six-tenths of 1% greater than that of "chemical nitrogen" (1.2505 g per liter). "Chemical nitrogen" had been synthesized through reaction of nitric oxide (NO) or nitrous oxide (N_2O, laughing gas) with hydrogen gas, heating of ammonium nitrite (NH_4NO_2), or reaction of urea (NH_2CONH_2) with sodium hypochlorite (NaOCl, pool disinfectant). The source of the discrepancy could be the presence of a light impurity, such as traces of residual hydrogen, in "chemical nitrogen" or, more likely, a heavy impurity in "atmospheric nitrogen." Rayleigh and Ramsay roughly estimated that this impurity might be present at a level of around 1%. It was hard to imagine that the very air they breathed could contain 1% of a hitherto-unknown substance[1]:

> The simplest explanation in many respects was to admit the existence of a second ingredient in air from which oxygen, moisture, and carbonic anhydride had already been removed. The proportional amount was not great. . . . But in accepting this explanation, even provisionally, we had to face the improbability that a gas surrounding us on all sides and present in enormous quantities could have remained so long unsuspected.

The meticulous work of Rayleigh and Ramsay led them to discover the gaseous element argon in 1894. They withheld announcement while they submitted their paper for the Smithsonian Institution's Hodgkin's Prize for the most important discovery related to atmospheric air.[2] They published their work in the *Philosophical Transactions of the Royal Society* during 1895 and the prize-winning paper was published by the Smithsonian in 1896.[1] Among their numerous careful experiments was the generation of "chemical nitrogen" from "atmospheric nitrogen" by removal of carbon dioxide and water from air using soda-lime and phosphoric anhydride and removal of oxygen through exposure to red-hot copper. The remaining "atmospheric nitrogen" was then ignited over magnesium at a "bright-red" heat to form powdery magnesium nitride (Mg_3N_2). Addition of water to the nitride produced ammonia (NH_3), which oxidized with calcium hypochlorite [$Ca(OCl)_2$] to produce "chemical nitrogen." Oxygen reacts rapidly with copper to form a salt. Nitrogen, being much less chemically reactive than oxygen, escapes red-hot copper unscathed. Magnesium is a much more reactive metal than copper. Indeed, it was unknown as a free metal until freed by Davy from its compounds in 1808 using a voltaic pile.

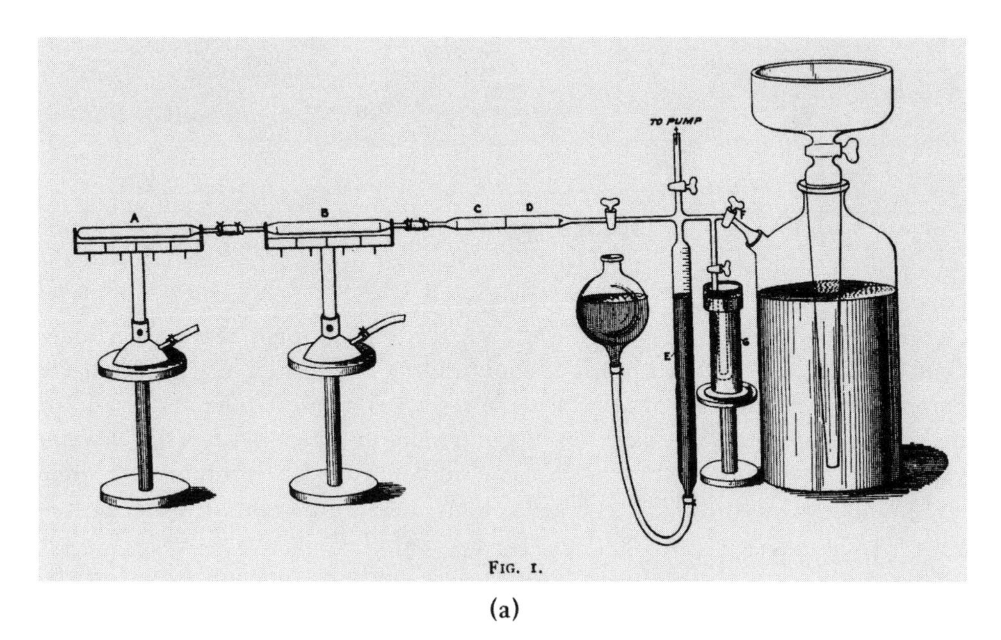

(a)

FIGURE 140 ■ (a) to (c) are described in the text. They are from the prize-winning essay published by Lord Rayleigh and William Ramsay (*Argon, a New Constituent of the Atmosphere*, Washington, D.C., 1896). Rayleigh had noted that atmospheric nitrogen is very slightly more dense than "chemical" nitrogen. After removing water and carbon dioxide from air, oxygen was removed with red-hot copper and then magnesium burned in the remaining nitrogen. The unreacted gas, comprising less than 1%, was mostly argon. *Illustration continued on following page*

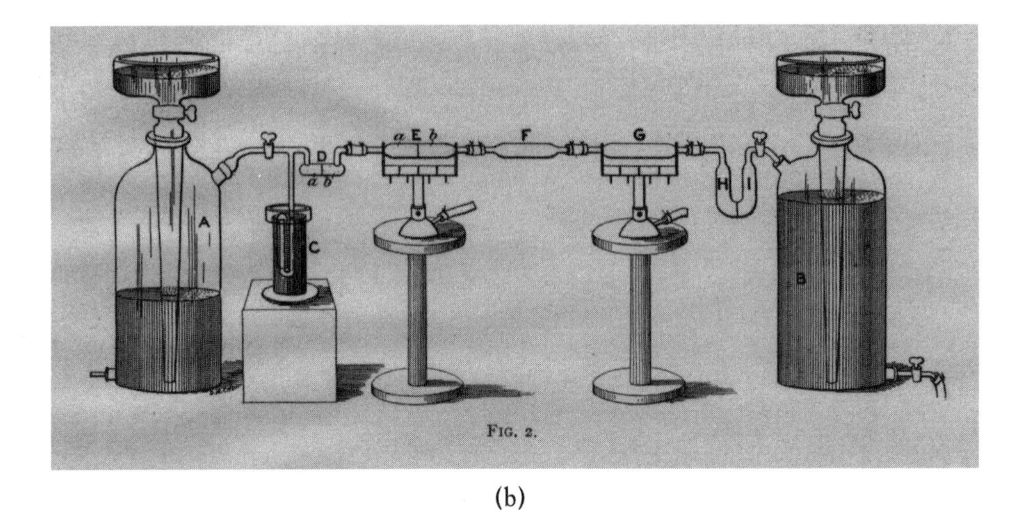

(b)

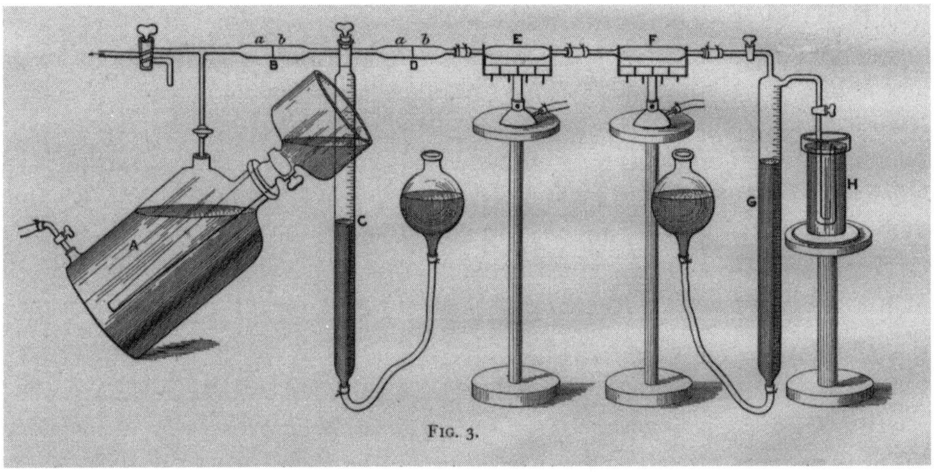

FIG. 3.

(c)

FIGURE 140 ■ *Continued*

The apparatus in Figure 140(a) (see the Smithsonian report[1]) includes combustion tube *A* filled with magnesium turnings and heated over a wide-flame burner and combustion tube *B* filled with copper oxide (to remove residual hydrogen gas generated by reaction of magnesium in tube *A* with residual water vapor) and also heated with a wide-flame burner. Tube *CD* contains soda-lime and phosphoric anhydride, *E* is a gas volume measuring vessel, *F* is connected with the "atmospheric nitrogen" gas holder, and G stores unabsorbed gas after each cycle. Figure 140(b) shows a larger-scale apparatus in which gas can be introduced via C into gas holder *A*. Tube *D* is filled with soda-lime [in Fig. 140(a)] and phosphoric anhydride [in Fig. 140(b)]; combustion tube *E*, heated with a wide flame, is half-filled with porous copper and half with granular copper oxide; tube *F* contained granular soda-lime and G contains magnesium turnings heated to bright red over a wide-flame burner; *H* contains phosphoric anhydride and *I* soda-lime. Nitrogen prepared by passing atmospheric air through red-hot copper is introduced via C into vessel *A*. Over the course of 10 days this nitrogen is passed slowly back and forth between *A* and *B*. Magnesium is replenished as needed. The remaining small residue of gas was transferred to the apparatus in Figure 140(c), which was designed to exclude atmospheric air in the remaining operations.

It was difficult to accurately determine the density of argon since there were impurities, mainly nitrogen, associated with it. Values were typically in the range of 1.75 to 1.82 g per liter, approximately 20 times that of hydrogen (H_2) gas. Since the molecular weight of hydrogen is 2.0 amu, then the "molecular" weight of argon should be about 40 amu.

Rayleigh and Ramsay characterized the new gas by observing its light spectrum: "The spectrum seen in this tube has nothing in common with that of nitrogen, nor indeed, so far as we know, with that of any known substance."[1] They tested the reactivity of this new element with about every nasty chemical they could and found it totally unreactive. They gave this new element the name argon derived from the Latin *a* (without) and *ergon* (work), meaning "idle."

And in an eloquent salute to Henry Cavendish, who first reported in 1785 the isolation of an unreactive gas comprising $^1/_{120}$ of the phlogisticated air, Rayleigh and Ramsay write[1]:

> Attempts to repeat Cavendish's experiment in Cavendish's manner have only increased the admiration with which we regard this wonderful investigation. Working on almost microscopical quantities of material and by operations extending over days and weeks, he thus established one of the most important facts in chemistry. And what is still more to the purpose, he raises as distinctly as we could do, and to a certain extent resolves, the question above suggested.

1. Lord Rayleigh and Professor William Ramsay, *Argon, A new Constituent of the Atmosphere*, Smithsonian Institution, Washington, D.C., 1896.
2. W.H. Brock, *The Norton History of Chemistry*, Norton, New York, 1993, pp. 331–340. This is an especially enjoyable and accessible discussion.

BUT ARGON IS A MONATOMIC GAS—AND THERE ARE OTHERS!

There is another amazing aspect in the discovery of argon beyond its total chemical inertness. Rayleigh and Ramsay reported measurements of the speed of sound in argon that indicated that the ratio of its heat capacity at constant pressure to that at constant volume (C_P/C_V) was too high for a diatomic molecule. The only other similar observation was for monatomic mercury (vapor) whose atomic weight was known since it forms compounds. At constant volume, heat added to a diatomic molecule such as N_2 goes into both movement of the molecule (translation) as well as vibration of the bond. In a monatomic substance there is no bond vibration and, thus, less capacity to absorb heat.

The finding that argon is a monatomic gas and has an atomic weight of 40 dealt a severe jolt to the established order.[2] First, if it was a diatomic molecule, its atomic weight would be about 20 (see above), thus fitting it confusingly well between fluorine (19) and sodium (23). However, a new monatomic substance with an atomic weight of 40 would not only require a new and totally unanticipated family in the Periodic Table, it coincided with the atomic weight of calcium and messed up the order that Mendeleev first employed to organize his table. These findings did indeed upset Mendeleev and his students.[2] Rayleigh and Ramsay themselves noted: "If argon be a single element then there is reason to doubt whether the periodic classification of elements is complete."[1] Their report[1] concluded: "We would suggest for this gas, assuming provisionally that it is not a mixture, the symbol A" (later changed to Ar).

At the end of the nineteenth century techniques were developed to liquefy air by cooling and expansion. The front page of the Sunday, December 30, 1900 issue of *The Brooklyn Daily Eagle* gives the following page-wide headline: "LIQUID AIR WILL OPEN UP A NEW WORLD OF WONDERS" and under it a subheadline: "Pictet, Foremost of Savants, Calls the Liquid the Elixir of Life, and Declares It Will Banish Poverty From the Earth."[3,4] Using similar techniques to condense air, in 1898 Ramsay discovered the related inert or "noble" gases neon (Ne), krypton (Kr), and xenon (Xe). Helium (He), as its name bears witness, was discovered on the sun in 1868 through its light spectrum measured during a solar eclipse. It was isolated by Ramsay in 1895 through heating uranium ores. For their studies, Rayleigh received the Nobel Prize in Physics in 1904 and Ramsay the 1904 Nobel Prize in Chemistry. In 1908, Ramsay isolated the last of the inert gases, radioactive radon (Rn) from radium-containing minerals.

In his enjoyable book *The Periodic Kingdom*,[5] P.W. Atkins describes the Periodic Table as a land of mountains, valleys, lakes, and shores. The noble gases are termed a strip of land on the eastern shore and Atkins notes that " . . . no other complete strip of land of the kingdom owes so much to a single person" —Ramsay.

1. Lord Rayleigh and Professor William Ramsay, *Argon, A New Constituent of the Atmosphere*, Smithsonian Institution, Washington, D.C., 1896.
2. W.H. Brock, *The Norton History of Chemistry*, Norton, New York, 1993, pp. 331–340. This is an especially enjoyable and accessible discussion.
3. Special *Newsday* reproduction of *The Brooklyn Daily Eagle*, Vol. 60, No. 360, Sunday, December 30, 1900.
4. Not to be too curmudgeonly about it, but note that in Brooklyn at least, the end of the century was properly celebrated and not snuck in at the end of 1899—mathematical authority still held sway over Madison Avenue if the latter indeed existed.
5. P.W. Atkins, *The Periodic Kingdom*, Basic Books, New York, 1995, pp. 53–54.

JUST HOW MANY DIFFERENT SUBSTANCES ARE IN ATMOSPHERIC AIR?

How many substances there are in air depends upon how low you will go (in measuring concentrations). At the percent (part-per-hundred or pph) level, there is only nitrogen (78.08%) and oxygen (20.95%).[1,2] If we stretch a bit and add argon (0.93%), over 99.9% of the dry atmosphere is accounted for by just three substances. Water concentrations can vary over five orders of magnitude and actually reach percent levels in tropical rain forests.[1] These percentages are volume/volume (v/v), and since equal numbers of gas molecules occupy equal volumes under the same pressure and temperature, that means that one thousand molecules of dry air will have on average 780 N_2 molecules, 210 O_2 molecules, and 9 argon atoms. Carbon dioxide is present at about 350 parts-per-million (ppm). Other gases at or near the low-ppm levels include Ne, He, methane (CH_4), and Kr giving a total of nine substances including water. At the parts-per-billion (ppb) level, we start adding hydrogen, carbon monoxide, sulfur dioxide, ammonia, and ozone. Below that, in the ppb to ppt (parts-per-trillion) range we encounter oxides of nitrogen and hundreds of organic vapors such as benzene, toluene, and tetrachloroethylene.[3] Indeed the number of expected organic air pollutants at the trace level numbers in the thousands.[4]

What is a part-per-billion? Imagine adding a drop of alcohol to a pool of water 6 ft deep × 12 ft wide × 18 ft long and stirring thoroughly. Alternatively, imagine a golf foursome compared to the world's total population.[5]

1. T.E. Graedel and P.J. Crutzen, *Atmospheric Change: An Earth System Perspective*, Freeman, New York, 1993, p. 8.
2. J.H. Seinfeld, *Atmospheric Chemistry and Physics of Air Pollution*, Wiley, New York, 1986, p. 8.
3. B.B. Kebbekus and S. Mitra, *Environmental Chemical Analysis*, Blackie Academic and Professional, London, 1998, pp. 229–230.
4. T.E. Graedel, D.T. Hawkins, and L.D. Claxton, *Atmospheric Chemical Compounds: Sources, Occurrence, and Bioassay*, Academic, Orlando, 1986.
5. Thanks to Professor Joel F. Liebman for this suggestion.

ATOMS OF THE CELESTIAL ETHER

Early hints of the wave nature of light included the seventeenth-century discovery of diffraction by Hooke and other manifestations of interference. It was obvious that dropping a rock into a pond created waves, and Boyle showed that air was necessary for the transmission of sound waves. Thus, it appeared that there had to be a medium for transmitting light waves and it was thought to be a kind of "universal ether"—present everywhere, yet imperceptable. During the 1880s, the physicists Michaelson and Morley disproved, experimentally, the existence of the ether. Nevertheless, the concept continued to influence many outstanding scientists for perhaps two more decades. In a book published in 1895 titled *Cod-Liver Oil and Chemistry*, the author Friedrich Möller explains the rotation of plane-polarized light, clockwise or counterclockwise, by invoking clockwise or counterclockwise rotation of a bond in the molecule producing clockwise or counterclockwise "wakes" in the ether.

Mendeleev was clearly a believer in the ether. His explanation was straightforwardly chemical and constructed from his Periodic Table and the newly discovered inert gases.[1] The 1904 English edition of Mendeleev's book *An Attempt Toward a Chemical Conception of the Ether* appeared when the Russian master was 70. He postulates that the ether is composed of atoms of an as-yet-unknown superlight inert gas. Clearly, the gas must be inert in order to penetrate all matter without being reacted or absorbed and clearly it must be superlight not to be perceived.

He fits the "ether element" into his Periodic Table in the manner shown in Figure 141. Mendeleev placed the inert gases in Group 0, to the left of hydrogen and the alkali metals. This places helium in Period 2 and leaves a gap to the left of hydrogen in Period 1. Our modern Periodic Tables place the inert gases in Group 18 (8A in some versions) and thus helium now sits in Period 1 for reasons theoretical as well as practical. Mendeleev postulated a new Group 0–Period 1 element, element y in the accompanying figure, which he calculated to have a relative atomic weight of 0.4 (hydrogen = 1.0) and notes that while this is clearly far too massive for atoms of the ether, it may correspond to unassigned lines in the solar spectrum (remember, helium was already known). He then postulates another new element x (see Fig. 141) in the Group 0–Period 0 space, which he reasons has a relative mass in the range 0.00000096–0.000000000055, the atom comprising the celestial ether.

This all-too-human attempt by Mendeleev to cram the ether concept into his Periodic Table illustrates our very human limitations in trying to fit our own world views to facts. Figure 142 depicts mid-nineteenth-century illustrations of dinosaurs. The bones were "crammed" into the shapes of bear-like or ox-like creatures because these were the largest land carnivores and herbivores then

Series	Zero Group	Group I	Group II	Group III	Group IV	Group V	Group VI	Group VII	Group VIII
0	x								
1	y	Hydrogen H=1·008							
2	Helium He=4·0	Lithium Li=7·03	Beryllium Be=9·1	Boron B=11·0	Carbon C=12·0	Nitrogen N=14·04	Oxygen O=16·00	Fluorine F=19·0	
3	Neon Ne=19·9	Sodium Na=23·05	Magnesium Mg=24·1	Aluminium Al=27·0	Silicon Si=28·4	Phosphorus P=31·0	Sulphur S=32·06	Chlorine Cl=35·45	
4	Argon Ar=38	Potassium K=39·1	Calcium Ca=40·1	Scandium Sc=44·1	Titanium Ti=48·1	Vanadium V=51·4	Chromium Cr=52·1	Manganese Mn=55·0	Iron Fe=55·9 Cobalt Co=59 Nickel Ni=59 (Cu)
5		Copper Cu=63·6	Zinc Zn=65·4	Gallium Ga=70·0	Germanium Ge=72·3	Arsenic As=75·0	Selenium Se=79	Bromine Br=79·95	
6	Krypton Kr=81·8	Rubidium Rb=85·4	Strontium Sr=87·6	Yttrium Y=89·0	Zirconium Zr=90·6	Niobium Nb=94·0	Molybdenum Mo=96·0	—	Ruthenium Ru=101·7 Rhodium Rh=103·0 Palladium Pd=106·5 (Ag)
7		Silver Ag=107·9	Cadmium Cd=112·4	Indium In=114·0	Tin Sn=119·0	Antimony Sb=120·0	Tellurium Te=127	Iodine I=127	
8	Xenon Xe=128	Caesium Cs=132·9	Barium Ba=137·4	Lanthanum La=139	Cerium Ce=140	—	—		— — — (—)
9		—	—		—	—	—	—	
10	—	—	—	Ytterbium Yb=173	—	Tantalum Ta=183	Tungsten W=184	—	Osmium Os=191 Iridium Ir=193 Platinum Pt=194·9 (Au)
11		Gold Au=197·2	Mercury Hg=200·0	Thallium Tl=204·1	Lead Pb=206·9	Bismuth Bi=208	—	—	
12	—	—	Radium Rd=224	—	Thorium Th=232	—	Uranium U=239		

FIGURE 141 ■ In *An Attempt Towards A Chemical Conception of the Ether* (London, 1904), the aged Mendeleev postulates that the "universal ether" is composed of unimaginably light inert gas atoms (x) in series zero–group zero of his Periodic Table. Below x, there would have to be another new inert gas (y) with an atomic mass of 0.4 (H = 1.0).

A MASSIVE ANTEDILUVIAN ANIMAL—THE MEGALOSAURUS.

IMMENSE PRE-HISTORIC ANIMALS—THE IGUANODON AND MEGALOSAURUS.

FIGURE 142 ■ The all-too-human attempt by Mendeleev to "cram" his periodic law into an explanation of the defunct ether theory is similar to the attempts by nineteenth-century paleontologists to "cram" the bones of dinosaurs into the shapes of bears and other known land animals.

known. Indeed, the planetary model of the atom, developed by Bohr in 1913 and later completely eclipsed, was probably based upon his desire for a unity in the universe and an analogy with the solar system.

1. A. Greenberg, *The Chemical Intelligencer*, April, 1995, pp. 31–36.

NON-ATOMUS

Nonindivisible! The Greek philosophers conceived of the smallest unit of matter as atomos (Latin *atomus*): indivisible. John Dalton had said: "Thou knows . . . no man can split the atom" (see earlier discussion of Dalton). However, toward the end of the nineteenth century, this view had to be completely modified.[1,2] In 1859, Julius Plucker discovered that the visible discharges in vacuum tubes could be deflected by a magnetic field. The term *cathode ray* was coined around 1883 and William Crookes established that they were negatively charged. Joseph John (J.J.) Thomson established the particulate nature of these emissions and he determined a charge-to-mass ratio $e/m = 1.2 \times 10^7$ emu/g; present value, 1.7×10^7 emu/g $= 5.1 \times 10^{17}$ esu/g, for his "corpuscles." The term *electron* was introduced by G.J. Stoney over Thomson's objections. It was also known at this time that the e/m value for the electron was about 1300 times that of the hydrogen ion (modern ratio ca. 2000).

In 1908 Robert Millikan (1923 Nobel Prize in Physics) first performed his famous oil droplet experiment in which he determined a unit charge of 4.77×10^{10} esu (later 4.80×10^{10}) esu. With the modern e/m value (1.7×10^7 esu/g), the mass of the electron was found to be only $^1/_{1837}$ that of the lightest atom, hydrogen.

The cathode-ray tubes were also found to eject positive ions in the opposite direction from the electrons. These *canal rays* were comprised of much more massive particles. J.J. Thomson (1906 Nobel Prize in Physics) used a magnetic field to bend the paths of these ions and record their collisions on film. He discovered that pure neon gas produced two masses, 20 and 22, due to isotopes. The term was coined by Frederick Soddy (1921 Nobel Prize in Chemistry) during his studies of radioactive elements having the same chemical but different radioactive properties.[3] The separation of positive ions using a magnetic field followed by recording them on a photographic plate is the basis of mass spectrometry, developed by Francis W. Aston (1922 Nobel Prize in Chemistry).[1]

1. J.R. Partington, *A History of Chemistry*, MacMillan, London, 1964, Vol. 4, pp. 929–934.
2. A.J. Ihde, *The Development of Modern Chemistry*, Harper & Row, New York, 1964, pp. 478–483; 486.
3. J.R. Partington, op. cit., pp. 941–947.

CRYSTALS CAN DIFFRACT X-RAYS

X-rays were discovered accidentally by William Röntgen in 1895.[1] He had a cathode-ray tube inside a cardboard box and nearby there was, by chance, a sheet of paper coated with phosphorescent material. When the tube was on, the phosphorescent material glowed in the dark. Röntgen found that the same penetrating radiation fogged photographic plates. He called the radiation x-rays and even took images of his own hand using them.[1] Röntgen won the first Nobel Prize in Physics (1901).

Light diffraction was a well-known and well-understood phenomenon by the end of the nineteenth century. It was known that if a transparent film is scored with lines separated by a distance close to the wavelength of light, interference (diffraction) occurs. For example, sodium light (wavelength = 0.0000589 cm or 589 nm) is diffracted by a grating having 7000 lines per centimeter (0.000143 cm spacing).[2] However, x-rays are not diffracted by such gratings despite the fact that they are electromagnetic radiation just like light. In 1912, Max von Laue (1879–1950) correctly hypothesized that the wavelengths of x-rays, thought to be about 10^{-8} or 10^{-9} cm (1×10^{-8} cm = 1 angstrom), might be comparable to the distances between atoms (and ions) in crystals.[3] He discovered that these crystalline lattices were capable of diffracting x-rays. In the upper part of Figure 143 we see depictions of the crystalline lattices

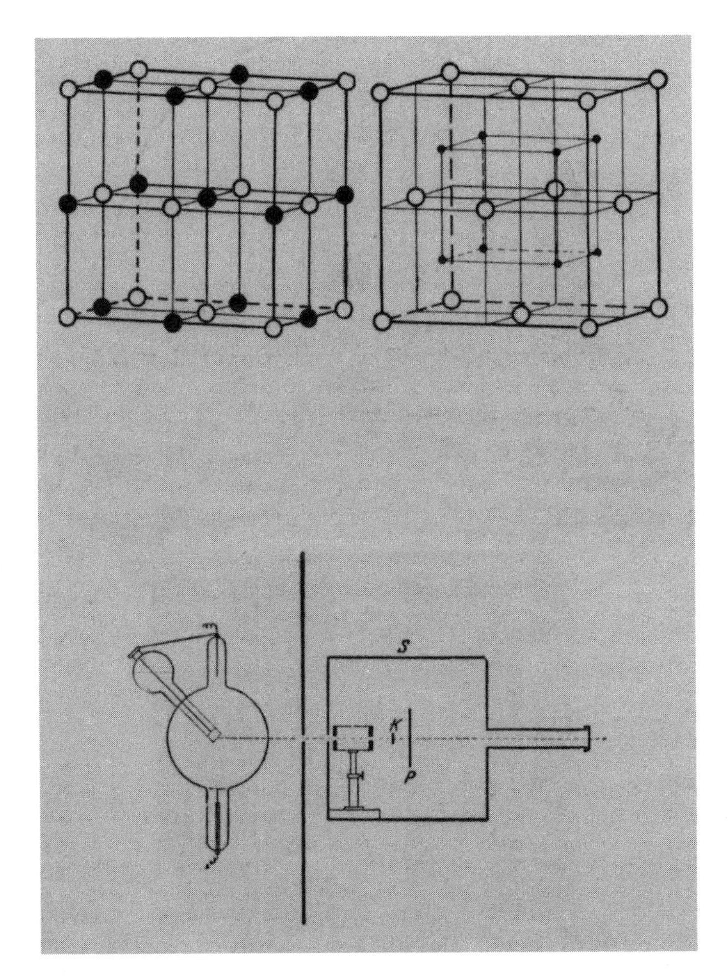

FIGURE 143 ■ Shortly after x-rays were discovered by Röntgen, Max von Laue postulated that their wavelengths were similar to the separations between atoms in ionic crystals such as rock salt and fluorspar (top). His x-ray unit is pictured at bottom (from Max Born, *The Constitution of Matter* (London, 1923).

of sodium chloride (rock salt) and calcium fluoride (fluorspar).[4] The lower half of Figure 143 depicts von Laue's x-ray apparatus: focused x-rays meet crystal C and then impinge on photographic plate *P*. The diffraction of the x-rays (theoretical construct, top of Fig. 144),[4] produces a pattern on the photographic plate (bottom of Fig. 144) that provides immediate clues to the crystal's symmetry. Von Laue won the 1914 Nobel Prize in Physics.

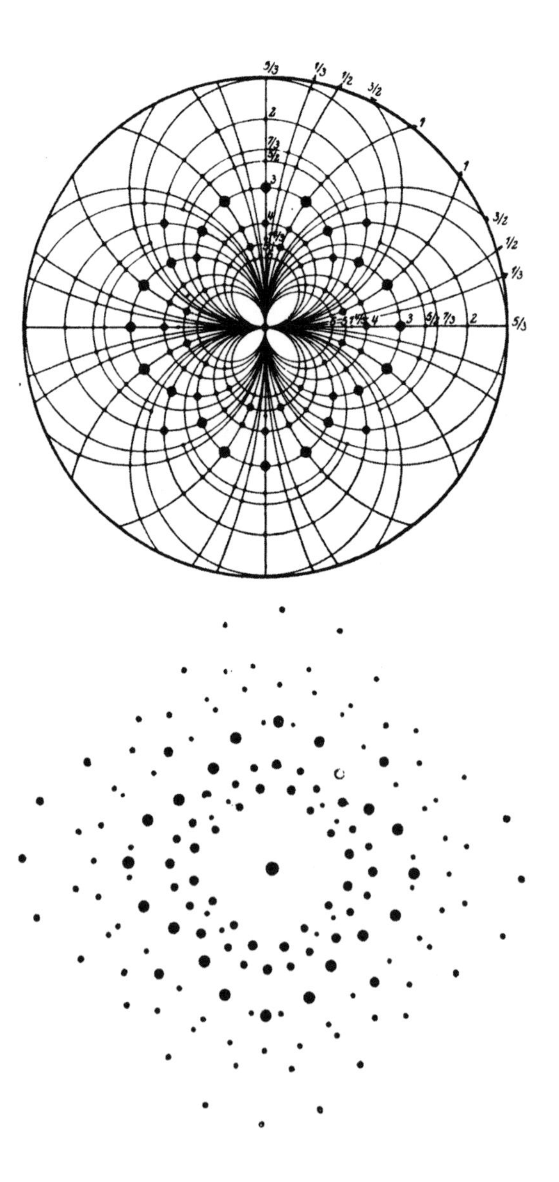

FIGURE 144 ■ Schematics of the x-ray pattern produced by von Laue's diffraction experiment (from Born, see Fig. 143).

1. J.R. Partington, *A History of Chemistry*, MacMillan, London, 1964, Vol. 4, pp. 934–935.
2. W.H. Bragg and W.L. Bragg, *X-Rays and Crystal Structure*, 4th ed., Bell, London, 1924, pp. 1–5.
3. A.J. Ihde, *The Development of Modern Chemistry*, Harper & Row, New York, 1964, pp. 483–486.
4. M. Born, *The Constitution of Matter*, Methuen, London, 1923, pp. 12–19.

TWO NOBEL PRIZES? NOT GOOD ENOUGH FOR THE ACADEMIE DES SCIENCES!

Stimulated by Röntgen's discovery of x-rays, Henri Becquerel (1852–1908) postulated a relationship between x-rays and fluorescence. He placed a variety of fluorescent crystalline samples in contact with photographic plates that were wrapped and well protected from sunlight. Upon exposing the samples to sunlight, he discovered that potassium uranyl sulfate caused fogging of the photographic plates. Seemingly, sunlight stimulated these compounds to release x-rays just as high-energy electrons kicked x-rays from anti-cathodes. However, Becquerel also made a surprising discovery. When the zinc uranyl sulfate–photographic film combination was kept in the dark, the film was also fogged. Becquerel had discovered radioactivity.[1,2]

The term *radioactive* was apparently[1] introduced by Marie and Pierre Curie in their paper in *Comptes Rendus*[3] in which they reported the discovery of the element polonium (see Fig. 145). Marya Sklodowska (1867–1934) came to Paris from Poland in 1891 to study mathematics and physics. Despite considerable privation she completed the equivalent of a Masters degree in physics at the Sorbonne in 1893 (top of her class) and a similar degree in mathematics in 1894. In that year she met Pierre Curie (1859–1906), a professor at the Municipal School of Industrial Physics and Chemistry.[4] She had plans to return to her beloved Poland and teach, and she rejected Pierre's proposals of marriage. She accepted his proposal when he offered to give up his research career and move with her to Poland.[4] Following their marriage in 1895, the couple decided to remain in Paris. Pierre finished his doctorate and his wife, Marie Sklodowska Curie, completed a license in teaching. They were given the opportunity to jointly pursue research at the Municipal School.

While Pierre performed research on piezoelectricity, Marie began her studies in the newly discovered field of radioactivity using her husband's electrometer as a detector. Madame Curie soon discovered that thorium (discovered by Berzelius in 1829) was radioactive like uranium, a finding made independently by Gerhardt Carl Schmidt. In 1898, she found that the ore pitchblende was much more radioactive than its uranium content (80% U_3O_8) would predict. She suspected the presence of an unknown and intensely radioactive element. At this point, Pierre joined Marie in her studies. Pitchblende was very expensive and the Curies were forced to use the insoluble waste material they received from a pitchblende mine in Bohemia.[4] In order to perform chemical separations on tons of material poor in pitchblende, they worked in an abandoned dissection shed of the Municipal School. Pierre's work centered on studies of radioactivity, while Marie's work concentrated on chemical separation and analysis. She comments

(175)

tube à potentiel très élevé, elle se charge négativement, ce qu'il est aisé de vérifier. De même, en touchant du doigt un tube de Crookes loin de la cathode, la paroi touchée devient cathode et il y a répulsion.

» Soit maintenant un tube à cathode plane centrée, de même diamètre que le tube. Les surfaces équipotentielles sont sensiblement planes et le faisceau est cylindrique. Vient-on à réduire le diamètre de la cathode, les surfaces de niveau se courbent et le faisceau est divergent. Si la cathode présente la forme d'un rectangle allongé, les rayons cathodiques doivent s'étaler en éventail dans un plan perpendiculaire à la plus grande dimension du rectangle, et c'est en effet ce qui a lieu.

» Supposons, au contraire, une cathode sphérique concave : à un vide peu avancé, les rayons émis forment un cône creux; menons un plan tangent à ce cône, le rayon contenu dans ce plan est repoussé d'une manière prépondérante par la partie de la cathode située du même côté de ce plan que le centre. De cette dissymétrie résulte une déviation du rayon qui tend à devenir parallèle à l'axe du cône. On peut également dire que les projectiles cathodiques, rencontrant obliquement les surfaces de niveau, se comportent comme des corps pesants lancés obliquement de haut en bas. De là cet allongement bien connu du foyer cathodique, d'autant plus marqué que le vide est plus avancé et le champ, par suite plus intense, près de la cathode. Plaçant au-devant de celle-ci un diaphragme à deux trous, on a deux faisceaux concourants, rectilignes à partir du diaphragme, se coupant cependant au delà du centre de courbure de la cathode; c'est donc surtout au voisinage de celle-ci que se produit l'inflexion des trajectoires, là précisément où le champ a son maximum d'intensité. »

PHYSICO-CHIMIE. — *Sur une substance nouvelle radio-active, contenue dans la pechblende* (¹). Note de M. **P. Curie** et de Mme **S. Curie**, présentée par M. Becquerel.

« Certains minéraux contenant de l'uranium et du thorium (pechblende, chalcolite, uranite) sont très actifs au point de vue de l'émission des rayons de Becquerel. Dans un travail antérieur, l'un de nous a montré que

(¹) Ce travail a été fait à l'École municipale de Physique et Chimie industrielles. Nous remercions tout particulièrement M. Bémont, chef des travaux de Chimie, pour les conseils et l'aide qu'il a bien voulu nous donner.

FIGURE 145 ■ First page of Pierre and Marie Curie's paper announcing the discovery of polonium in pitchblende and inventing the word *radioactive* (*Comptes Rendus*, **127**: 175, 1898).

that "Sometimes I had to spend a whole day mixing a boiling mass with a heavy iron rod nearly as large as myself. I would be broken with fatigue at the end of the day."[4] In one chemical fraction laboriously derived from the impure pitchblende, Marie Curie discovered in July, 1898 a new element, polonium, named after her native land (Fig. 145 shows the title page of this article).

However, another chemical fraction that contained barium and other alkaline earth salts exhibited intense radioactivity. When Madame Curie had purified this fraction to a point where the specific radioactivity was 60 times that of uranium, a new spectral line was detected in the fraction. As sensitive as the spectroscope (developed[5] by Robert Wilhelm Bunsen and Gustav Robert Kirchoff around 1860) was in its detection of emitted light, the electrometer was even more sensitive to the detection of radioactivity. Further fractionation to a level of 900 times the specific radioactivity was accompanied by a corresponding increase in the intensity of the new spectral line. This gave the Curies the assurance to report the new chemical element, radium, in the *Comptes Rendus*, in December, 1898.[1,2,4] It was only in July, 1902 that further separation provided pure radium. Several tons of pitchblende waste had been employed to yield 0.1 g of pure radium chloride.[1] Using the chemical analogy with its alkaline earth contaminant barium, very much in the manner of Mendeleev, the Curies assumed that the chloride was $RaCl_2$ and assigned its atomic weight at 225, thus leaving yawning gaps in the Periodic Table. Marie Curie presented her doctoral thesis in 1902 and it was published in 1903 (*Recherches sur les Substances Radioactives*).[1]

The Curies and Becquerel shared the Nobel Prize for Physics in 1903. The French Academy of Sciences had nominated Pierre Curie and Henri Becquerel for the Prize but Swedish scientist Magnus Gosta Mittag-Leffler was able to add Marie to the nomination.[4] Pierre was appointed to the faculty at the University of Paris in 1904 while Marie was promoted to Professor at the women teacher's college in Sevres.[4] Already suffering from the effects of radiation poisoning, Pierre died in a street accident in 1906. Marie was then appointed to the faculty of the University of Paris—the first woman on its faculty in its 650-year history.[4] Incredibly, in 1911 she failed to be elected to the French Academy of Sciences, but later in the year she received the Nobel Prize in Chemistry—the only person to win two Nobels until Linus Pauling did so in 1963. Although she had received only two nominations, one was by the Swedish chemist and 1903 Nobel Laureate Svante Arrhenius, who was an enlightened advocate for women in science.[4]

Marie Curie's story is very dramatic and the discussion of her by the Rayner-Canhams[4] is succinct, sensitive and balanced. During World War I, Marie Curie stopped her research and she and daughter Irene (born in 1897; Eve was born in 1904) served as x-ray technicians with mobile units in the battlefield. Marie began investigations of the medical applications of radiation including cancer therapy at about this time. Irene Joliot-Curie[6] and her husband Frederic Joliot-Curie would eventually share the 1935 Nobel Prize in Chemistry for their discovery of artificial radioactivity. Irene's intense left-wing political activities furnished at least one excuse for rejection of her nomination to the French Academy of Sciences. The Rayner-Canhams note that although the evidence of radiation poisoning and cancers among her co-workers was clear, Marie Curie resisted the obvious conclusions about the health hazards. Daughter Irene died at 59 of leukemia.[6] Marie died of leukemia at age 67.[4] The Rayner-Canhams note the profound influence of Marie Curie in attracting a kind of "critical mass" of intellectually gifted women into nuclear chemistry and physics. One of these, Marguerite Perey discovered element 87 (francium) and became, in 1962, the first woman to be elected to the French Academy of Sciences.[6] She died of cancer at the age of 65.[6] The Rayner-Canhams further note the development of "critical masses" (my term) of women scientists in crystallography[7] as well as biochemistry.[8] The impact of these newly established and formidable "old-girl" networks in chemistry will be an interesting topic for future sociologists of science. For the record we note that the National Academy of Sciences (U.S.) was formed in 1863 and had an initial membership of 50. The first woman was admitted in 1925—Dr. Florence R. Sabin, Professor of Histology, Johns Hopkins University. As of April 27, 1999, there were 2,222 members of whom 132 are women.[9]

1. J.R. Partington, *A History of Chemistry*, MacMillan, London, 1964, Vol. 4, pp. 936–939.
2. A.J. Ihde, *The Development of Modern Chemistry*, Harper & Row, New York, 1964, pp. 487–490.
3. M. Curie and P. Curie, *Comptes Rendus*, **127**:175, 1898.
4. M. Rayner-Canham and G. Rayner-Canham, *Women In Chemistry: Their Changing Roles From Alchemical Times To The Mid-Twentieth Century*, American Chemical Society and Chemical Heritage Foundation, Washington, D.C. and Philadelphia, 1998, pp. 97–107.
5. A.J. Ihde, op. cit., pp. 233–235.
6. M. Rayner-Canham, op. cit., pp. 112–116.
7. M. Rayner-Canham, op. cit., pp. 67–91.
8. M. Rayner-Canham, op. cit., pp. 135–164.
9. Public Information Office, National Academy of Sciences.

IT'S THE ATOMIC *NUMBER*, DMITRY!

The first explicit use of atomic number is attributed to John Newlands, who arranged his 1864 table of elements by "the number of the element" in the order of their "equivalents" using Cannizzaro's system.[1] At the time, Professor George Carey Foster "humorously enquired of Mr. Newlands whether he had ever examined the elements according to the order of their initial letters."[1]

The fact that atoms have their identities locked inside their nuclei was only discovered at the beginning of the twentieth century. The Curies first postulated that radiation emitted from uranium and other radioactive substances was particulate in nature.[2] In 1906, Rutherford and Geiger determined a value for the charge-to-mass ratio of the α particle that was one-half that for the hydrogen ion (H^+). Therefore, the α particle could have been either H_2^+ or He^{2+}.[2] The latter was confirmed in 1911 and was, of course, consistent with the emission of helium gas from radioactive nuclei.[2] In 1909, Geiger and Marsden, working in Rutherford's laboratory found that many α particles pass through 0.01-mm-thick gold leaf with little deflection while only a few suffer major deflections or rebounds. Similar results had been obtained by Rutherford and Geiger a year earlier.[2] These and related studies using Wilson's new cloud chamber led Rutherford to conclude in 1911 that atoms were mostly empty space with a tiny, positively charged *nucleus* (term he introduced in 1912) at the center.[2,3]

The measurement of deflection angles using the cloud chamber led Geiger and Marsden to the conclusion that the positive charge in the nucleus (in whole-number multiples of the charge on an electron) tended to be about half the atomic weight.[2,3] A. Van den Broek, in 1913, suggested that the nuclear charge, in electron-charge units, is equal to the ordinal number (1, 2, 3, . . .) of the element in the Periodic Table.[2]

It was Henry G.J. Moseley (1887–1915) who, in 1913, used the term *atomic number* and established its significance.[2,3] Moseley made a study of the vibrational frequencies of certain x-rays (the *K* series) emitted from different metallic anticathodes. In Figure 146 (right), we see a decent-looking correlation between the square root of the frequencies of the *K* radiations with the atomic weights of the corresponding elements. However, the correlation with the atomic number (Fig. 146, left) was virtually perfect. Clearly, the atomic number was more than a counting device. Ultimately, it explained certain troubling anomalies—the reversal in placement between tellurium and iodine that worried Mendeleev and the apparent anomaly that the recently discovered argon (which almost equaled calcium in atomic weight) had to be placed before the lighter potassium. It verified the placement of cobalt before nickel on the basis of chemical properties despite the inversion of their atomic weights and confirmed gaps in the Periodic Table for as-yet-undiscovered metals.[3] Moseley was drafted during World War I and was killed at the age of 28 in the battle of Gallipoli.[2,3]

Starting around 1920, it was assumed that the difference between the atomic mass and the atomic number was due to protons *combined in the nucleus* with electrons. Thus, chlorine-35 would have 17 protons in the nucleus, 18 protons combined with 18 nuclear electrons with 17 electrons outside the nu-

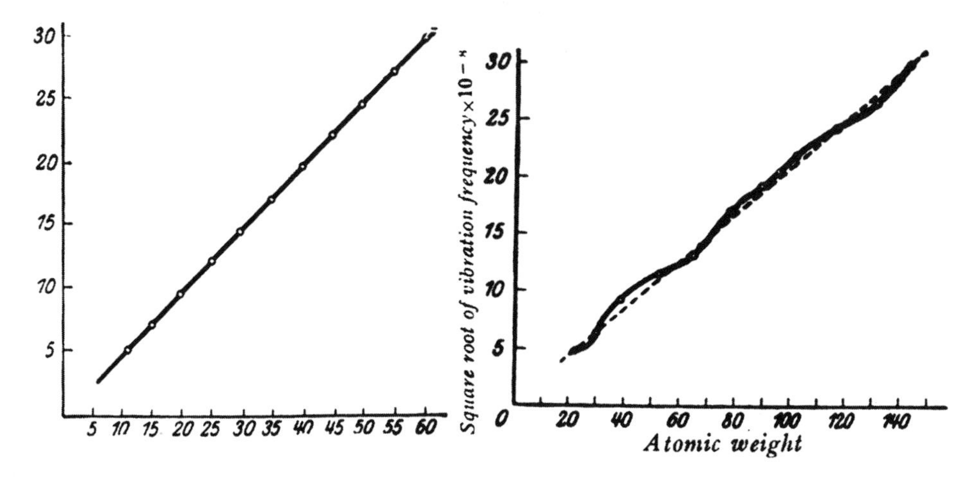

FIGURE 146 ■ The fundamental basis of the Periodic Table is the Atomic Number and not the Atomic Weight. The square root of the frequency of emitted x-rays from different metallic cathodes is imperfectly related to Atomic Mass but directly proportional to Atomic Number. This immediately explained certain anomalies in the Periodic Table. Henry G.J. Moseley, who made this critical discovery, was drafted in World War I and died at Gallipoli at the age of 28 (figure from Born; see Fig. 143).

cleus.[4] This picture changed when Chadwick discovered the neutron in 1932. But remember, a free neutron decomposes to a proton and an electron (plus an antineutrino).

1. J.R. Partington, *A History of Chemistry*, MacMillan, London, 1964, Vol. 4, pp. 887–888.
2. J.R. Partington, op. cit., pp. 942–953.
3. A.J. Ihde, *The Development of Modern Chemistry*, Harper & Row, New York, 1964, pp. 485–486.
4. J.R. Partington, *Everyday Chemistry*, MacMillan, London, 1929, pp. 245–249.

X-RAYS MEASURE THE DISTANCES BETWEEN ATOMS OR IONS

While Max von Laue used crystals to perform an experiment with x-rays, William H. Bragg (1862–1942) and his son William Lawrence Bragg (1890–1971) used x-rays to determine the structures of crystals. In 1912 and 1913 the Braggs developed and applied the diffraction equation that bears their name:

$$nL = 2d \sin \theta$$

where $n = 1, 2, 3, \ldots$ L is the wavelength of the x-rays; d is the distance between layers of atoms (ions), and θ is the angle of incidence to the surface.

Figure 147(a) from the Braggs' text[1] demonstrates the reinforcement of x-ray waves that obey Braggs' Law. In Figure 147(b),[1] we see a schematic of their x-ray apparatus in which single crystals (or powders) were placed on a rotating table so that reflections could be collected from all angles.

Not only did the Braggs' x-ray diffraction apparatus allow these critical measurements of distances, they helped confirm the reality of ions since, as determined by J.J. Thomson, the intensity of scattering was proportional to the number of electrons (furnishing, in effect, an additional confirmation of atomic numbers).[2,3]

X-ray crystallography soon became the most important "optic" for structural chemistry in the solid state. It laid the basis for Linus Pauling's crystallographic studies that led to the synthesis of the principles expounded in his 1939 text, *The Nature of the Chemical Bond.* These principles were applied by him to solve the α-helical structure of proteins through simple use of homegrown molecular models. J.D. Watson and Francis Crick used Pauling's model-building approach, combined with x-ray data to beat him at his own game and arrive at the structure of DNA. When I was a graduate student in the late 1960s, the complete solution of a crystalline structure by x-ray data was a relatively rare event. It was then used primarily for structural chemistry studies in which researchers desired accurate bond lengths, bond angles, and other related data. Improved instrumentation and especially the incredibly increased power of computers have now made x-ray crystallography a fairly routine tool for structure confirmation of fairly large molecules that form good crystals. Large globular molecules are still, however, immense challenges.

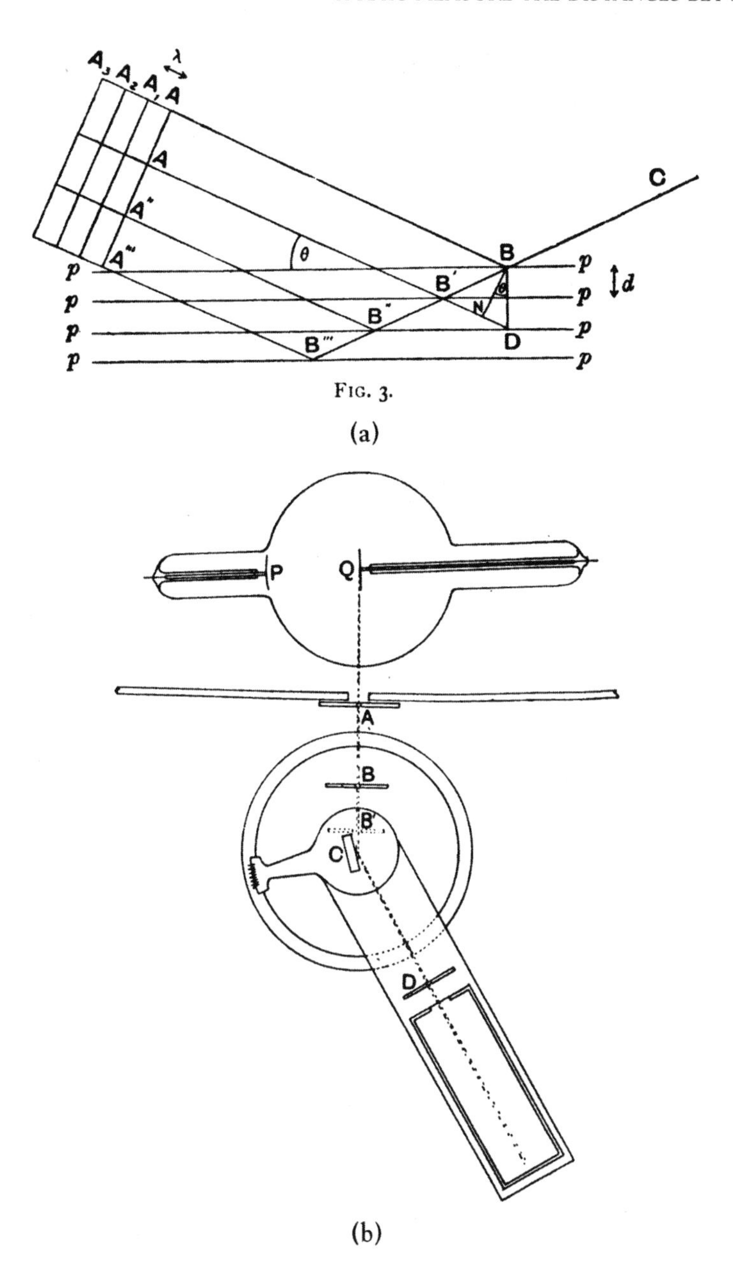

FIGURE 147 ■ William H. Bragg and his son William L. Bragg reversed von Laue's experiment and used x-rays to measure the distances between ions or atoms in crystals (see text). (a) Depicts the conditions for constructive interference of x-rays termed *Braggs' Law*; (b) schematic of the Braggs x-ray apparatus (W.H. Bragg and W.L. Bragg, *X-Rays And Crystal Structure*, 4th ed., London, 1924).

1. W.H. Bragg and W.L. Bragg, *X-Rays And Crystal Structure*, 4th ed., Bell, London, 1924.
2. J.R. Partington, *A History of Chemistry*, MacMillan, London, 1964, Vol. 4, pp. 934–936.
3. W.H. Brock, *The Norton History of Chemistry*, Norton, New York, 1993, pp. 393; 492–494.

WHERE DID WE DIG UP THE MOLE?

The modern concept of valence explains the early concept of chemical "equivalents" first introduced by Cavendish around 1767.[1] Thus, 36.5 g of acid of salt (HCl) gas neutralizes 56.0 g of potash (KOH); 49.0 g of acid of vitriol (H_2SO_4) or 32.7 g of acid of phosphorus (H_3PO_4) equivalently neutralizes the same 56.0 g of potash. The ratio of equivalent weights for $HCl/H_2SO_4/H_3PO_4$ is always 1.00/1.34/0.90. Similarly, 53.0 g of soda (Na_2CO_3) neutralizes 36.5 g of HCl gas: the ratio of equivalent weights for KOH/Na_2CO_3 is always 1.06. The same 36.5 g of HCl neutralizes 29.1 g of "milk of magnesia" [$Mg(OH)_2$; note that $Ca(OH)_2$ was once called "milk of lime"] to produce 47.6 g of $MgCl_2$ and 18.0 g of water. If we place platinum electrodes into 47.6 g of molten $MgCl_2$, 12.1 g of magnesium will electroplate, and 35.5 g of chlorine gas [12.2 liters at standard temperature and pressure (25°C and 1.00 atm)] will be released. Thus, the ratio of equivalent weights of Cl_2/Mg is always 2.93.

Equivalent masses (and the related concept of "normality") have gradually disappeared from modern chemistry texts in favor of a definition based directly on numbers of "particles" (atoms, molecules, ions, electrons), and this is truly ironic.

The term *mole* was first introduced by Wilhelm Ostwald in 1901.[1] It is derived from the Latin for "mass, hump, or pile"[1] (the term *molecule*, introduced by Pierre Gassendi[2] in the early seventeenth century has the same root; presumably it means a mass of atoms). Specifically, Ostwald used the term to represent the formula weight of a substance in grams: 36.5 g of HCl is one mole. The formal definition of the mole adopted by the Fourteenth Conference Générale des Poids at Mésures in 1971 is: "the amount of a substance of a system that contains as many elementary entities as there are atoms in 0.012 kilograms of carbon-12."[1] The rich irony is that Ostwald fiercely resisted the atomic concept at the time Boltzmann committed suicide in 1906 but his mole is now defined explicitly in terms of atoms.

The number of atoms in 0.012 kg of carbon-12 is Avogadro's Number ($6.02213670 \times 10^{23}$). Perrin's 1908 experiments on dust particles and particles of gamboge and mastic in water gave a value of about 6×10^{23}. Once Millikan had determined the charge of an electron (modern physical value, $q = 1.6021773 \times 10^{-19}$ coulombs or C) and this was combined with the modern value for the faraday (1 F = 96,485.31 C, the total charge in one mole of electrons), another completely independent determination was available for Avogadro's Number. Here's another: 1 g of radium yields 11.6×10^{17} α particles in one year and these produce 0.043 liters of helium gas at standard temperature and pressure (STP).[1] Indeed, the Rayleigh scattering that causes our sky to be blue allows calculation of Avogadro's Number. The current accepted value is based upon density, atomic mass, and x-ray diffraction measurements of pure crystalline silicon. Its uncertainty[1] is only 3.5×10^{17}! Let's remember Avogadro's Number as "six-point-oh-two-and-twenty-three-oh-oh-oh's" (like "Pennsylvania-6-5-oh-oh-oh" for aging Glen Miller buffs including lapsed hippies whose memories of Glen Miller are only prenatal).

1. J.J. Lagowski (ed.), *MacMillan Encyclopedia of Chemistry*, Simon & Schuster, MacMillan, New York, 1997, Vol. 1, pp. 198–199; Vol. 3, pp. 951–955.
2. J.R. Partington, *A History of Chemistry*, MacMillan, London, 1961, Vol. 2, p. 462.

XENON IS SLIGHTLY IGNOBLE AND KRYPTON IS NOT INVINCIBLE

The inertness of the noble gases as well as Richard Abegg's law of valence and countervalence[1] were important leads to understanding of valence and bonding. In 1916 Walther Kossel proposed that atoms used their valence-shell electrons to form bonds and that they tried to attain the electronic structure of the rare gas preceding them (electropositive elements) or immediately following them (electronegative elements). Kossel's theory is depicted in Figure 148.[2] Gilbert N. Lewis formulated the octet rule in his article "The Atom And The Molecule."[3] In Figure 149(a) we see his representations of the valence electrons of the first complete row of elements in families IA to VIIA.[3] The noble gas neon occurs at the end of this period and has each corner of the cube "occupied" by an electron. Thus, inertness corresponds to a completed octet and this "filling" of the valence shell explains an atom's valence. The unreactivity of noble gases

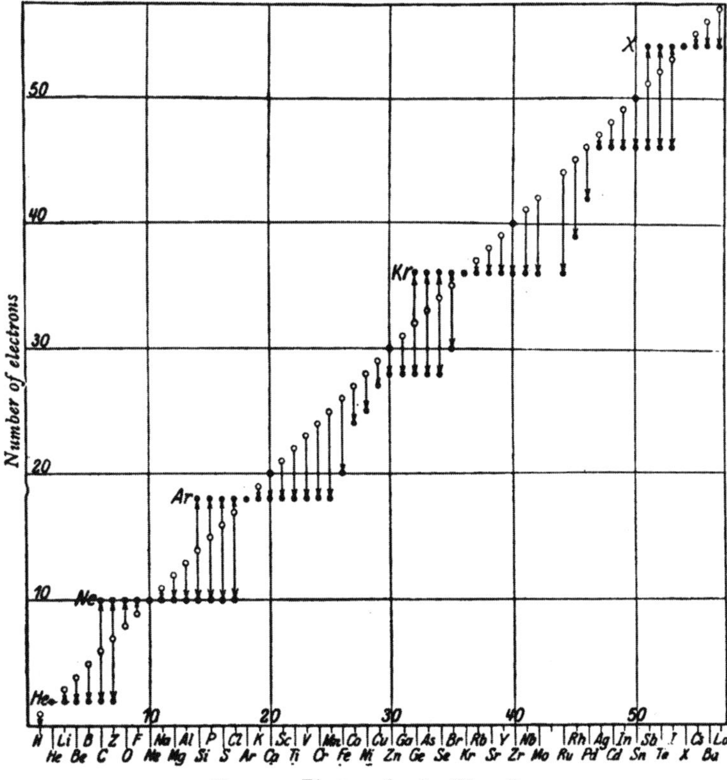

FIG. 20.—Electro-valencies (Kossel).

FIGURE 148 ■ Walther Kossel's theory in which atoms adopt the valence shell of the nearest inert gas by loss or gain of electrons (from Born, see Fig. 143).

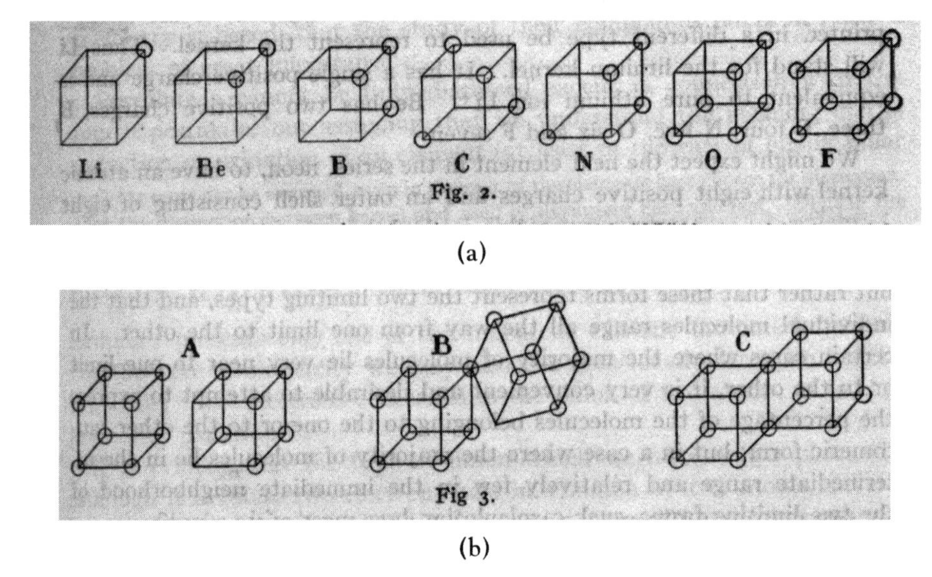

FIGURE 149 ■ The original Gilbert N. Lewis dot structures (*Journal of the American Chemical Society*, **38**:762, 1916).

had become an article of faith, verified forcefully when Henri Moissan turned his newly found "Tasmanian devil" fluorine loose on a sample of argon sent by Ramsay in 1894. The result: Nothing![4]

However, in 1962 Neil Bartlett discovered that molecular oxygen (O_2) is oxidized by (loses an electron to) hexafluoroplatinum (PtF_6) to give the new compound $O_2^+PtF_6^-$.[5] He realized that the oxygen molecule holds its electron about as tightly as a xenon atom does. Therefore, Xe just might also lose an electron to PtF_6. The resulting red crystalline solid was originally thought to be $Xe^+PtF_6^-$,[4] but it is now thought to be $[XeF^+][Pt_2F_{11}^-]$.[6]

In any case, the conceptual threshold had been crossed and a family of fluorine-containing xenon compounds is now known. For example, a 1:5 mixture of Xe and F_2, heated in a nickel vessel, produces XeF_4 (melting point 117°C).[6] XeF_6, formed by Xe and F_2 at high temperature and pressure, attacks quartz and reacts violently with water to produce XeO_3, itself a high explosive.[6] Obviously, this is not "The Friendly World of Chemistry Neighborhood."

Krypton reacts with F_2 under an electric discharge at −183°C to form a solid (KrF_2) that decomposes slowly at room temperature.[6] There are also salts of KrF^+, such as $KrF^+SbF_6^-$.[6] Although radon loses electrons much more easily than xenon, its most stable isotope has a half-life of 3.8 days, and not much chemistry is done although compounds thought to be RnF_2, $RnF^+TaF_6^-$, and possibly RnO_3 are known.[6] As to the rumors concerning a reputed green ore of krypton—doubtful.

Figure 149(b) shows "steps" in the sharing of an edge (two electrons) to form a single bond between two cubic iodine atoms. If two atoms share a face formed by fusing the two cubes, they share four electrons and form a double bond. Tetrahedral bonding in methane (CH_4) was explained by sharing of two opposite edges on the top of cubic carbon with two cubic hydrogens and, similarly, sharing the two alternate edges with hydrogens on the bottom of the cube. While van't Hoff explained triple bonds through sharing faces of two tetrahedra,

Lewis's picture is a bit more strained. In 1919 Irving Langmuir slightly modified Lewis's picture and extended coverage to the transition metals. Figure 150 depicts oxides of nitrogen, the two allotropes of oxygen as well as possible isomers of hydrogen peroxide (never found), and oxides of phosphorus. It is interesting

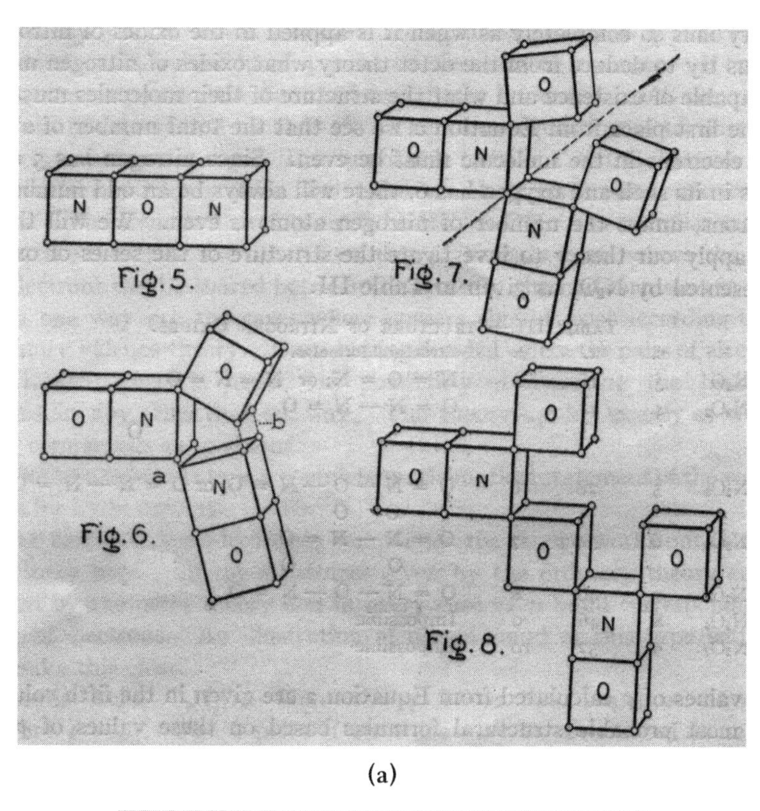

(a)

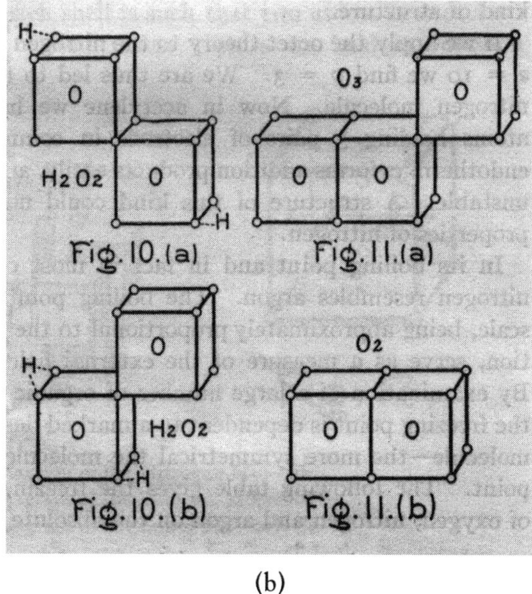

(b)

FIGURE 150 ■ Irving Langmuir's modification of the Lewis structures [*Journal of the American Chemical Society*, **41**, No. 6 and No. 10 (1919) and **42** (No. 2) (1920)].

to note that Langmuir handles hydrogen's completed valence shell "duet" by having it bridge along the edge of a cube.

In the 1960s *quadruple* bonds between metallic atoms were discovered in species such as $Re_2Cl_8^{2-}$.[7] There would be no obvious way to explain quadruple bonds with cubic atoms or tetrahedra.

Lewis's paper proposed the bookkeeping dot structures (e.g., H:H) that bear his name. How does one convey to an introductory chemistry student just how ridiculously simple and powerful Lewis structures are for prediction?

1. Lewis (Ref. 3) states Abegg's Law as: "the total difference between the maximum negative and positive values or polar numbers of an element is frequently eight and is in no case more than eight."
2. M. Born, *The Constitution of Matter*, Methuen, London, 1923, pp. 21–23.
3. G.N. Lewis, *Journal of the American Chemical Society*, **38**:762–785, 1916.
4. W.H. Brock, *The Norton History of Chemistry*, Norton, New York, 1993, p. 337.
5. See N. Bartlett, *American Scientist*, **51**:114, 1963.
6. A. Cotton, G. Wilkinson, C.A. Murillo and M. Bochmann, *Advanced Inorganic Chemistry*, 6th ed., Wiley, New York, 1999, pp. 588–597.
7. F.A. Cotton, *Accounts of Chemical Research*, **2**:242, 1969.

THE ATOM AS SOLAR SYSTEM

The line spectra obtained by heating elements and refracting the light through a prism was employed by Bunsen to identify salts. His gas-powered burner was first used to obtain colorless flame to study light emissions of these salts—not for heating flasks.[1] The spectroscope designed by Bunsen and Kirchoff immediately led to the discovery of cesium in 1860 and rubidium in 1861.[1] In 1868, the emission spectrum of another new element, helium, was discovered in the spectrum of the solar chromosphere.[2] But what was the origin of line spectra—light having very precise frequencies (or wavelengths) unique to each element? What was the origin of the photoelectric effect: A small quantity of high-energy (high-frequency) light waves could kick an electron off of a metal surface, but a huge quantity of light of a lower frequency could not? Apparently the *quality* of the energy, not its quantity was the issue. These phenomena were addressed by Max Planck (1858–1947), who developed quantum theory around the year 1900.[3] The simple equation he advanced, $E = h\nu$, indicated that the frequency of light emitted by an excited atom, for example, was directly proportional to the energy decrease of the emitting atom (h is Planck's constant).

Shortly after the Rutherford model of the atom was established, it was obvious to wonder where the electrons were. In 1913, Niels Bohr[3] (1885–1962) used Planck's quantum theory, combined with the line spectra (visible, ultraviolet, infrared) of hydrogen, to postulate the circular planetary model of the atom. If negative electrons were orbiting the positively charged nucleus, classical physics required them to spiral into the nucleus. Bohr postulated that electrons could only have certain discrete energies (occupy only certain circular orbits) and never "in-between" values. These orbits corresponded to quantum numbers n =

1, 2, 3, ... The model was revolutionary and even subversive. Where were the electrons when they moved between orbits? They could never be found in the "in-between." The model beautifully explained the spectrum of the hydrogen atom and the helium ion (He^+), and failed for all other atoms. Arnold Sommerfeld modified Bohr's orbits to allow both circular and elliptical orbits.[3] He explained the fine structure of the spectrum of hydrogen by adding a second quantum number for angular momentum. Now there was occasional reference to *orbitals*—really *suborbits*. Sommerfeld's theory enjoyed success in explaining H and He^+ and other atomic spectra. Figure 151, from a 1923 book by Born,[4] depicts the Bohr model for H, He, and He^+ and its extension to He_2. Figures 152 and 153, from Smith's 1924 book,[5] depict the "Rococo era" of the "old" quantum theory a few "hours before the dawn" of quantum mechanics in 1926. The pretty image of an electron spiraling on the surface of a 4*s* orbital would be seen to violate Heisenberg's Uncertainty Principle. The picture of atoms that emerged in 1926 would almost seem to be more suited to abstract art than "hard" science.

1. A.J. Ihde, *The Development of Modern Chemistry*, Harper & Row, New York, 1964, pp. 231–235.
2. A.J. Ihde, op. cit., p. 373.
3. A.J. Ihde, op. cit., pp. 499–507.
4. M. Born, *The Constitution of Matter*, Methuen, London, 1923, pp. 24–32.
5. J.D. Main Smith, *Chemistry and Atomic Structure*, D. Van Nostrand, New York, 1924, pp. 160–176.

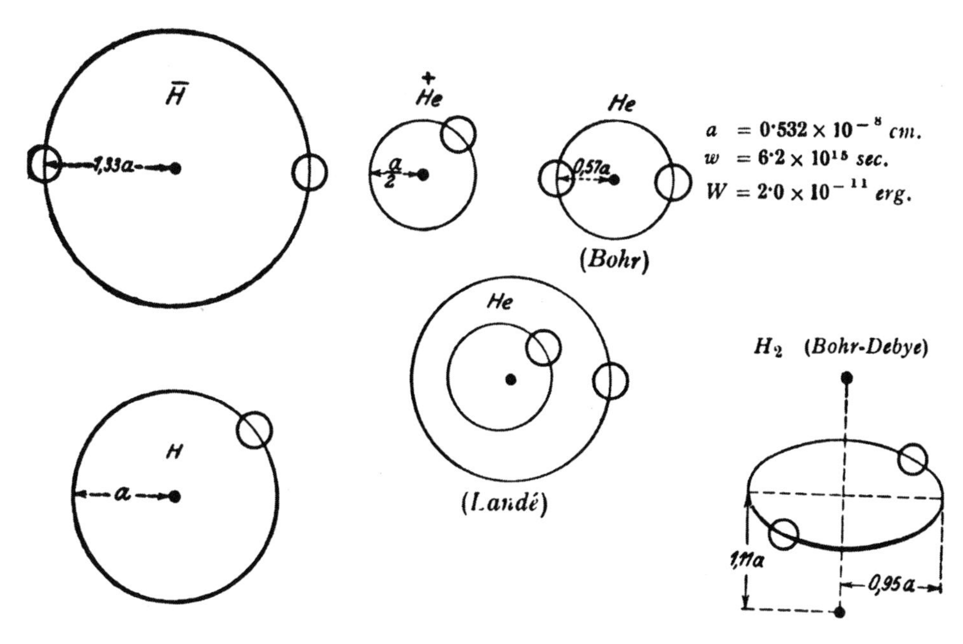

FIGURE 151 ■ Variations of the Bohr model of the atom that was really much more "subversive" than it looks. If an electron is forbidden to exist between orbits, how does it pass from one orbit to the next? (from Born; see Fig. 143).

The Dynamic Atom 171

DIAGRAM XV

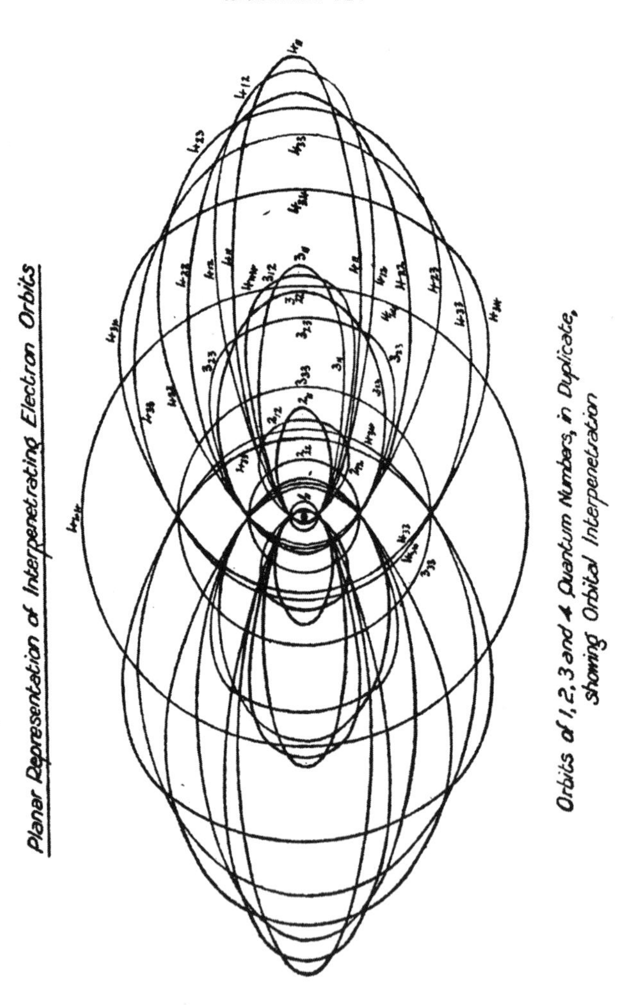

FIGURE 152 ■ The "Rococo" era of the "old" quantum theory comes to an end (J.D. Main Smith, *Chemistry and Atomic Structure*, New York, 1924).

DIAGRAM XVI

Spatial Representation of Electron Orbital Domain

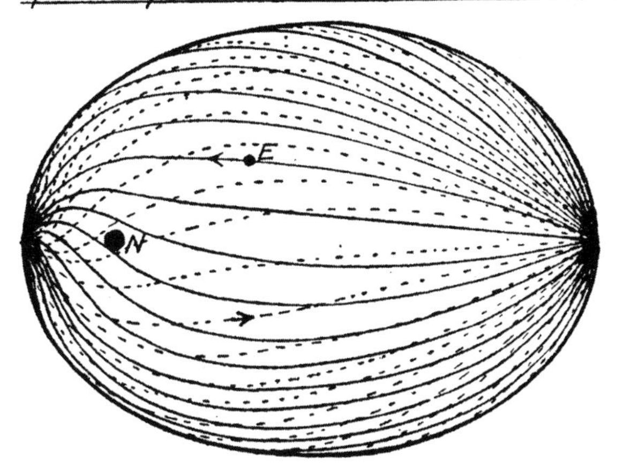

Precession Paths of a 4s Electron moving on the Surface of an Imaginary Solid Ellipsoid of Revolution with Nucleus at the Focus

FIGURE 153 ■ An electron spiraling down a 4*s* orbital before the Heisenberg Uncertainty Principle smudged the picture (from Main Smith, see Fig. 152).

'TIS A GIFT TO BE SIMPLE

Simple "counting rules" are useful, powerful, and hint at underlying structure. When Gregor Mendel reported the laws of heredity in 1865, the "two-ness" of his results were incredibly simple and powerful, yet made little initial impact. The source was the as-yet-unknown genes and the ultimate origin: the double helix of DNA. What emerged from quantum mechanics in 1926 were four quantum numbers (n, l, m_l and m_s). Allowed values for these quantum numbers specified the energy, orbital ("domain"), and spin for every electron in an atom. The periodicities of the properties of the elements were manifestations of the quantum numbers. The transition metals corresponded to filling 3*d*, 4*d*, and 5*d* orbitals; the lanthanides to filling 4*f* orbitals; the actinides to filling 5*f* orbitals. The octet rule, which explained why H_2 and F_2 have single bonds, why N_2 has a triple bond, and why sodium chloride has Na^+ and Cl^- ions, while magnesium oxide has Mg^{2+} and O^{2-} ions, is consistent with quantum mechanics. There are lots of other "counting rules" in chemistry with quantum mechanics at the core. Nyholm and Gillespie's valence-shell electron pair repulsion (VSEPR) theory is incredibly good at predicting molecular geometries (CO_2 is linear, H_2O is bent). All one does is count electron pairs, obtained from straightforward Lewis octets or "expanded octets" (e.g., PF_5, SF_6) around the central atom. The stability of benzene is understood by Hückel's $4n + 2$ rule. The Woodward–Hoffmann rules follow similar $4n + 2$ and $4n$ alternation with the ability to predict thermal chemistry and photochemistry as further alternatives. I do not mean to imply that quantum mechanics (or chemistry) is easy. But the occurrence and power of simple counting rules in chemistry continues to amaze and delight me.

TRANSMUTING QUANTUM MECHANICS INTO CHEMISTRY

The contributions of Linus Pauling (1901–1994) to twentieth-century chemistry are arguably as fundamental as those of Lavoisier to the late eighteenth and early nineteenth centuries. Pauling received his Ph.D. under the guidance of Roscoe Dickinson, California Institute of Technology's first Ph.D. in chemistry, who literally took him by the hand and taught him crystallography.[1] Arthur Amos Noyes had been recruited from MIT to direct the Gates Chemical Laboratory just three years before Pauling's arrival as a student at the newly energized Caltech. Following the completion of Pauling's Ph.D., Noyes wished to retain the brilliant young man on his faculty and fretted over his growing friendship with Gilbert N. Lewis, the Department Chair at Berkeley.[2] The solution was to send him to Europe (immediately if not sooner). He arranged a dinner for the 24-year-old Pauling with the Director of the Guggenheim Foundation (a friend of Noyes). A fellowship for work at the great quantum theory centers of Europe followed. Still, there remained a waiting period and Pauling's planned visit to Berkeley. Noyes encouraged an early departure: "If the Paulings left early," he proposed, "then they would have time for stopovers in Madeira, Algiers, and Gibralter before docking in Naples, then a few weeks for touring Italy." Italy! Noyes spoke glowingly of the glories of Rome, the fabulous ruins at Paestum. "I'll give you enough money to pay the fare in Europe," he said, "and support you from the end of March until the beginning of the Guggenheim Fellowship."[2] Noyes retained his "franchise player."

Arriving in Munich during the Spring of 1926, Pauling immediately contacted Arnold Sommerfeld. He would later spend time with Niels Bohr in Copenhagen. However, the "old" quantum theory underlying the Bohr–Sommerfeld atom was just starting to crumble in late 1925 and Pauling bore witness to the work of physicists Louis De Broglie, Erwin Schrödinger, Wolfgang Pauli, Paul Dirac, Max Born, Walther Heitler, and Fritz London. At one point, Pauling excitedly presented his ideas on the power of the Bohr–Sommerfeld model to Pauli. "Not interesting" was the terse response.[3] But Pauling learned the new quantum mechanics and the application of the Schrödinger equation and made them accessible to chemists.

The Bible of mid-twentieth century chemistry was Pauling's *The Nature of the Chemical Bond* (Ithaca, 1939; 2nd ed., 1940; 3rd ed., 1960). In *The Double Helix*, J.D. Watson writes: "The book I poked open the most was Francis' copy of *The Nature of the Chemical Bond.* Increasingly often, when Francis needed to look up a crucial bond length, it would turn up on the quarter bench of lab space that John had given to me for experimental work. Somewhere in Pauling's masterpiece I hoped the real secret would lie . . . ."[4] Pauling's book was based on a series of articles, "The Nature of the Chemical Bond," that were published starting in 1931. Figure 154 shows the title page of the first article in the series. So much of what we teach in the first year of chemistry is presented in these works.

Although the title has an almost magical sound to it, the nature of the chemical bond was truly the domain Pauling began to explore. He formulated the concept of *hybridization* to explain how localized atomic orbitals best overlap to form two-electron bonds. The Kossel–Lewis–Langmuir picture explained ionic and covalent bonding in terms of the octet rule. An interesting question was whether the transition from covalent to ionic bonding (the nature of the chemical bond) was smooth and continuous or abrupt. In his work, Pauling examined the abrupt change in melting points of the second-row series of fluorides:[5] NaF (995°C); MgF_2 (1263°C); AlF_3 (1257°C); SiF_4 (−90°C); PF_5 (−94°C); SF_6 (−51°C). The seemingly obvious conclusion is that the first three are ionic (electrons cleanly transferred, not shared) and the next three are covalent (electrons shared). However, Pauling noted that structure is the key here

April, 1931 THE NATURE OF THE CHEMICAL BOND 1367

[CONTRIBUTION FROM GATES CHEMICAL LABORATORY, CALIFORNIA INSTITUTE TECHNOLOGY, NO. 280]

THE NATURE OF THE CHEMICAL BOND. APPLICATION OF RESULTS OBTAINED FROM THE QUANTUM MECHANICS AND FROM A THEORY OF PARAMAGNETIC SUSCEPTIBILITY TO THE STRUCTURE OF MOLECULES

BY LINUS PAULING

RECEIVED FEBRUARY 17, 1931 PUBLISHED APRIL 6, 1931

During the last four years the problem of the nature of the chemical bond has been attacked by theoretical physicists, especially Heitler and London, by the application of the quantum mechanics. This work has led to an approximate theoretical calculation of the energy of formation and of other properties of very simple molecules, such as H_2, and has also provided a formal justification of the rules set up in 1916 by G. N. Lewis for his electron-pair bond. In the following paper it will be shown that many more results of chemical significance can be obtained from the quantum mechanical equations, permitting the formulation of an extensive and powerful set of rules for the electron-pair bond supplementing those of Lewis. These rules provide information regarding the relative strengths of bonds formed by different atoms, the angles between bonds, free rotation or lack of free rotation about bond axes, the relation between the quantum numbers of bonding electrons and the number and spatial arrangement of the bonds, etc. A complete theory of the magnetic moments of molecules and complex ions is also developed, and it is shown that for many compounds involving elements of the transition groups this theory together with the rules for electron-pair bonds leads to a unique assignment of electron structures as well as a definite determination of the type of bonds involved.[1]

I. The Electron-Pair Bond

The Interaction of Simple Atoms.—The discussion of the wave equation for the hydrogen molecule by Heitler and London,[2] Sugiura,[3] and Wang[4] showed that two normal hydrogen atoms can interact in either of two ways, one of which gives rise to repulsion with no molecule formation, the other

[1] A preliminary announcement of some of these results was made three years ago [Linus Pauling, *Proc. Nat. Acad. Sci.*, **14**, 359 (1928)]. Two of the results (90° bond angles for *p* eigenfunctions, and the existence, but not the stability, of tetrahedral eigenfunctions) have been independently discovered by Professor J. C. Slater and announced at meetings of the National Academy of Sciences (Washington, April, 1930) and the American Physical Society (Cleveland, December, 1930).

[2] W. Heitler and F. London, *Z. Physik*, **44**, 455 (1927).

[3] Y. Sugiura, *ibid.*, **45**, 484 (1927).

[4] S. C. Wang, *Phys. Rev.*, **31**, 579 (1928).

FIGURE 154 ▪ The first article in the series *The Nature of the Chemical Bond* by Linus Pauling (*Journal of the American Chemical Society*, 53:1367, 1931). These articles formed the foundation of his book of the same title (Ithaca, 1939), in turn the core of twentieth-century structural chemistry.

and that while AlF_3 is polymeric, SiF_4 exists as individual molecules. He concluded that the Al–F and Si–F bonds were both polar covalent and not that dissimilar in nature. Pauling further explored what were termed one-electron bonds (thought to be present in diborane B_2H_6, the "nonclassical" structure reported in 1951 by Hedberg and Schomaker, using electron diffraction; its three-center bonding explained by Lipscomb) and three-electron bonds in species such as nitric oxide (NO). He developed the concept of *electronegativity* to quantitate the transition in nature from pure covalent to pure ionic bonding. His concept of *resonance* rationalized the transition from the highly polar covalent bond in hydrogen fluoride (comparable contributions from the H^+F^- and H–F resonance contributors) to the less polar HI (less H^+I^- contribution). It also furnished the explanation for the 70-year-old quandary of the relationship of benzene's structure to its reactivity. Much as two tuning forks embedded in the same wooden block exchange vibrations—one vibrates and then transfers its vibration to the other and the exchange reverses—so too can benzene be represented as two equivalent structures "in resonance." This is only an analogy and benzene is thought of as a resonance "hybrid" of the two limiting classical Lewis-type ("canonical") structures. It is worth mentioning here that the rival molecular orbital approached championed by Robert Mulliken has, with the aid of computers, probably become the more powerful technique in present-day research.

Pauling's audacious scientific career included the use of first principles and molecular models to intuit the structure of the protein α-keratin. He also was the first to characterize the basis for a disease at the molecular level—sickle-cell anemia—the result of a substitution of one amino acid for another in hemoglobin. Pauling was awarded the Nobel Prize in Chemistry in 1954.

Pauling's political activities, characterized as left-wing during the McCarthy era of the 1950s, caused him difficulties at Caltech as well as with the State Department. Its denial of a passport caused him to miss a 1952 meeting of the Royal Society in which critical information about DNA was exchanged.[6] Ultimately, his political activities were critical in obtaining agreement on a ban in atmospheric testing of nuclear weapons and he was awarded the 1962 Nobel Peace Prize on October 1, 1963—the date of the test-ban treaty. It is not unreasonable to consider Pauling to be one of the parents, along with Rachel Carson, of the environmental movement. Pauling's resonance theory was considered "revisionist" in Stalin's Soviet Union and he was vilified by staunch Communists. Anybody who can simultaneously upset Communists and McCarthyites must be doing something right!

1. T. Hager, *Force of Nature: The Life of Linus Pauling*, Simon & Schuster, New York, 1995, pp. 82–85, 88–91.
2. T. Hager, op. cit., pp. 107–109.
3. T. Hager, op. cit., pp. 116–117.
4. J.D. Watson, *The Double Helix*, A Norton Critical Edition, edited by G.S. Stent, W.W. Norton & Co., New York, 1980.
5. L. Pauling, *The Nature of the Chemical Bond*, 3rd ed., Cornell University Press, Ithaca, 1960, pp. 71–73.
6. T. Hager, op. cit., pp. 400–407.

MERCURY CAN BE TRANSMUTED TO GOLD

Transmutation happens! And mercury *can* be transmuted to gold—but not by chemistry or alchemy. The *Strong Force* that binds an atomic nucleus is on the order of millions of electron volts (MeV) per nuclear particle (proton or neutron). A neutron isolated from a nucleus has a half-life of a mere 17 minutes before disintegrating into a proton and an electron. Since the mass of the neutron equals the mass of a proton plus 2.5 electrons, the lost mass of 1.5 electrons is equivalent (Einstein's $E = mc^2$) to 0.78 MeV.[1] Now, chemistry involves the gain or loss of electrons only and thus chemistry happens with energies on the order of a few eV at most—roughly a millionth of the nuclear binding force. The nuclei in the carbon, hydrogen, nitrogen, and oxygen atoms doze peacefully when TNT explodes and all chemical hell breaks loose.

Radioactivity is emitted from atomic nuclei that are unstable and spontaneously change their structure. In 1896, Henri Becquerel first discovered radioactivity when he placed a piece of zinc uranyl sulfate wrapped in paper on a photographic plate. Two years later, Marie and Pierre Curie discovered two highly radioactive elements, polonium and radium, in pitchblende.[2] α particles, the nuclei of helium atoms, were among the radiations emitted by these substances which were spontaneously transmuting. Indeed, since the earth had billions of years ago lost its original complement of light, inert helium, all helium in our present environment comes from radioactive decay. The sun makes its own helium fresh every day (and night) by fusing hydrogen atoms. The first man-made transmutation was achieved by Rutherford in 1919 when he bombarded nitrogen (^{14}N) with α particles and made oxygen (^{17}O).[3] In 1932, Chadwick observed the neutron, thus explaining the existence of isotopes and largely unifying knowledge about the nucleus.

The neutron plays a pivotal role in manmade transmutations. In the words of Bronowski:[4] "At twilight on the sixth day of Creation, so say the Hebrew commentators to the Old Testament, God made for man a number of tools that gave him also the gift of creation. If the commentators were alive today, they would write "God made the neutron." Is it far-fetched to consider the neutron to be the Stone of the Philosophers (and atom smashers to be athanors—the furnaces of the Philosophic Egg)? Frankly, yes. But, in 1941, fast neutrons were used to transmute mercury into a tiny quantity of gold.[5] Was the age-old dream realized? Would a modern day version of the Roman Emperor Diocletian have to burn all of the notebooks and journal articles and destroy the atom smashers in order to protect the world's currency? Well, probably not. It is likely that an ounce of such gold would cost more than the net worth of the planet. Also, the gold so obtained is radioactive[6] and lives for only a few days at most.[5] But, we are not always logical when it comes to gold. In the words of Black Elk, a holy man of the Oglala Lakota–Sioux on the Pine Ridge Reservation in South Dakota:[7]

> Afterward I learned that it was Pahuska[8] who had led his soldiers into the Black Hills that summer to see what he could find. He had no right to go there, because all that country was ours. Also the Wasichus[8] had made a

treaty with Red Cloud (1868) that said it would be ours as long as grass should grow and water flow. Later I learned too that Pahuska had found there much of the yellow metal that makes the Wasichus crazy; and that is what made the bad trouble, just as it did before, when the hundred were rubbed out.

Our people knew there was yellow metal in little chunks up there; but they did not bother with it, because it was not good for anything.

1. *Encyclopedia Brittanica*, 15th ed., Chicago, 1986, Vol. 14, p. 332.
2. J.R. Partington, *A History of Chemistry*, MacMillan, London, 1964, Vol. 4, pp. 936–947, 953–955.
3. P.W. Atkins and L.L. Jones, *Chemistry: Molecules, Matter and Change*, 3rd ed., Freeman, New York, 1997, pp. 858–860.
4. J. Bronowski, *The Ascent of Man*, Little, Brown, Boston, 1973, p. 341.
5. R. Sherr, K.T. Bainbridge, and H.H. Anderson, *The Physical Review*, **60**:473–479, 1941.
6. In the 1964 James Bond movie *Goldfinger*, the arch-villain Auric Goldfinger tries to detonate a nuclear weapon inside Fort Knox to make the U.S. gold supply radioactive in order to increase the value of his own gold horde. Apparently, novelist Ian Fleming knew there is only one stable isotope of gold.
7. P. Riley (ed.), *Growing Up Native American*, William Morrow, New York, 1993, p. 99. Thanks to Professor Susan Gardner for this suggestion.
8. *Pahuska* is "Long Hair"—General George Armstrong Custer; *Wasichus*, the term for white settlers and soldiers, translates as "greedy ones." (Thanks to Professor Susan Gardner for the suggestion and background for this topic.)

MODERN ALCHEMISTS SEEK ATLANTIS

In Figure 155 we see a 1944 formulation of the Periodic Table by Glenn T. Seaborg.[1] We are commonly told that there are 92 naturally occurring elements. Logically, this would seem to end with uranium (atomic number 92), and it is true that uranium is the highest atomic number element found naturally in any significant amount and that only ultratrace quantities of neptunium and plutonium occur in uranium ores. However, there are gaps: for example, at element 43. That element, technetium (Tc) was the first synthetic element, although it was later discovered that exceedingly minute (trace) quantities occur naturally due to uranium decay.[2,3] Perrier and Ségré succeeded in 1937 by bombarding molybdenum (Mo) with deuterons (nuclei of deuterium). The half-life of ^{97}Tc is 2.6 million years. Today, the Tc-99m (m = metastable) ($t_{1/2} \sim 6$ hr) isotope is used for heart imaging.[3] Element 87 is francium, synthesized from actinium by Marguerite Perey in 1939.[2] Its most stable isotope, ^{223}Fr, has a half-life of 21.8 minutes. It is also found in ultratrace (2×10^{-18} ppm) quantities in uranium ores since new francium is made as the "old" decays leaving a minute steady-state concentration.[3] Its properties, though little studied, resemble those of rubidium and cesium. Element 85, astatine, was produced by bombarding bismuth with accelerated α particles.[2] Its most stable isotope, ^{210}As, has a half-life of 8.3 hours and is also found in ultratrace quantities in uranium ores.[3] The largest quantity made of astatine (50 billionths of a gram) allowed a limited amount of study: It is concentrated in the thyroid like iodine and AtI_2^- is even more stable

1 H 1.008																1 H 1.008	2 He 4.003
3 Li 6.940	4 Be 9.02											5 B 10.82	6 C 12.010	7 N 14.008	8 O 16.000	9 F 19.00	10 Ne 20.183
11 Na 22.997	12 Mg 24.32	13 Al 26.97										13 Al 26.97	14 Si 28.06	15 P 30.98	16 S 32.06	17 Cl 35.457	18 A 39.944
19 K 39.096	20 Ca 40.08	21 Sc 45.10	22 Ti 47.90	23 V 50.95	24 Cr 52.01	25 Mn 54.93	26 Fe 55.85	27 Co 58.94	28 Ni 58.69	29 Cu 63.57	30 Zn 65.38	31 Ga 69.72	32 Ge 72.60	33 As 74.91	34 Se 78.96	35 Br 79.916	36 Kr 83.7
37 Rb 85.48	38 Sr 87.63	39 Y 88.92	40 Zr 91.22	41 Cb 92.91	42 Mo 95.95	43	44 Ru 101.7	45 Rh 102.91	46 Pd 106.7	47 Ag 107.880	48 Cd 112.41	49 In 114.76	50 Sn 118.70	51 Sb 121.76	52 Te 127.61	53 I 126.92	54 Xe 131.3
55 Cs 132.91	56 Ba 137.36	57 LA 138.92 58-71 SEE La SERIES	72 Hf 178.6	73 Ta 180.88	74 W 183.92	75 Re 186.31	76 Os 190.2	77 Ir 193.1	78 Pt 195.23	79 Au 197.2	80 Hg 200.61	81 Tl 204.39	82 Pb 207.21	83 Bi 209.00	84 Po	85	86 Rn 222
87	88 Ra	89 AC SEE Ac SERIES	90 Th	91 Pa	92 U	93 Np	94 Pu	95	96								

LANTHANIDE SERIES	57 La 138.92	58 Ce 140.13	59 Pr 140.92	60 Nd 144.27	61	62 Sm 150.43	63 Eu 152.0	64 Gd 156.9	65 Tb 159.2	66 Dy 162.46	67 Ho 163.5	68 Er 167.2	69 Tm 169.4	70 Yb 173.04	71 Lu 174.99
ACTINIDE SERIES	89 Ac	90 Th 232.12	91 Pa 231	92 U 238.07	93 Np 237	94 Pu	95	96							

1 H																	2 He
3 Li	4 Be											5 B	6 C	7 N	8 O	9 F	10 Ne
11 Na	12 Mg											13 Al	14 Si	15 P	16 S	17 Cl	18 Ar
19 K	20 Ca	21 Sc	22 Ti	23 V	24 Cr	25 Mn	26 Fe	27 Co	28 Ni	29 Cu	30 Zn	31 Ga	32 Ge	33 As	34 Se	35 Br	36 Kr
37 Rb	38 Sr	39 Y	40 Zr	41 Nb	42 Mo	43 Tc	44 Ru	45 Rh	46 Pd	47 Ag	48 Cd	49 In	50 Sn	51 Sb	52 Te	53 I	54 Xe
55 Cs	56 Ba	57 La	72 Hf	73 Ta	74 W	75 Re	76 Os	77 Ir	78 Pt	79 Au	80 Hg	81 Tl	82 Pb	83 Bi	84 Po	85 At	86 Rn
87 Fr	88 Ra	89 Ac	*104* Rf	*105* Db	*106* Sg	*107* Bh	*108* Hs	*109* Mt	110	111	**112**	(113)	**114**	(115)	**116**	(117)	(118)
(119)	(120)	(121)	(154)	(155)	(156)	(157)	(158)	(159)	(160)	(161)	(162)	(163)	(164)	(165)	(166)	(167)	(168)

LANTHANIDES	58 Ce	59 Pr	60 Nd	61 Pm	62 Sm	63 Eu	64 Gd	65 Tb	66 Dy	67 Ho	68 Er	69 Tm	70 Yb	71 Lu
ACTINIDES	90 Th	91 Pa	92 U	93 Np	94 Pu	95 Am	96 Cm	97 Bk	98 Cf	99 Es	100 Fm	*101* Md	*102* No	*103* Lr
SUPER-ACTINIDES	(122)	(123)	(124)	(125)	(126)									(153)

FIGURE 155 ■ The top figure is the Periodic Table published by Glenn Seaborg in 1944 and the bottom is a futuristic table (slightly modified from his diagram in *Accounts of Chemical Research*, **28**:257, 1995). Seaborg died in March, 1999 following a stroke some six months earlier. Sadly, he was unaware of a Russian group achieving the anticipated "Isle of Stability" with element 114 in January 1999 and elements 116 and 118 announced by Lawrence Berkeley National Laboratory in June 1999 (courtesy American Chemical Society). However, the claimed discovery of 118 has since been withdrawn.

than the commonplace I_3^-.[3] Element 61, promethium, was conclusively reported in 1947 as a trace by-product of uranium fission ($<1 \times 10^{-11}$ ppm).[2,3] The isotope reported (^{147}Pm) has a half-life of only 2.6 years. Subsequently, ^{145}Pm was found to have a half-life of 17.7 years.[3]

So, it seems that of these 92 "natural" elements, only 88 can be considered naturally occurring since the above four are transient species, newly formed by radioactive decay. Neptunium and plutonium can similarly be found in ultratrace quantities due to *de novo* synthesis coupled to rapid decay. We could stretch a point by noting that all helium on our planet is also formed *de novo*. However, although these fresh helium atoms are lost into space, the nuclei are totally stable.

The true stars of Seaborg's 1944 Periodic Table are the transuranium elements neptunium (Np) and plutonium (Pu) as well as elements 89 to 92 (actinium, thorium, protactinium, and uranium). Neptunium was synthesized by McMillan and Abelson at Berkeley in 1940.[1] In late 1940 and early 1941 McMillan, Kennedy, Wahl, and Seaborg made ^{238}Pu through bombardment of uranium with deuterons in early 1941, and ^{239}Pu was obtained by bombarding uranium with neutrons.[1] It was Seaborg who, in 1944, proposed a new series of compounds for the Periodic Table—the actinides—analogous to the rare earths or lanthanides. In his book *The Periodic Kingdom*, Atkins describes the lanthanides as the northern shore of an island off the south coast of the Periodic Kingdom.[4] Seaborg, thus, discovered the southern shore of this island.

The transmutations described here are nuclear physics and not chemistry (or alchemy). But, if we play with our earlier metaphor and liken the neutron to the Philosopher's Stone, then ^{239}Pu could be likened to gold as both a blessing and a curse. It was the fuel in the atomic bomb used on Nagasaki that ended World War II while killing and maiming a city's population. The incredible stockpile assembled during the Cold War now leaves the earth with hundreds of tons of this scary yet useful substance: the curse of King Midas on the one hand, a source of energy on the other.

In Figure 155 we also see a modified version of a futuristic Periodic Table published by Seaborg in 1995.[5] The southern shore of the coastal island is completed through lawrencium (Lr). Seaborg used Mendeleevian logic[5] to predict the properties of element 101 as "*eka*-thulium," just as Mendeleev predicted an *eka*-silicon (germanium) below silicon. Appropriately, it is named Mendelevium (Md). There was, unfortunately, a controversy about naming elements with atomic numbers over 100, which was finally settled in 1997:[6] 101, mendelevium (Md); 102, nobelium (No); 103, lawrencium (Lw); 104, rutherfordium (Rf); 105, dubnium (Db); 106, seaborgium (Sg); 107, nielsbohrium (Bh); 108, hassium (Hs); 109, meitnerium (Mt). Naming 106 for Glenn Seaborg was a particularly significant gesture honoring his massive contributions. Elements 107 to 109 have half-lives of milliseconds.[5] Fittingly, element 106 was the last of the series to have a lifetime (tens of seconds) to permit chemical study. Perhaps there *will* be seaborgic sulfate or calcium seaborgate.[5] Ten years after the discovery of hassium in 1984, elements 110, 111, and 112 were identified in that order. Elements 110 to 112 have half-lives on the order of microseconds.[5] These findings are consistent with the nuclear shell-structure theory of Maria Goeppart-Mayer and Hans Jensen, who shared the Nobel Prize in Physics in 1963.[2]

Göppart-Mayer and Jensen's theory predicted the existence of "Islands of Stability" among the superheavy elements. Atkins calls these Atlantis.[4] An audacious experiment by the Joint Institute for Nuclear Research in Dubna, Russia electrified the scientific world in early 1999 with a cautious announcement of the synthesis of element 114 (atomic mass 289) with a half-life (α-decay) of 20 seconds by bombarding ^{244}Pu with ^{48}Ca ions.[7] Albert Ghiorso said: "This is the most exciting event in our lives."[7] The neutron number (N = 175) approached the "magic" closed neutron shell (N = 184). Stronger proof for element 114 came with the synthesis of a second isotope, $^{287}114$ (N = 173) with a half-life (α-decay) of about 5 seconds.[8] Indeed, $^{283}112$ (N = 171) has a half-life of about 1.5 minutes (spontaneous fission), which is 3×10^5 longer than the first reported isotope ($^{277}112$; N = 165).[8] And, in June, 1999, the Lawrence Berkeley National Laboratory reported element 118 (mass 293), by bombardment of lead with krypton ions, as well as its daughter nuclide 116 (mass 289)—both showed some stability (lifetimes on the order of hundreds of microseconds).[9] Sadly, Glenn T. Seaborg suffered a crippling stroke in August, 1998 and died in early 1999, unaware of the landing on Atlantis. Even more sadly, the claim for element 118 was withdrawn among allegations of error and even fraud.[10]

1. G.T. Seaborg, *Chemical and Engineering News*, **23** (23), December 10, 1945.
2. J.R. Partington, *A History of Chemistry*, MacMillan, London, 1964, Vol. 4, pp. 953–955.
3. H. Rossotti, *Diverse Atoms: Profiles of the Chemical Elements*, Oxford University Press, Oxford, 1998.
4. P.W. Atkins, *The Periodic Kingdom*, Basic Books, New York, 1995, pp. 25–26, 56.
5. G.T. Seaborg, *Accounts of Chemical Research*, Vol. **28**:257–264, 1995.
6. *Chemical and Engineering News*, Vol. **75**, September 8:10, 1997.
7. R. Stone, *Science*, **283**:474, 1999.
8. Yu.Ts. Oganesian, A.V. Yeremin, A.G. Popeko, S.L. Bogomolov, G.L. Buklanov, M.L. Chelnokov, V.I. Chepigin, B.N. Gikal, V.A. Gorshkov, G.G. Gulbekian, M.G. Itkis, A.P. Kabachenko, A.Yu Laurentev, O.N. Malyshev, J. Rohac, R.N. Sagaidak, S. Hofmann, S. Saro, G. Giardina and K. Morita, *Nature*, Vol. 400, p. 242 (1999).
9. R. Dagani, *Chem. Eng. News*, Vol. 77 (June 14, 1999), p. 6.
10. R. Monastersky, *The Chronicle of Higher Education*, August 16, 2002, pp. A16–A21.

THE CHEMISTRY OF GOLD IS NOBLE BUT NOT SIMPLE[1]

Now here is an interesting point: it was obvious to the ancients that gold was "noble" in the same sense that we think of the Group 8A gases (helium, neon, etc.) as "noble"—namely, unreactive. It did not tarnish like its valuable cousin silver. Silver tarnish is due to atmospheric hydrogen sulfide that forms a coating of black silver sulfide (Ag_2S) and it could be heated repeatedly with no change. It did not dissolve in hydrochloric acid or nitric acid. It did "dissolve" in *aqua regia*, a mixture of 3:1 HCl/HNO_3, but evaporation of the solution and intense heating recovered the gold unchanged, unlike the baser metals where a calx remained. However, gold chemistry is not obvious, even today.

We know that gold does exhibit reactivity (but so does xenon).[2,3] For example, the "dissolution" in *aqua regia* is really a chemical reaction in which HNO_3 and HCl act synergistically (as a team). The oxidation of elemental gold to Au(III) can only happen because of the stability of the $AuCl_4^-$ ion:

$$Au + HCl + HNO_3 \rightarrow AuCl_4^- + NO_2 + H_3O^+$$

$$Au + HCl + Cl_2 + H_2O \rightarrow H_3O^+[AuCl_4^-]\ (H_2O)_3$$

where the final product is chloroauric acid. Similarly, in the presence of dilute cyanide solutions, it will oxidize in air at room temperature to form stable $Au(CN)_2^-$ ions. This reaction is used to extract gold from ores.

The explanation of gold's nobility is *not* obvious.[2] In modern terms, we note that the outermost electrons in the atoms of the coinage metals copper, silver and gold are $4s^1$, $5s^1$, and $6s^1$. Thus, they appear at first to be close cousins of the decidedly *ignoble* alkali metals potassium, rubidium, and cesium (also $4s^1$, $5s^1$, and $6s^1$, repectively)—cesium would "date" almost anybody. The alkali metals described "underlie" their outermost electrons with a completed subshell structure—the octets of the preceding noble gas. These octets shield the outermost electrons (ns^1) from nuclear attraction and make them easy to ionize (lose) and, thus, render the alkali metals reactive. Cesium has its outermost electron furthest from the nucleus and is most reactive. In contrast, the coinage metals "underlie" their ns^1 electrons with a completed 18-electron shell that would also imply great stability. However, the *d* electrons are not particularly effective in shielding the ns^1 electrons, which are thus strongly attracted to the nucleus and hard to ionize.[2] In further contrast to the alkali metals, the order of reactivity is smallest to largest. Copper is most reactive, and gold the least reactive. Apparently, relativistic physics is required to explain the behavior of the $6s^1$ electron in gold.[3] So it appears that, in this instance at least, chemistry has been rescued by "the triumphal chariot of physics."

1. I thank my wife, Susan Greenberg, for suggesting this essay.
2. B.E. Douglas, D.H. McDaniel, and J.J. Alexander, *Concepts and Models of Inorganic Chemistry*, 3rd ed., Wiley, New York, 1994, pp. 724–725.
3. F.A. Cotton and G. Wilkinson, *Advanced Inorganic Chemistry*, 5th ed., Wiley, New York, 1988, pp. 937–939.

THE "PERFECT BIOLOGICAL PRINCIPLE"

> We wish to suggest a structure for the salt of deoxyribose nucleic acid (D.N.A.). This structure has novel features which are of considerable biological interest.

So reads the first paragraph of the ground-breaking communication by James D. Watson and Francis H.C. Crick in *Nature* reporting their double-helical structure for DNA.[1] The third paragraph from the end says:

> It has not escaped our notice that the specific pairing we have postulated immediately suggests a possible copying mechanism for the genetic material.

In his now-classic personal narrative, *The Double Helix*,[2] J.D. Watson imagines this understated eloquence of the paper-to-be. The book is a wonderfully idiosyncratic history, from Watson's perspective, of the race to discover the structure of DNA. It shows lay readers that scientists are human, for better or worse.

The thesis in Watson's narrative is that understanding the function of DNA may hint at its structure. Hopefully, the structure of DNA will be "beautiful" and make its function self-evident. In the book, Watson recalls Francis Crick's postulation, after sharing a few beers, of a "perfect biological principle" —the perfect self-replication of the gene.[3] It is this kind of overarching interest in function and a willingness to "play" with molecular models that give Watson and Crick an advantage in the search. Of continuing historical debate is their use of the x-ray crystallographic data obtained by Rosalind Franklin without her permission or knowledge. There remain to this day troubling questions[4,5] in spite of the acknowledgement of her data in the *Nature* paper and the fact that the next two papers in the issue were authored by Wilkins, A.R. Stokes, and H.R. Wilson followed by Franklin and R.G. Gosling. It is clear that Franklin correctly

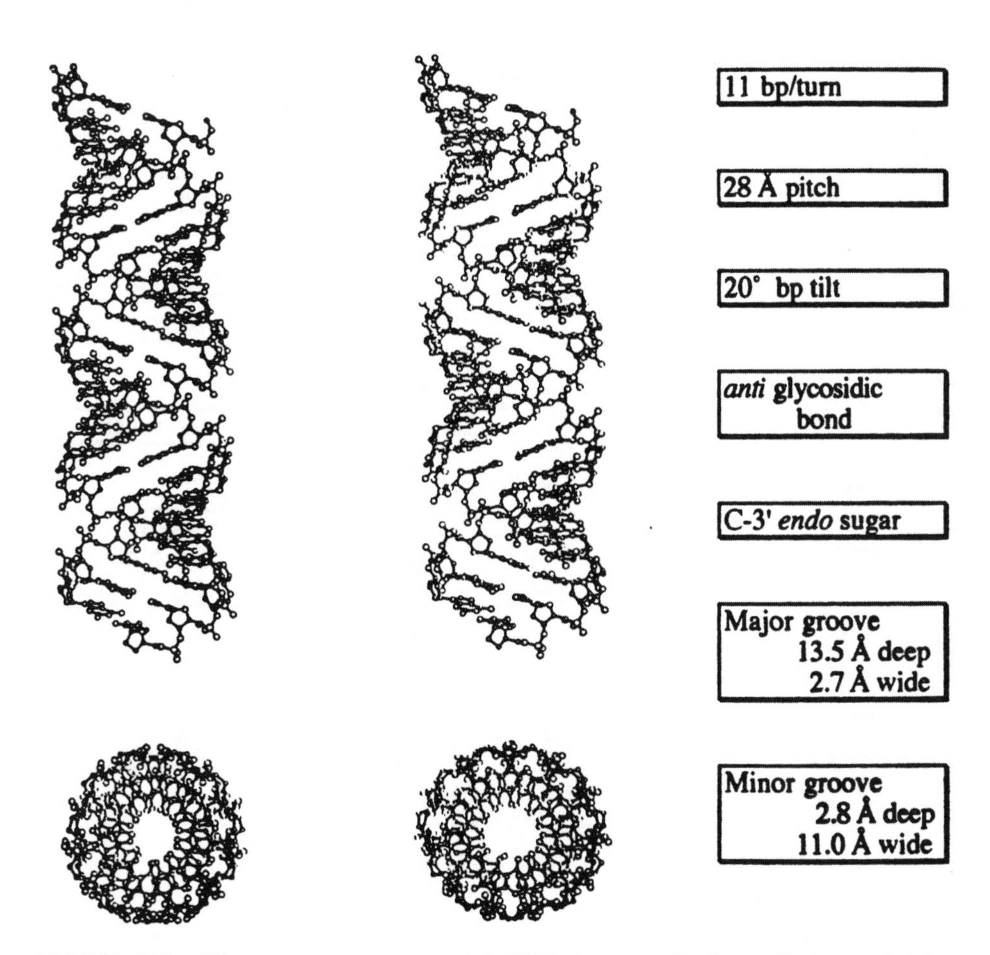

FIGURE 156 ■ The compact structure of A-DNA (courtesy Professor Catherine J. Murphy based on structures in Arnott and Chandrasekaran, *Proceedings of the Second SUNYA Conversation in the Discipline Biomolecular Stereodynamics*, R. Sarma (ed.), Vol. 1, Adenine Press, 1981, pp. 99–122; courtesy Adenine Press).

concluded that the phosphates were on the outside of the helix. Furthermore, she understood that the data indicated helicity.[4,5] Her approach was a rigorous solution of the structure based upon straightforward data analysis although she had used molecular models in the past.[4,5]

Linus Pauling had solved the structure of α-keratin using the principles of bonding in his *Nature of the Chemical Bond* to construct models and take a short-cut to the laborious and incredibly complex interpretation of the x-ray data. Watson and Crick succeeded at beating him at his own game. There is a deli-

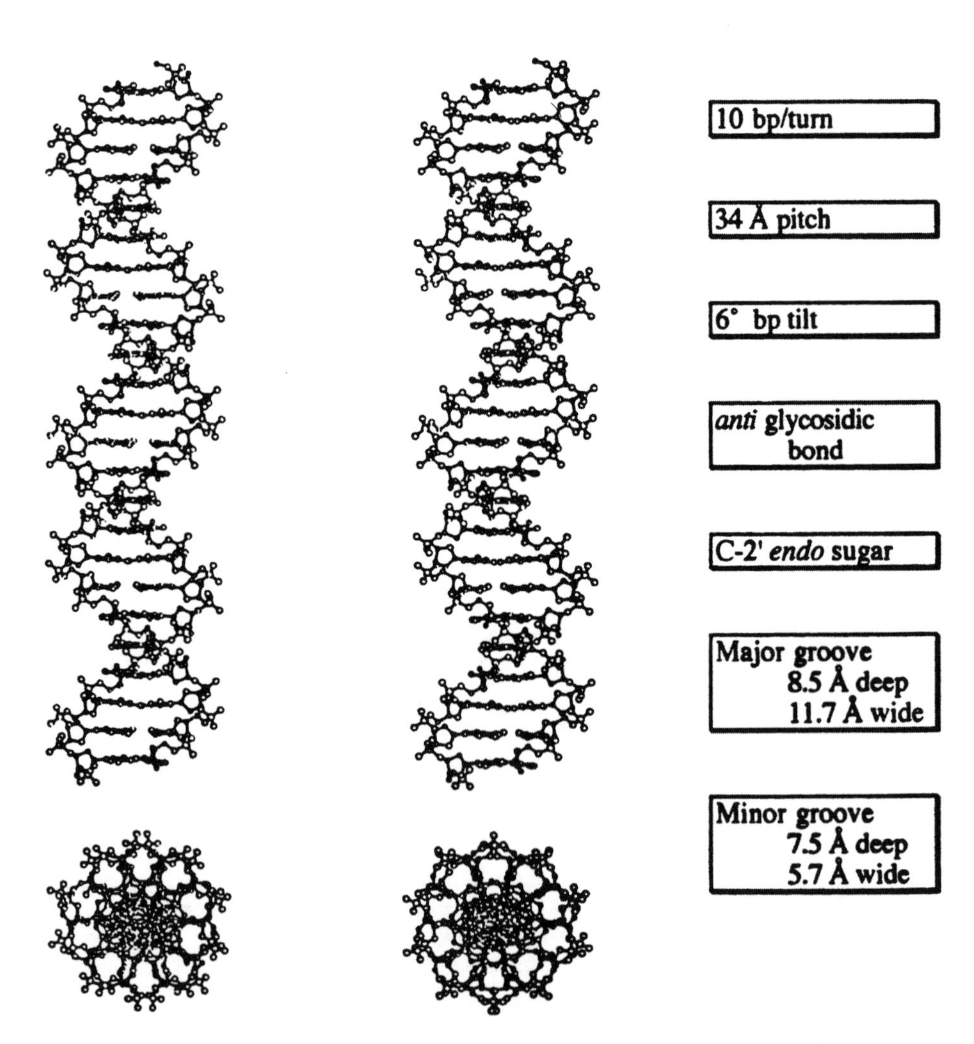

FIGURE 157 ■ The elongated B structure of DNA which provided Rosalind Franklin with a clear x-ray diffraction Diagram indicating helical structure (courtesy Professor Catherine J. Murphy based on structures in Arnott and Chandrasekaran, *Proceedings of the Second SUNYA Conversation in the Discipline Biomolecular Stereodynamics*, R. Sarma (ed.), Vol. 1, Adenine Press, 1981, pp. 99–122; courtesy Adenine Press).

cious moment in Watson's book when Peter Pauling, Linus' son who is also at Cambridge, informs Watson and Crick that his father has solved the structure and that it is a triple helix.[6] Watson grabs the manuscript from the younger Pauling's coat pocket and he and Crick read it with trepidation. To their surprise and delight, Linus has goofed. The phosphates are protonated as if phosphoric acid were not an acid. Their next concern is that he would be chagrined by his mistake, redouble his determination, solve the problem, and win *their* Nobel! (The year is 1953. Pauling would win the Nobel Prize in Chemistry in 1954.)

Figure 156 shows the compact structure of A-DNA. Franklin discovered the technique of moistening A-DNA, which forms the more hydrated and elongated B-DNA (Figure 157) whose x-ray pattern spoke so eloquently of its helical structure. Watson, Crick, and Wilkins shared the Nobel Prize in Medicine in 1962. After Cambridge, Rosalind Franklin joined the efforts of J.D. Bernal at Birkbeck College in London, where she was given charge of her own research group. She was an effective group leader and became a world-renowned expert in the crystallography of viruses. Her work established that viruses are hollow-cored. She was diagnosed with ovarian cancer in 1956 and, during her final months, performed studies on the incredibly dangerous polio virus. She died in 1958 at the age of 37.[4,5]

In 1976, BBC released a film titled *The Race For The Double Helix*. It was a very intelligent film that was essentially a dramatization of Watson's book although the interpretations were not identical. Jeff ("Jurassic Park") Goldblum played James D. Watson beautifully. I have carefully searched for a video of this film and it is not available for purchase. I wish I knew why.[7]

1. J.D. Watson and F.H.C. Crick, *Nature*, **171**:737–738, 1953.
2. J.D. Watson, *The Double Helix*, Simon & Schuster, New York, 1968.
3. J.D. Watson, op. cit., p. 126.
4. A. Sayre, *Rosalind Franklin and DNA*, Norton, New York, 1975.
5. M. Rayner-Canham and G. Rayner-Canham, *Women In Chemistry: Their Changing Roles from Alchemical Times to The Mid-Twentieth Century*, American Chemical Society and Chemical Heritage Foundation, Washington, D.C. and Philadelphia, 1998, pp. 82–90.
6. J.D. Watson, op. cit., pp. 157–163.
7. Now available: Films for the Humanities and Sciences, Box 2053, Princeton, NJ 08543-2053.

NANOSCOPIC "HEAVENS"

In the movie "Fantastic Voyage" actress Raquel Welch is among a group of scientists and doctors tasked to remove an inoperable brain tumor from a VIP. The team enters a submarinelike vessel which is then reduced to microscopic dimensions and injected into the patient's bloodstream. A moment of elevated drama occurs when Ms. Welch, outside of the vessel, is attacked by blobby antibodies and the men vie for the honor of removing them from her bodysuit. Needless to say, after some tense moments, the tumor operation eventually succeeds and the movie ends happily.

Microscopic refers to objects that can be detected in common optical microscopes—they are microns (micrometers = 10^{-6} m) in dimension. Individual atoms are ångstroms (10^{-10} m or 10^{-8} cm) in size. Large enough clusters of atoms form molecules or aggregates of molecules (such as viruses) that are tens of ångstroms or nanometers (1 nm = 10^{-9} m) in scale. What if we could make computers, machines and even robots out of nanoscale parts? Clearly, Nature has already mastered nanotechnology, why can't we?

Figure 158 depicts two molecules that were each synthesized by merely mixing equal quantities of a *linear* bifunctional molecule (the "edges" of the squares end in nitrogen atoms) and an *angular* bifunctional molecule that makes a 90° bend [the "corners" of the square are centered on metallic (M) atoms].[1] This synthesis is depicted in equation (b) of Figure 159. Figures 159 and 160 show other possibilities [equations (a) to (h)] for joining bifunctional linear (l) and/or bifunctional angular (a) molecules to form regular polygons.[1] If one of the two molecules is trifunctional and it is combined with a bifunctional molecule, the result is a regular polygon [equations (j) to (m) in Figure 160].

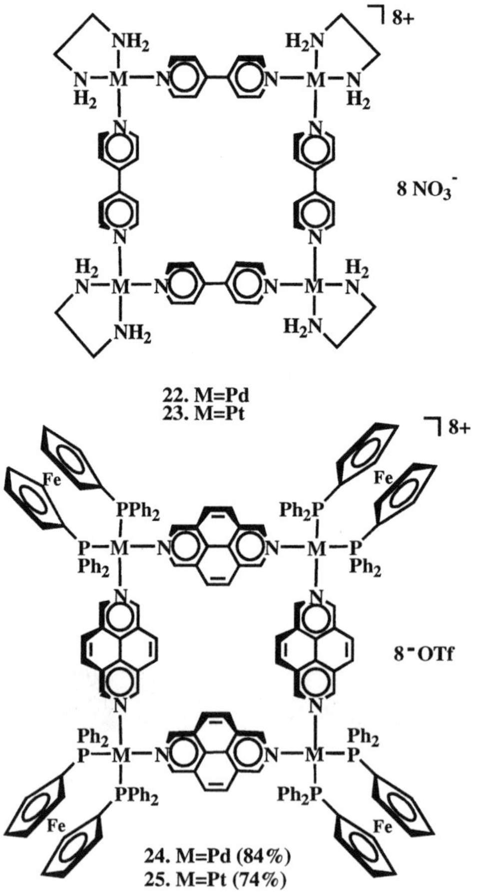

22. M=Pd
23. M=Pt

24. M=Pd (84%)
25. M=Pt (74%)

Chart 2.

FIGURE 158 ■ Chemical squares joined by bifunctional linear molecules and bifunctional angular (90°) molecules joined by dative bonds about 20% as strong as covalent bonds. Nature allows the four molecules that join to form each of these structures to form–break–reform until they "get it right" (P.J. Stang and B. Olenyuk, *Accounts of Chemical Research*, Vol. **30**:502, 1997) (courtesy American Chemical Society).

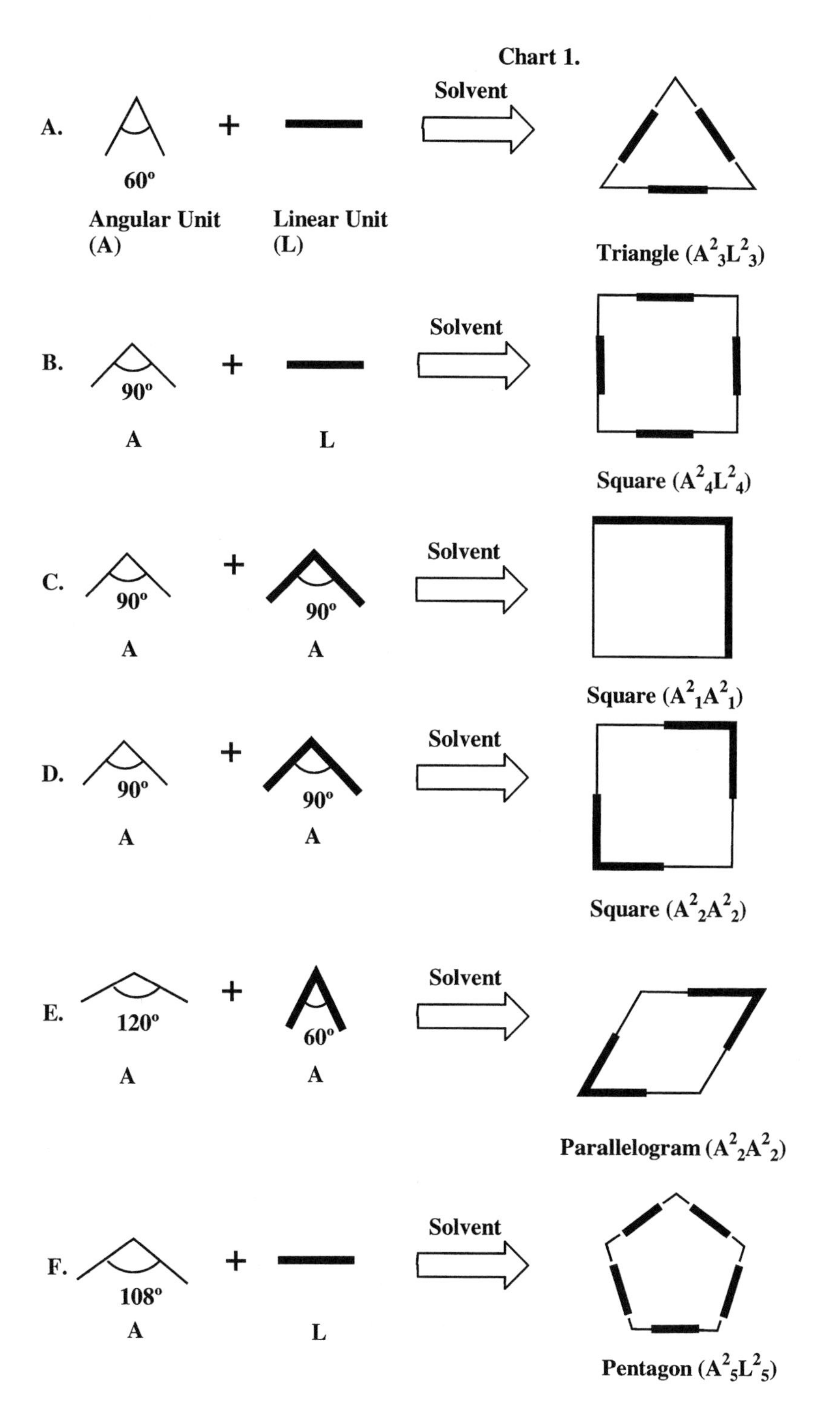

FIGURE 159 ■ Molecules that join to form polygons (see Stang and Olenyuk, Fig. 158; courtesy American Chemical Society).

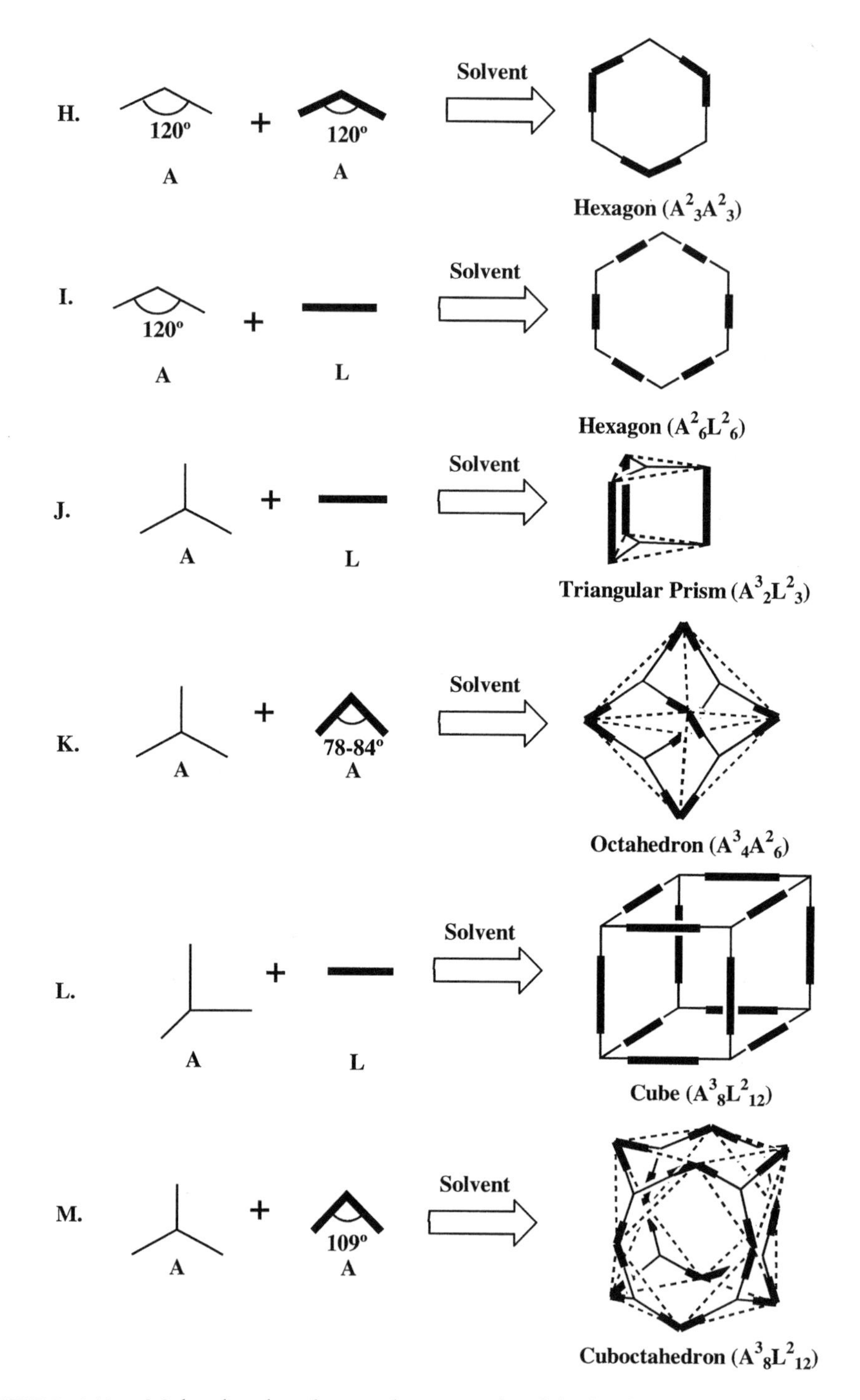

FIGURE 160 ■ Molecules that form polygons and polyhedra (see Stang and Olenyuk, Fig. 158; courtesy American Chemical Society).

Using this approach (see Fig. 161), *planar* trifunctional molecule **1** was merely mixed with *angular* bifunctional molecule **3** in a 2:3 molar ratio in methylene chloride solution. In 10 minutes a virtually perfect reaction yielding pure cuboctahedron **5** was complete.[2] The dimension of the huge molecule is about 5 nm (or 50 ångstroms). Furthermore, this remarkable approach was successful

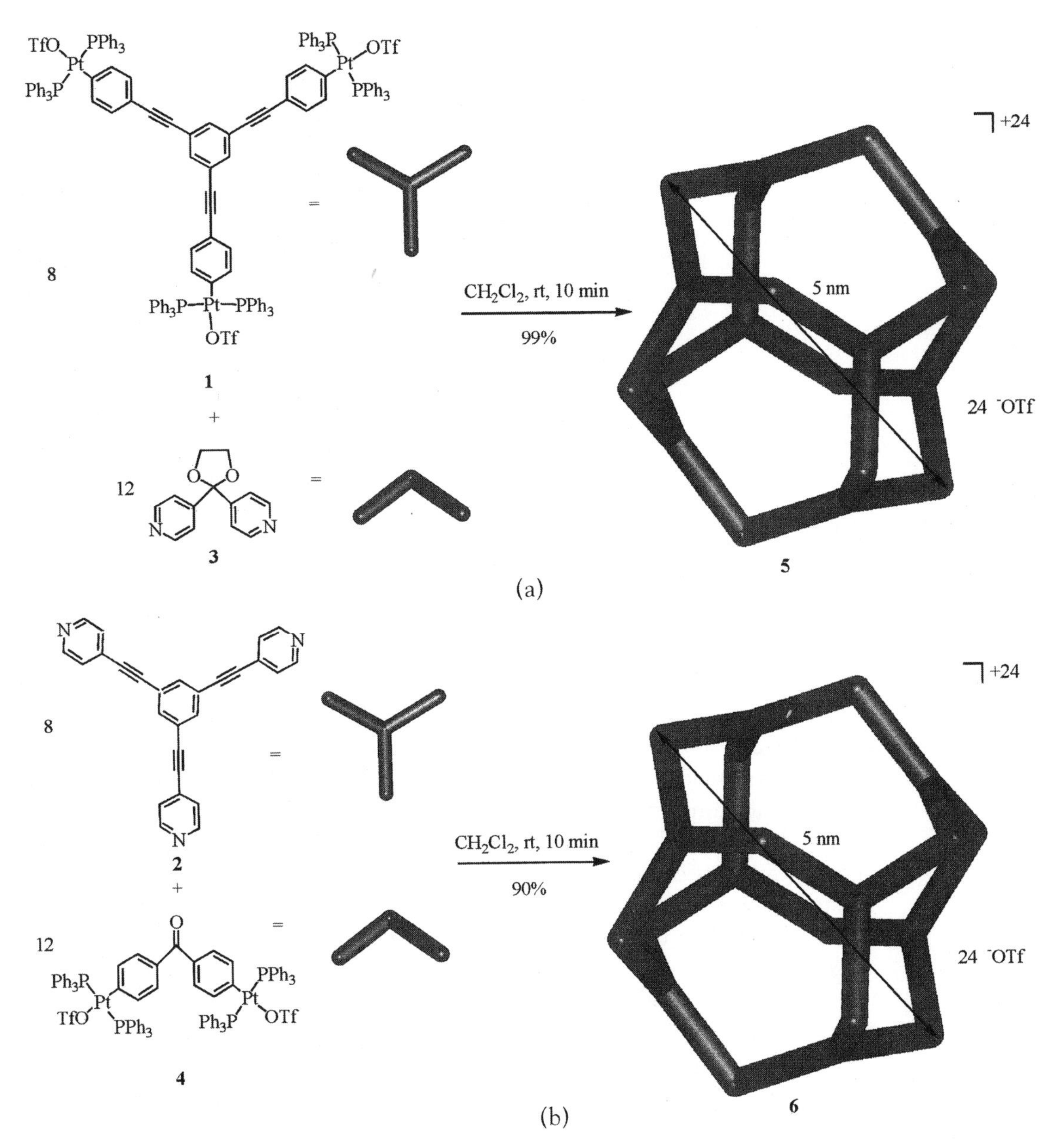

FIGURE 161 ■ The formation of a nanoscopic cuboctahedron (5 nanometers across) in 99% yield in 10 minutes using the scheme outlined in Figure 160 (B. Olenyuk, J.A. Whiteford, A. Fechtenkotter, and P.J. Stang, *Nature*, Vol. **398**:794, 1999; courtesy *Nature*; the author thanks Peter J. Stang for this figure).

in making a nanoscopic dodecahedron of 5880 atoms (see Figure 162) by merely mixing a 2:3 molar ratio of *nonplanar* trifunctional molecules and linear bifunctional molecules;[3] its formula: $C_{2900}H_{2300}N_{60}P_{120}S_{60}O_{200}F_{180}Pt_{60}$.

We have now come full circle over the course of 2500 years. The ancient Pythagoreans envisioned a mathematical basis to matter and the four earthly elements and the fifth, heavenly element (the "ether") were represented by the five Platonic solids (see Kepler's *Harmonices Mundi*, Fig. 3). Platonic solids held together by strong covalent bonds have been known for some time. In white phosphorus (P_4), the atoms occupy the corners of a tetrahedron.[4] How *did* the

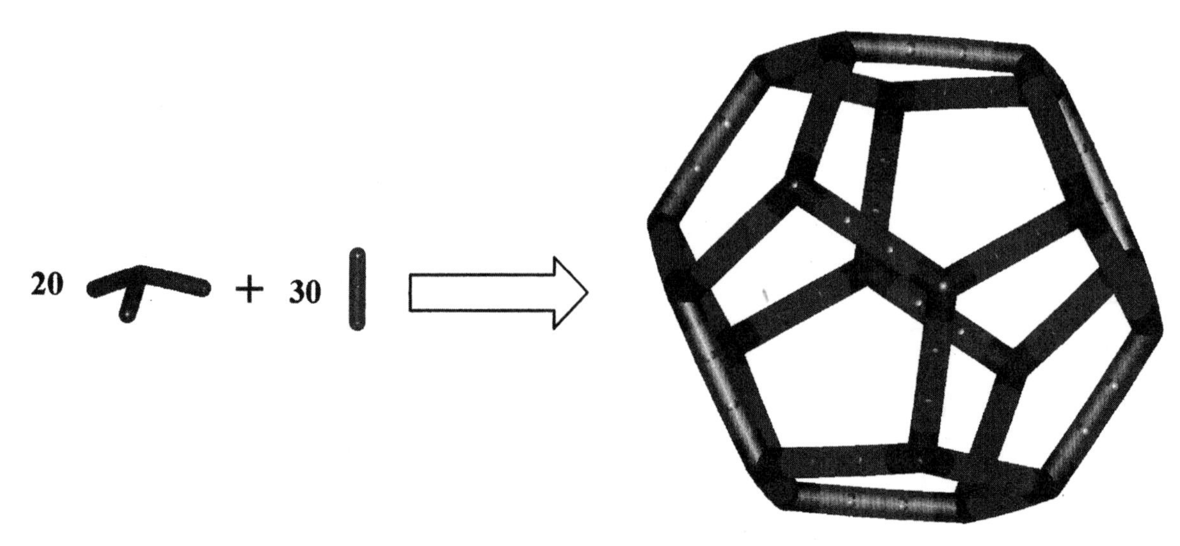

FIGURE 162 ■ The formation of a nanoscopic dodecahedron ("heaven," see Fig. 3) using the conceptual approach shown of Stang and Olenyuk.[3] It is the largest abiological system made by self-assembly.[3]

ancients know that the tetrahedron was appropriate for fire and also for fiery white phosphorus? Starting almost 40 years ago, clever organic chemists laboriously "tricked" nature, seemingly "thwarted" entropy, and assembled cubes, dodecahedra, and tetrahedra of covalently attached carbons.[5] Nature, however, had its own tricks in mind and in the late 1980s, the soccer ball of carbon atoms (C_{60}), "buckminsterfullerene" or "buckyball," a truncated icosahedron, was discovered in soot.[6] The truncated icosahedron ("soccer ball") and the cuboctahedron are two examples of the 13 Archimedean semiregular solids. The five Platonic solids all have one type of polygonal face—triangles for an icosahedron, for example. The truncated icosahedron, in contrast, has both pentagonal and hexagonal faces. Although octahedra and icosahedra are not stable structures for carbon, the former are well represented by molecules containing transition metals such as rhenium and cobalt and the latter by elemental boron and a variety of boron-containing molecules and ions.[4]

How does nature (and we chemists are part of the natural world) assemble large orderly nanostructures like viruses and nanoscopic dodecahedra? First, it prefabricates complex units such as proteins using the genetic code. Exacting chemical synthesis is required for the synthetic structural units shown in Figures 158 to 162. These units then self-assemble using weak forces such as van der Waals interactions, dipole–dipole forces, and hydrogen bonds to organize spontaneously and self-order into an optimal structure. In the case of the nanostructures described in Figures 158 to 162, ligand-metal "dative" bonds that are perhaps only 20% as strong as covalent bonds are employed. If strong covalent bonds are formed in a chemical reaction, the final product may well depend upon the initial reaction conditions (e.g., temperature, pressure) since sometimes the product formed fastest will prevail, sometimes the most stable product will prevail and sometimes mixtures of the two (if they are indeed different) will be found. This is precisely the issue that confounded Berthollet (Fig. 100) on the eve of the Atomic Theory. By contrast, structures held together by weak bonds will associate–dissociate–reassociate (build–repair, anneal) until the best structure is formed and the entire process may well be complete in minutes. In short, Nature will find a way.

1. P.J. Stang and B. Olenyuk, *Accounts of Chemical Research*, **30**:502–518, 1997.
2. B. Olenyuk, J.A. Whiteford, A. Fechtenkotter, and P.J. Stang, *Nature*, **398**:794–796, 1999.
3. B. Olenyuk and P.J. Stang, *Journal of the American Chemical Society*, **121**, 10, 434, 1999. See also: *Chemical & Engineering News*, November 15, 1999, p. 11.
4. F.A. Cotton and G. Wilkinson, *Inorganic Chemistry*, 5th ed., Wiley, New York, pp. 18–21.
5. A. Greenberg and J.F. Liebman, *Strained Organic Molecules*, Academic, New York, 1978.
6. R.F. Curl and R.E. Smalley, *Scientific American*, October, 1992, p. 54.

MOVING MATTER ATOM-BY-ATOM

Chemistry textbooks inform us that John Dalton formulated Atomic Theory in 1803 and imply that atoms were accepted from then on. Actually, such acceptance was far from universal and late-nineteenth-century books such as Brodie's *The Calculus of Chemical Operations* (London, 1866, 1877) and Hunt's *A New Basis for Chemistry: A Chemical Philosophy* (Boston, 1887), although antiatomic in nature, were not written by cranks or "nutters." The eminent physicist Ernst Mach and the famous chemist Wilhelm Ostwald resisted the reality of atoms into the beginning of the twentieth century. Jacob Bronowski strongly implies that the suicide in 1906 of Ludwig Boltzmann, who successfully explained heat as atomic and molecular motion, stemmed in part from his failure to totally convince the scientific community that atoms are real.[1]

However, at just about the same time, Albert Einstein developed a mathematical theory of the movement of microscopic particles in liquids (Brownian movement first analyzed by R. Brown in 1828) that modeled them as gas molecules.[2] In 1908, Jean Perrin explained the Brownian motion of microscopic particles in liquids and tobacco smoke, and used his data to make an excellent estimate of Avogadro's Number.[2] His book *Les Atomes* (Paris, 1913; London, 1916) laid out the case for the absolute reality of atoms and brought together a number of different ways of determining Avogadro's Number. These studies gained him the 1926 Nobel Prize in Physics.

Roughly 80 years after Boltzmann died by his own hand, we are imaging atoms, picking them up, moving them, and depositing them one at a time. Ernst Ruska, Gerd Binnig, and Heinrich Rohrer shared the 1986 Nobel Prize in Physics for their invention of the scanning tunneling microscope (STM). The STM "skates" a metallic tip of atomic dimensions to near atomic distances from surfaces of atoms or molecules. At these close distances, there is "crosstalk" between the electrons that "tunnel" between the two populations of atoms. The STM senses the miniscule changes in pressure required to keep a constant current and thus traces images of the atoms. Under certain conditions, an "energy trap" can be created under the STM probe tip that will allow the capture of an individual atom and its transfer across a surface. Figure 163 is a computer-generated model of an STM tip moving a xenon atom.[3]

Is the image in Figure 164 an extraterrestrial landscape, a fluted pie crust, the eye of a chameleon or the work of an abstract artist? Incredibly, it is an STM image of a "quantum corral" formed by moving 48 iron atoms one-by-one into a circle.[3] The ripples in the center are a standing wave produced by surface electrons confined by the circle of atoms and "provide a striking demonstration of the wave-particle nature of the electron."[3]

And what does that mean? In the 1920s Louis DeBroglie described electrons as both particles and waves because they have precise mass, go "splat-splat-splat" (or "click-click-click") into Geiger counters yet show interference like radio and light waves. It is one thing to say "particle-waves" and quite another to really *picture them*. Try it. Our problem is that electrons are outside of both our direct senses and experiences. As Bronowski notes, twentieth-century physics introduced abstraction and uncertainty and the need for what he describes as "tolerance" in modeling nature.[4] The nineteenth-century satire *Flatland* by Shakespearean scholar Edwin A. Abbott illustrates our limitations.[5]

A sphere, resident of the three-dimensional world of "Spaceland," visits the two-dimensional world of "Flatland" where he meets a square Flatlander. The square perceives the sphere in limited ways but only starts to truly understand his own limits in perception when the two visit one-dimensional "Lineland." Ironically, the square quite innocently turns the tables on the seemingly omniscient sphere as follows:[6]

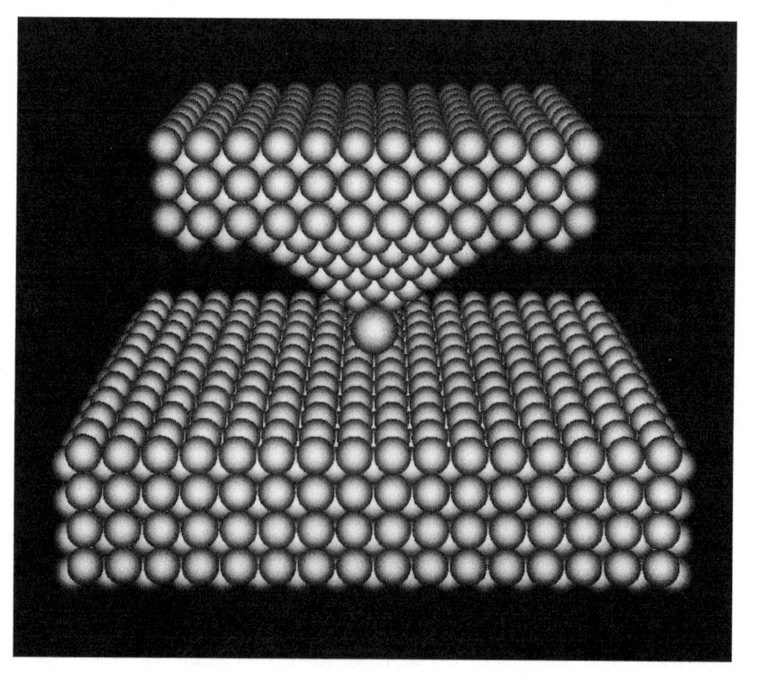

FIGURE 163 ■ Schematic of the scanning tunneling microscope (STM) the tip of which is of atomic dimension (P. Avouris, *Accounts of Chemical Research*, **28**:95, 1995; courtesy American Chemical Society; the author thanks Dr. Phaedon Avouris, IBM Research Division, for this figure).

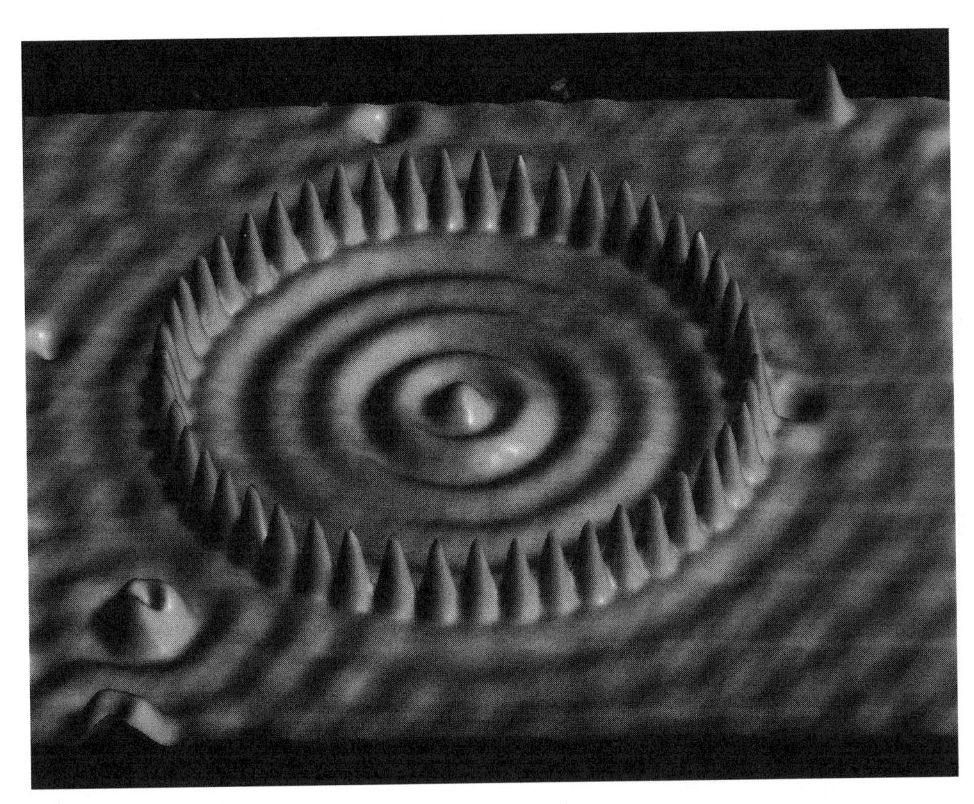

FIGURE 164 ■ STM image of the "quantum corral" consisting of 48 iron atoms placed one at a time. The image shows the particle–wave nature of electrons (P. Avouris, *Accounts of Chemical Research*, **28**:95, 1995; courtesy American Chemical Society; the author thanks Dr. Phaedon Avouris, IBM Research Division, for this figure).

Square: But my Lord has shewn me the intestines of all my countrymen in the Land of Two Dimensions by taking me with him into the Land of Three. What therefore more easy now than to take his servant on a second journey into the blessed Land of the Fourth Dimension . . . ?
Sphere: But where is this Land of Four Dimensions?
Square: I know not: but doubtless my Teacher knows.
Sphere: Not I. There is no such land. The very idea of it is inconceivable.

Incidentally, the sphere and the square finally visit zero-dimensional "Pointland" where they hear the sole resident singing hymns of self-praise:[6] "It fills all Space and what It fills, It is. What It thinks, that It utters; and what It utters, that It hears; and It itself is Thinker, Utterer, Hearer, Thought, Word, Audition; It is the One and yet the All in All. Ah, the happiness, ah, the happiness of Being." Have you ever met this type of person? Such self-satisfaction and isolation are inimical to all human endeavors including science.

Our mental images of matter continue to evolve. In late 1999, a group of scientists coupled x-ray and neutron diffraction techniques with quantum mechanical calculations to physically "see" the shape of an electron orbital.[7,8] The technique involved comparing an experimental electron density distribution with a calculated electron density distribution and plotting a difference density map. "We were just amazed when it first came up on the screen," exclaimed one of the scientists.[9] As we continue to probe the innermost secrets of chemical bonding, first explained in part by Bohr's solar system atom almost 90 years ago, I am reminded of the closing movements of Gustav Holst's symphonic opus *The Planets*. The mysterious outermost planets are evoked in the music which gradually disappears into the void leaving an open-ended sense of wonder—a metaphor for the very human curiosity that urges scientific exploration.

1. J. Bronowski, *The Ascent of Man*, Little, Brown, Boston, 1973, pp. 347–351.
2. J.R. Partington, *A History of Chemistry*, MacMillan, London, 1964, Vol. 4, pp. 744–746.
3. P. Avouris, *Accounts of Chemical Research*, **28**: 95–102, 1995. I am grateful to Dr. Phaedon Avouris, T.J. Watson Research Center, IBM Research Division, for kindly supplying original of Figures 163 and 164.
4. J. Bronowski, op. cit., Chap. 11.
5. E.A. Abbott, *Flatland: A Romance of Many Dimensions* (with Foreward by Isaac Asimov), Barnes & Nobles, New York, 1983.
6. E.A. Abbott, op. cit., pp. 102–103, 109–110.
7. J.M. Zuo, M. Kim, M. O'Keefe, and J.H. Spence, *Science*, **401**:49–52, 1999.
8. C.J. Humphreys, *Science*, **401**:21–22, 1999.
9. M. Jacoby, *Chemical and Engineering News*, September 6, 1999, p. 8. See also M.W. Browne, *New York Times*, September 7, 1999, pp. D1–D2.

SECTION IX
POST-SCRIPT

ENDING IN IMAGERY

We end this book as it was started—with metaphors that suggest a unity between matter, nature, and the human spirit and conclude with one short poem, "The Poplar," and excerpts from two longer poems by Seamus Heaney,[1] the Irish poet who was awarded the 1995 Nobel Prize in Literature.

"The Poplar"

Wind shakes the big poplar, quicksilvering
The whole tree in a single sweep.
What bright scale fell off and left this needle quivering?
What loaded balances have come to grief?

The shimmering of quicksilver (mercury) and the suggestion of the balance provide an image that "records a moment of beauty—and questions what natural balance might have been upset to produce it."[2]

"The Gravel Walks" (an excerpt)

Hoard and praise the verity of gravel.
Gems for the undeluded. Milt of earth.
Its plain, champing song against the shovel
Soundtests and soundblasts words like "honest worth."

The sound and feeling of gravel is likened to the noblest human values.

"To A Dutch Potter in Ireland" (an excerpt)

And if glazes, as you say, bring down the sun,
Your potter's wheel is bringing up the earth.
Hosannah ex infernis. Burning wells.

Hosannah in clean sand and kaolin
And, 'now that the rye crop waves beside the ruins',
In ash-pits, oxides, shards and chlorophylls.

1. S. Heaney, *The Spirit Level*, Farrar, Straus, and Giroux, New York, 1996. (We thank Farrar, Straus and Giroux for permission to reprint this material.)
2. R. Tillinghast, *New York Times Book Review*, July 21, 1996, p. 6.

INDEX